CRAIG WELLS

P9-DFQ-272

STATICS

Also by J. L. Meriam

DYNAMICS

STATICS

SECOND EDITION

J. L. Meriam

PROFESSOR OF ENGINEERING MECHANICS
SCHOOL OF ENGINEERING
DUKE UNIVERSITY

John Wiley & Sons, Inc.

NEW YORK   LONDON   SYDNEY   TORONTO

Copyright © 1966, 1971 by John Wiley & Sons, Inc.

All rights reserved. Published simultaneously in Canada.

No part of this book may be reproduced by any means, nor transmitted, nor translated into a machine language without the written permission of the publisher.

Library of Congress Catalogue Card Number: 71-136719

ISBN 471 59595 0

Printed in the United States of America

10 9 8 7

PREFACE

To the Student

The challenge and responsibility of modern engineering practice demand a high level of creative activity which, in turn, requires the support of strong analytical capability. The subject of engineering mechanics, which includes statics and dynamics, constitutes one of the cornerstones of analytical capability, and all engineers should have a basic background in this field of study.

Today's student of engineering becomes tomorrow's practicing engineer who must, through the exercise of his creative imagination and his professional knowledge, successfully combine theory and practice in the development of new structures, machines, devices, and processes which provide benefit to man. This process of modern creative design depends on the ability to visualize new configurations in terms of real materials and processes and the physical laws which govern them. Maximum progress to support the development of this design capability will be made when engineering theory is learned within the context of engineering reality so that the significance of theory can be perceived as it is being studied. This book is written with the foregoing view in mind, and it is hoped that the student will find interest and stimulation in the many problems which are taken from a wide variety of contemporary engineering situations to provide realistic and significant applications of the theory.

The purpose of the study of mechanics is to predict through calculation the behavior of engineering components and systems involving force and motion. Successful prediction in engineering design requires the careful formulation of problems with the aid of a dual thought process of physical understanding and mathematical reasoning. The process of formulating a problem is one of constructing a mathematical model which incorporates appropriate physical assumptions and mathematical approximations and approaches the actual situation with sufficient accuracy for the purpose at hand. Indeed, this process of matching the symbolic model to its physical prototype is, undoubtedly, one of the most valuable experiences of engineering study, and the problems which are included are intended to provide a comprehensive opportunity to develop this ability.

Success in analysis depends to a surprisingly large degree on a well-disciplined method of attack from hypothesis to conclusion where a straight path of rigorous application of principles has been followed. The student is urged to develop ability to represent his work in a clear, logical, and neat manner. The basic training in mechanics is a most excellent place for early development of this disciplined approach which is so necessary in most engineering work which follows.

More material is contained in this text than is covered in the usual first course in engineering mechanics, so that the book also includes an introduction to more advanced topics in mechanics and can serve as a future reference for basic principles. The more advanced topics are identified along the margins of the pages to alert the reader. To aid the student in his initial study of each topic this second edition of the text includes an expanded collection of introductory problems and problems of intermediate difficulty. Problems that offer a special challenge because of their difficulty are marked with a black triangle and a red problem number at the end of each set.

I extend encouragement to all students of mechanics, and I hope that this book will provide substantial assistance to them in acquiring a strong and meaningful engineering background.

July 1971

J. L. Meriam

PREFACE

To the Instructor

In recent times the strong trend to increase the analytical capability in engineering has resulted in increased emphasis on the mathematical generalities in mechanics. When adequate emphasis on physical understanding and engineering application is preserved, then the trend is of great benefit in extending capabilities for the analytical description of difficult problems. On the other hand, when primary attention is focused on the mathematical framework of mechanics with secondary attention to physical reality and engineering application, then the trend is of questionable benefit. Instruction in engineering mechanics has as its basic purpose the development of capacity to predict the effects of force and motion as an aid in carrying out the creative design process of engineering. Therefore the primary focus should be on the engineering significance of physical quantities with the mathematical structure acting in its role as servant. When this basic purpose is kept in mind, a proper balance between theory and application can be realized.

In this same connection there is often the temptation for the instructor of mechanics who has reached a high level of theoretical ability to forget the frame of reference of his students and present the subject with an overemphasis on generalization. There is considerable danger in this approach for the first basic course, since students lack the background necessary to cope with excessive early generality and they are also deprived of experiencing some of the historical and natural development of the subject..

A further consideration of philosophy is the strong need to provide an environment of challenging engineering reality as a means of developing the motives for learning mechanics. A solid background of analytical capability can be established in no better way than by creating a genuine interest and a compelling engineering need for the effective use of theory.

Statics is an engineering text and is written with these views in mind. Effort has been made to present the theory rigorously, concisely, and with a generality commensurate with the background of basic calculus assumed of the reader. For students who do not have a background in vector analysis, the necessary concepts and explanation are introduced in the text as needed. A more formal introduction and summary of the algebra and calculus of vectors as used in mechanics are included in Appendix B for convenient reference. It is my firm belief that facility with vector analysis is developed best within the context of its meaningful application in mechanics, and this view has guided the treatment of vectors in this book.

For two-dimensional analysis, the scalar-geometric method is generally employed as the simplest and most direct description. For three-dimensional problems, vector notation is more frequently used as the most direct and appropriate description. Tensors are introduced briefly in the discussion of internal stresses, which is one of the optional and more advanced topics in the text. The exclusive use of scalar notation, of vector notation, or of tensor index notation is rejected in favor of the choice of mathematical tool which is most appropriate for the situation at hand. In the author's view it is far more important in the basic course in mechanics to preserve and strengthen dependence on geometrical visualization and physical understanding than it is to emphasize the extensive or exclusive use of a tensor notation which reduces geometry essentially to a notational manipulation. The creative ideas that find greatest use in those branches of engineering which are supported by mechanics are born and develop more through the visualization of physical configurations than through the manipulation of notation in analysis.

The presentation and the problems in *Statics* have not been intentionally structured to provide exercises in the use of the computer. If this were the case, the role of the computer would be largely artificial. However, the student who has access to computer facilities should be encouraged to use them for the solution of occasional problems where a machine solution offers a distinct advantage.

Statics is intended for use primarily by first- and second-year engineering students. Included, however, are introductions to more advanced topics such as the criteria for the adequacy of equilibrium constraints, the differential equations of equilibrium for internal stresses in a continuum, and stability criteria for two-degree-of-freedom systems. These more advanced topics are optional and can serve as a basis for further study and future reference. *Statics* includes more material than can be easily covered in an introductory course of the usual length. It is organized in such a way that considerable flexibility is available in the choice of topics for assignments, so that the book may be adapted to a variety of course structures. The instructor should exercise caution in attempting to cover too much material in a limited time at the expense of thoroughness and understanding.

In this second edition of *Statics,* I have greatly expanded the problem sets by substantially increasing the number of introductory problems and problems of intermediate difficulty in order to help the beginning student gain initial confidence and understanding of the basic principles and methods of statics. Also, the number of simple, introductory sample problems has been increased, along with the expansion of explanations in a number of the topics. Deleted from the second edition is the material on mass moments of inertia which has been shifted to the companion volume on *Dynamics,* where it finds direct use. Continued strong emphasis is placed on the presentation of a large number of interesting and practical problems drawn from a broad range of engineering applications. Numerous examples have been taken from the subject of space mechanics and other contemporary developments. The problems are arranged approximately in order of increasing difficulty. The assessment of difficulty, of course, depends not only on the recognition of theory but also on the obstacles encountered in constructing the idealized model and in carrying out the required mathematics, and these factors vary considerably among individuals.

The emphasis placed in the first edition on clear and detailed illustrations in an effort to establish a sense of engineering reality has been continued in the present edition. This effort is consistent with my firm belief that experience with the formulation of problems that incorporate a high degree of reality, including a choice of the approach for their solution, is perhaps the most important aspect of the study of engineering mechanics. With this approach, theory takes on a significance that it cannot possibly have when the student encounters primarily idealized, and hence performulated, problems. Special attention has been given to the format of the present edition through the identification of optional topics and difficult problems. Color has been introduced, which greatly facilitates the recognition of external forces and provides a new dimension to the function and appearance of the book.

I am pleased to give continued recognition to Dr. A. L. Hale of Bell Telephone Laboratories for his valuable assistance in reviewing the manuscript and in offering numerous helpful suggestions. Dr. Hale rendered similar assistance in my previous books on mechanics, and it is a genuine pleasure to have his continued interest and contribution in this new volume. Acknowledgment is also given to the critical reviews and numerous helpful suggestions of Professor Paul Jones of the University of Illinois and Professor Andrew Pytel of The Pennsylvania State University during the preparation of this second edition. Also I am grateful for the high standards and professional contributions of the staff of John Wiley & Sons in the planning and production of this book. Finally, I acknowledge continued encouragement, patience, and assistance from my wife, Julia, during the many hours required to prepare this manuscript.

Durham, North Carolina
July 1971

J. L. Meriam

CONTENTS

* Symbol ■ indicates that the article contains topics of a somewhat advanced or specialized nature.

GUIDE TO THE USE OF STATICS

1 *Principal equations* are identified by a red triangle to the left and a red equation number to the right, such as

▶ $$\Sigma\mathbf{F} = 0 \qquad \Sigma\mathbf{M} = 0 \tag{13}$$

2 *Advanced and specialized topics* included in the text for optional study are preceded by a row of triangles

▼   ▼   ▼   ▼   ▼

and are identified by a gray band along the outer margin of the page.

3 Sample Problems

are set off from the remainder of the text for ready identification by horizontal red rules and by a vertical red rule along the outer margin of the page.

4 *Problems* in the problem sets are

numbered consecutively by chapter,

arranged generally in order of increasing difficulty,

identified by a black triangle and red number (◀ **2/43** for example) when they incorporate special challenge or difficulty.

5 *Force vectors* are represented on the diagrams by heavy red arrows to focus attention on their unique significance and to distinguish them from other lines or vectors.

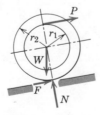

Color is also used selectively to highlight or clarify other geometric elements in the figures.

STATICS

1 PRINCIPLES OF STATICS

1 Mechanics. Mechanics is that physical science which deals with the state of rest or motion of bodies under the action of forces. No one subject plays a greater role in engineering analysis than does mechanics. The early history of this subject is synonymous with the very beginnings of engineering. Modern research and development in the fields of vibrations, stability and strength of structures and machines, rocket and spacecraft design, automatic control, engine performance, fluid flow, electrical machines and apparatus, and molecular, atomic, and subatomic behavior are highly dependent upon the basic principles of mechanics. A thorough understanding of this subject is an absolute prerequisite for work in these and many other fields.

Mechanics is the oldest of the physical sciences. The earliest recorded writings in this field are those of Archimedes (287–212 B.C.) which concern the principle of the lever and the principle of buoyancy. Substantial progress awaited the formulation of the laws of vector combination of forces by Stevinus (1548–1620), who also formulated most of the principles of statics. The first investigation of a dynamic problem is credited to Galileo (1564–1642) in connection with his experiments with falling stones. The accurate formulation of the laws of motion, as well as the law of gravitation, was made by Newton (1642–1727), who also conceived the idea of the infinitesimal in mathematical analysis. Substantial contributions to the development of mechanics were also made by da Vinci, Varignon, D'Alembert, Lagrange, Laplace, and others.

The subject of mechanics is logically divided into two parts, *statics,* which concerns the equilibrium of bodies under the action of forces, and *dynamics,* which concerns the motion of bodies. Dynamics in turn includes *kinematics,* which is the study of the motion of bodies without reference to the forces that cause the motion, and *kinetics,* which relates the forces and the resulting motions.

2 Basic Concepts. Certain definitions and concepts are basic to the study of mechanics, and they should be understood at the outset.

Space. Space is the geometric region in which events take place. In this book the word *space* will be used to refer to a three-dimensional region. It is not uncommon, however, to refer to motion along a straight line or in a plane as occurring in one- or two-dimensional space, respectively. The concept of *n*-dimensional space is an abstract device for describing relations among *n* quantities.

1

Reference Frame. Position in space is determined relative to some geometric reference system by means of linear and angular measurements. The basic frame of reference for the laws of Newtonian mechanics is the *primary inertial system* or *astronomical frame of reference,* which is an imaginary set of rectangular axes that are assumed to have no translation or rotation in space. Measurements show that the laws of Newtonian mechanics are valid for this reference system as long as any velocities involved are negligible compared with the speed of light.* Measurements made with respect to this reference are said to be *absolute,* and this reference system is considered "fixed" in space. A reference frame attached to the surface of the earth has a somewhat complicated motion in the primary system, and a correction to the basic equations of mechanics must be applied for measurements made relative to the earth's reference frame. In the calculation of rocket and space flight trajectories, for example, the absolute motion of the earth becomes an important parameter. For most engineering problems of machines and structures which remain on the earth's surface, the corrections are extremely small and may be neglected. For these problems the laws of mechanics may be applied directly for measurements made relative to the earth, and in a practical sense such measurements will be referred to as *absolute.*

Time. Time is a measure of the succession of events and is considered an absolute quantity in Newtonian mechanics. The unit of time is the second, which is a convenient fraction of the 24-hour day.

Force. Force is the action of one body on another. A force tends to move a body in the direction of its action upon the body. The properties of force are discussed in detail in Chapter 2.

Matter. Matter is substance which occupies space. A *body* is matter bounded by a closed surface.

Inertia. Inertia is the property of matter causing a resistance to change in motion.

Mass. Mass is the quantitive measure of inertia. Mass is also a property of every body which is always accompanied by mutual attraction to other bodies.

Particle. A body of negligible dimensions is called a particle. In the mathematical sense a particle is a body whose dimensions approach zero, so that it may be analyzed as a point mass. Frequently a particle is chosen as a differential element of a body. Also, when the dimensions of a body are irrelevant to the description of its position or its motion, the body may be treated as a particle.

Rigid body. A body that has no relative deformation between its parts is said to be a rigid body. This is an ideal condition, since all real bodies change shape to a certain extent when subjected to forces. When such changes in shape are negligible compared with the overall dimensions of the body or

---

* For velocities of the same order as the speed of light, 186,000 mi/sec, the theory of relativity must be applied.

with the changes of position of the body as a whole, the assumption of rigidity is permissible. For a rigid body, then, the difference in configurations between its initial and its deformed states is neglected. As an example of the assumption of rigidity, for an airplane flying through turbulent air, the flexural movement of a few inches of its wing tip in relation to the body of the aircraft is clearly of no consequence to the average distribution of aerodynamic forces on its wings or to the specification of the motion of the airplane as a whole in its flight path. For these considerations, then, the treatment of the airplane as a rigid body offers no complication.

Deformable body. When the effects of externally applied forces on the internal stresses and strains of a body are to be examined, then the deformation characteristics of the body must be considered. For this purpose the body is viewed as deformable. The internal forces induced by the flexural movement of the wings of the airplane, for example, are of critical concern in the structural design of the airplane which, for this purpose, cannot be treated as a rigid body.

3 Scalars and Vectors. The quantities dealt with in statics are of two kinds, scalars and vectors. Scalar quantities are those with which a magnitude alone is associated. Examples of scalar quantities in mechanics are time, volume, density, speed, energy, and mass. Vector quantities, on the other hand, possess direction as well as magnitude and must obey the parallelogram law of addition as described in this article. Examples of vectors are displacement, velocity, acceleration, force, moment, and momentum.

Physical quantities that are vectors fall into one of three classifications, free, sliding, or fixed.

A *free vector* is one whose action is not confined to or associated with a unique line in space. For example, if a body moves without rotation, then the movement or displacement of any point in the body may be taken as a vector, and this vector will describe equally well the direction and magnitude of the displacement of every point in the body. Hence the displacement of such a body may be represented by a free vector.

A *sliding vector* is one for which a unique line in space must be maintained along which the quantity acts. When we deal with the external action of a force on a rigid body, the force may be applied at any point along its line of action without changing its effect on the body as a whole* and hence may be considered a sliding vector.

A *fixed vector* is one for which a unique point of application is specified, and therefore the vector occupies a particular position in space. The action of a force on a deformable or nonrigid body must be specified by a fixed vector at the point of application of the force. In this problem the forces and movements internal to the body will be dependent on the point of application of the force as well as its line of action.

---

* This is the so-called *principle of transmissibility,* which is discussed in Art. 9 of Chapter 2.

A vector quantity **V** is represented by a line segment, Fig. 1, having the direction of the vector and having an arrowhead to indicate the sense. The length of the directed line segment represents to some convenient scale the magnitude $|\mathbf{V}|$ of the vector and is written with lightface type V. Boldface type is used for vector quantities whenever the directional aspect of the vector is a part of its representation. When writing vector equations it is important to preserve the mathematical distinction between vectors and scalars. It is recommended that in all handwritten work a distinguishing mark be used for each vector quantity, such as an underline, $\underline{V}$, or an arrow over the symbol, $\vec{V}$, to take the place of boldface type in print. The direction of the vector **V** may be measured by an angle θ from some known reference direction as indicated. The negative of **V** is a vector $-\mathbf{V}$ directed in the sense opposite to **V** as shown.

In addition to possessing the properties of magnitude and direction, vectors must also obey the parallelogram law of combination. This law requires that two vectors $\mathbf{V}_1$ and $\mathbf{V}_2$, treated as free vectors, Fig. 2a, may be replaced by their equivalent **V** which is the diagonal of the parallelogram formed by $\mathbf{V}_1$ and $\mathbf{V}_2$ as its two sides, as shown in Fig. 2b. This combination or vector sum is represented by the vector equation

$$\mathbf{V} = \mathbf{V}_1 + \mathbf{V}_2$$

where the plus sign used in conjunction with the vector quantities (boldface type) means *vector* and not *scalar* addition. The scalar sum of the magnitudes of the two vectors is written in the usual way as $V_1 + V_2$, and it is clear from the geometry of the parallelogram that $V \neq V_1 + V_2$.

The two vectors $\mathbf{V}_1$ and $\mathbf{V}_2$, again treated as free vectors, may also be added head-to-tail by the triangle law, as shown in Fig. 2c, to obtain the identical vector sum **V**. It is clear from the diagram that the order of addition of the vectors does not affect their sum, so that $\mathbf{V}_1 + \mathbf{V}_2 = \mathbf{V}_2 + \mathbf{V}_1$.

The difference $\mathbf{V}_1 - \mathbf{V}_2$ between the two vectors is easily obtained by adding $-\mathbf{V}_2$ to $\mathbf{V}_1$ as shown in Fig. 3 where either the parallelogram or triangle procedure may be used. The difference $\mathbf{V}'$ between the two vectors is expressed by the vector equation

$$\mathbf{V}' = \mathbf{V}_1 - \mathbf{V}_2$$

where the minus sign is used to denote *vector subtraction*.

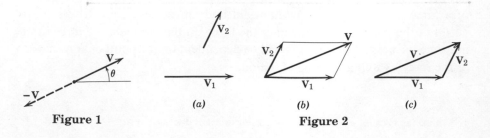

Figure 1

(a)   (b)   (c)

Figure 2

Any two or more vectors whose sum equals a certain vector $\mathbf{V}$ are said to be the *components* of that vector. Hence the vectors $\mathbf{V}_1$ and $\mathbf{V}_2$ in Fig. 4a are the components of $\mathbf{V}$ in the directions 1 and 2, respectively. It is usually more convenient to deal with vector components that are mutually perpendicular, and these are called *rectangular components*. The vectors $\mathbf{V}_x$ and $\mathbf{V}_y$ in Fig. 4b are the x- and y-components, respectively, of $\mathbf{V}$. Likewise, in Fig. 4c, $\mathbf{V}_{x'}$ and $\mathbf{V}_{y'}$ are the x'- and y'-components of $\mathbf{V}$. When expressed in rectangular components, the direction of the vector with respect to, say, the x-axis is clearly specified by

$$\theta = \tan^{-1} \frac{V_y}{V_x}$$

For some problems, particularly three-dimensional ones, it is convenient to express the rectangular components of $\mathbf{V}$ in terms of unit vectors $\mathbf{i}$, $\mathbf{j}$, $\mathbf{k}$, which are vectors in the x-, y-, and z-directions, respectively, with magnitudes of unity. The vector sum* of the components is written

$$\mathbf{V} = \mathbf{i}V_x + \mathbf{j}V_y + \mathbf{k}V_z$$

If l, m, n are the direction cosines of $\mathbf{V}$ with respect to the x-, y-, and z-axes, the components are seen to have the magnitudes

$$V_x = lV \qquad V_y = mV \qquad V_z = nV$$

with $$V^2 = V_x{}^2 + V_y{}^2 + V_z{}^2$$

Note also that $l^2 + m^2 + n^2 = 1$.

4 Newton's Laws. Sir Isaac Newton was the first to state correctly the basic laws governing the motion of a particle and to demonstrate their validity.†

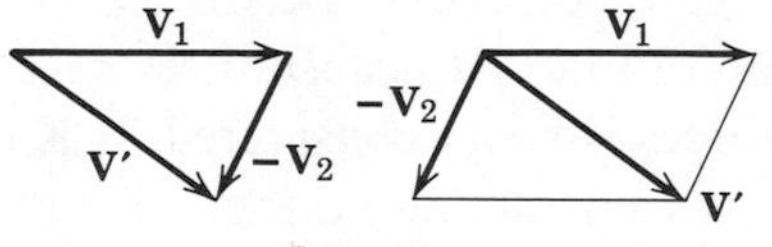

Figure 3

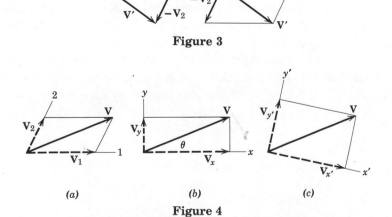

(a)   *(b)*   *(c)*

Figure 4

---

* See Fig. B1, Appendix B.

† Newton's original formulations may be found in the translation of his *Principia* (1687) revised by F. Cajori, University of California Press, 1934.

Slightly reworded to use modern terminology, these laws are:

Law I. A particle remains at rest or continues to move in a straight line with a uniform velocity if there is no unbalanced force acting on it.

Law II. The acceleration of a particle is proportional to the resultant force acting on it and is in the direction of this force.*

Law III. The forces of action and reaction between interacting bodies are equal in magnitude, opposite in direction, and collinear.

The correctness of these laws has been verified by innumerable accurate physical measurements. Newton's second law forms the basis for most of the analysis in dynamics. As applied to a particle of mass m it may be stated as

$$\mathbf{F} = m\mathbf{a} \tag{1}$$

where $\mathbf{F}$ is the resultant force acting on the particle and $\mathbf{a}$ is the resulting acceleration. This equation is a *vector* equation since the direction of $\mathbf{F}$ must be equal to the direction of $\mathbf{a}$ in addition to the equality in magnitudes of $\mathbf{F}$ and $m\mathbf{a}$. Newton's first law contains the principle of the equilibrium of forces, which is the main topic of concern in statics. Actually this law is a consequence of the second law, since there is no acceleration when the force is zero, and the particle either is at rest or moves with a constant velocity. The first law adds nothing new to the description of motion but is included since it was a part of Newton's classical statements.

The third law is basic to our understanding of force. It states that forces always occur in pairs of equal and opposite forces. Thus the downward force exerted on the desk by the pencil is accompanied by an upward force of equal magnitude exerted on the pencil by the desk. This principle holds for all forces, variable or constant, regardless of their source and holds at every instant of time during which the forces are applied. Lack of careful attention to this basic law is the cause of frequent error by the beginner. In analyzing bodies under the action of forces it is absolutely necessary to be clear about which of the pair of forces is being considered. It is necessary first of all to *isolate* the body under consideration and then to consider only the one force of the pair which acts *on* the body in question.

In addition to formulating the laws of motion for a particle Newton was also responsible for stating the law that governs the mutual attraction between bodies. This *law of gravitation* is expressed by the equation

$$F = K\frac{m_1 m_2}{r^2} \tag{2}$$

where F = the mutual force of attraction between two particles
      K = a universal constant known as the constant of gravitation
m_1, m_2 = the masses of the two particles
      r = the distance between the centers of the particles

---

*To some it is preferable to interpret Newton's second law as meaning that the resultant force acting on a particle is proportional to the time rate of change of momentum of the particle and that this rate of change is in the direction of the force. Both formulations are equally correct.

The mutual forces F obey the law of action and reaction, since they are equal and opposite and are directed along the line joining the centers of the particles. Experiment yields the value of $K = 6.67(10^{-8})$ cm³/(gm sec²) for the gravitational constant. Gravitational forces exist between every pair of bodies. On the surface of the earth the only gravitational force of appreciable magnitude is the force due to the earth's attraction. Thus each of two iron spheres 4 in. in diameter is attracted to the earth with a force of 8.90 lb which is called its *weight*. On the other hand the force of mutual attraction between them if they are just touching is 0.0000000234 lb. This force is clearly negligible compared with the earth's attraction of 8.90 lb, and consequently the gravitational attraction of the earth is the only gravitational force of any appreciable magnitude which need be considered for experiments conducted on the earth's surface.

The weight of a body is the force of attraction of the body to the earth and depends on the position of the body relative to the earth. An object weighing 10 lb at the earth's surface will weigh 9.99500 lb at an altitude of 1 mi, 9.803 lb at an altitude of 40 mi, and 2.50 lb at an altitude of 4000 mi or a height approximately equal to the radius of the earth. It is at once apparent that the variation in the gravitational attraction of high-altitude rockets and spacecraft becomes a major consideration.

Every object that is allowed to fall in a vacuum at a given location on the earth's surface will have the same acceleration g, as can be seen by combining Eqs. 1 and 2 and canceling the term representing the mass of the falling object. This combination gives

$$g = \frac{Km_0}{r^2}$$

where m_0 is the mass of the earth and r is the radius of the earth.* The mass m_0 and mean radius r of the earth have been found by experiment to be $5.98(10^{27})$ gm and $6.38(10^8)$ cm, respectively. These values together with the value for K already cited may be substituted into the expression for g to give

$$g = 980 \text{ cm/sec}^2 \qquad \text{or} \qquad g = 32.2 \text{ ft/sec}^2$$

An accurate determination of the acceleration due to gravity measured relative to the earth must account for the fact that the earth is essentially a rotating oblate spheroid with flattening at the poles. The value of g relative to the earth's surface has been found equal to 32.09 ft/sec² at the equator, 32.17 ft/sec² at a latitude of 45 deg, and 32.26 ft/sec² at the poles. The proximity of large land masses will also influence the local value of g by a small but detectable amount. It is sufficiently accurate in almost all engineering calculations involving experiments on or near the surface of the earth to use the value of 32.2 ft/sec² for g.

The mass m of a body may be calculated from the results of the simple gravitational experiment. If the gravitational force or *weight* has a magnitude

---

* It can be proved that for this purpose the earth may be considered a particle with its entire mass concentrated at its center.

W, then, since the body falls with an acceleration g, Eq. 1 gives

▶ $$W = mg \quad \text{or} \quad m = \frac{W}{g} \tag{3}$$

5 **Units.** There are a number of systems of units used in relating force, mass, and acceleration. Four of these systems are defined in the table that follows.

Systems of Units

Type of System (fundamental quantities)	Gravitational (length, force, time)		Absolute (length, mass, time)	
Name of System	British or FPS	MKS	British or FPS	CGS
length L force F time T mass M	foot (ft) pound (lb) second (sec) lb-ft^{-1}-sec^2	meter (m) kilogram (kg) second (sec) kg-m^{-1}-sec^2	foot (ft) poundal (pdl) second (sec) pound (lb)	centimeter (cm) dyne second (sec) gram (gm)
System in use by	Engineers in English-speaking countries	Engineers in non-English-speaking countries	Physicists (occasionally)	Physicists everywhere

Engineers use a gravitational system in which length, force, and time are considered fundamental quantities and the units of mass are derived. Physicists use an absolute system in which length, mass, and time are considered fundamental and the units of force are derived. Either system, of course, may be used with the same results. The engineer prefers to use force as a fundamental quantity because most of his experiments involve direct measurement of force. The British or FPS gravitational system is the one used in this book. The most commonly used unit of mass in this system is the *slug*, which is the mass of a body that weighs 32.2 lb at the earth's surface. More generally, however, mass is expressed directly in terms of the fundamental units lb-ft^{-1}-sec^2.

6 **Accuracy, Limits, and Approximations.** The number of significant figures shown in an answer should be no greater than the number of figures which can be justified by the accuracy of the given data. Hence the cross-sectional area of a square bar whose side, 0.24 in., say, was measured to the nearest hundredth of an inch should be written as 0.058 in.^{2} and not as 0.0576 in.^{2}, as would be indicated if the numbers were multiplied out.

When calculations involve small differences in large quantities, greater accuracy in the data is required to achieve a given accuracy in the results. Hence it is necessary to know the numbers 4.2503 and 4.2391 to an accuracy of five significant figures in order that their difference 0.0112 be expressed to three-figure accuracy. It is often difficult in somewhat lengthy computations to know at the outset the number of significant figures needed in the original data to ensure a certain accuracy in the answer.

Slide-rule accuracy, usually three significant figures, is considered satisfactory for the majority of engineering calculations. The decimal point should be located by a rough longhand approximation, which also serves as a check against large slide-rule error.

The *order* of differential quantities is the subject of frequent misunderstanding. Higher-order differentials may always be neglected compared with lower-order differentials when the mathematical limit is approached. As an example the element of volume ΔV of a right circular cone of altitude h and base radius r may be taken to be a circular slice a distance x from the vertex and of thickness Δx. It can be verified that the complete expression for the volume of the element may be written as

$$\Delta V = \frac{\pi r^2}{h^2}\left[x^2\,\Delta x + x\,(\Delta x)^2 + \tfrac{1}{3}\,(\Delta x)^3\right]$$

It should be recognized that, when passing to the limit in going from ΔV to dV and from Δx to dx, the terms in $(\Delta x)^2$ and $(\Delta x)^3$ drop out, leaving merely

$$dV = \frac{\pi r^2}{h^2}\,x^2\,dx$$

which is an exact expression in the limit.

In using trigonometric functions of differential quantities it is well to call attention to the following relations, which are true in the mathematical limit:

$$\sin d\theta = \tan d\theta = d\theta$$
$$\cos d\theta = 1$$

The angle $d\theta$ is, of course, expressed in radian measure. When dealing with small but finite angles it is often convenient to replace the sine by the tangent or either fuction by the angle itself. These approximations, $\sin\theta = \theta$ and $\tan\theta = \theta$, amount to retaining only the first term in the series expansions for the sine and tangent. If a closer approximation is desired, the first two terms may be retained, which gives $\sin\theta = \theta - \theta^3/6$ and $\tan\theta = \theta + \theta^3/3$. As an example of the first approximation, for an angle of 1 deg,

$$\sin 1° = 0.0174524 \qquad \text{and} \qquad 1° \text{ is } 0.0174533 \text{ radian}$$

The error in replacing the sine by the angle for 1 deg is only 0.005 per cent. For 5 deg the error is 0.13 per cent, and for 10 deg the error is still only 0.51 per cent. Similarly, for small angles the cosine may be approximated by the first two terms in its series expansion, which gives $\cos\theta = 1 - \theta^2/2$.

A few of the mathematical relations which are useful in mechanics are listed in Table C3, Appendix C.

7 Description of Statics Problems. The study of statics is directed toward the quantitative description of forces that act on engineering structures in equilibrium. Mathematics establishes the relations between the various quantities involved and makes it possible to predict effects from these relations. A dual thought process is required in formulating this description. It is necessary to think in terms of the physical situation and in terms of the corresponding mathematical description. Analysis of every problem will require the repeated transition of thought between the physical and the mathematical. Without question, one of the greatest difficulties encountered by the student is the inability to make this transition of thought freely. He should recognize that the mathematical formulation of a physical problem represents an ideal limiting description, or model, which approximates but never quite matches the actual physical situation.

In the course of constructing the idealized mathematical model for any given engineering problem certain approximations will always be involved. Some of these approximations may be mathematical, whereas others will be physical. For instance, it is often necessary to neglect small distances, angles, or forces compared with large distances, angles, or forces. A force which is actually distributed over a small area of the body upon which it acts may be considered as a concentrated force if the dimensions of the area involved are small compared with other pertinent dimensions. The weight of a steel cable per foot of its length may be neglected if the tension in the cable is many times greater than its total weight, whereas the cable weight may not be neglected if the problem calls for a determination of the deflection or sag of a suspended cable due to its weight. Thus the degree of assumption involved depends on what information is desired and on the accuracy required. The student should be constantly alert to the various assumptions called for in the formulation of real problems. The ability to understand and make use of the appropriate assumptions in the course of the formulation and solution of engineering problems is certainly one of the most important characteristics of a successful engineer. One of the major aims of this book is to provide a maximum of opportunity to develop this ability through the formulation and analysis of many practical problems involving the principles of statics.

Graphics is an important analytical tool and serves in three capacities. First, it makes possible the representation of a physical system on paper by means of a sketch or diagram. Geometrical representation is vital to physical interpretation and aids greatly in visualizing the three-dimensional aspects of many problems. Second, graphics often affords a means of solving physical relations where a direct mathematical solution would be awkward or difficult. Graphical solutions not only provide practical means for obtaining results, but they also aid greatly in making the transition of thought between the physical situation and the mathematical expression because both are represented simultaneously. A third use of graphics is in the display of results on charts or graphs which become a valuable aid to representation.

An effective method of attack on statics problems, as in all engineering problems, is essential. The development of good habits in formulating problems and in representing their solutions will prove to be an invaluable asset. Each solution should proceed with a logical sequence of steps from hypothesis to conclusion, and its representation should include a clear statement of the following parts, each clearly identified:

1. Given data
2. Results desired
3. Necessary diagrams
4. Calculations
5. Answers and conclusions

In addition it is well to incorporate a series of checks on the calculations at intermediate points in the solution. The reasonableness of numerical magnitudes should be observed, and the accuracy and dimensional homogeneity of terms should be frequently checked. It is also important that the arrangement of work be neat and orderly. Careless solutions that cannot be easily read by others are of little or no value. The discipline involved in adherence to good form will in itself be an invaluable aid to the development of the abilities for formulation and analysis. Many problems that at first may seem difficult and complicated become clear and straightforward once they are begun with a logical and disciplined method of attack.

The subject of statics is based on surprisingly few fundamental concepts and involves mainly the application of these basic relations to a variety of situations. In this application the *method* of analysis is all-important. In solving a problem it is essential that the laws which apply be carefully fixed in mind and that these principles be applied literally and exactly. In applying the principles which define the requirements for forces acting on a body it is essential that the body in question be *isolated* from all other bodies so that complete and accurate account of all forces which act on this body may be taken. This *isolation* should exist mentally as well as be represented on paper. The diagram of such an isolated body with the representation of *all* external forces acting on it is called a *free-body diagram*. It has long been established that the *free-body diagram method* is the key to the understanding of mechanics. This is so because the *isolation* of a body is the tool by which *cause* and *effect* are clearly separated and by which attention to the literal application of a principle is accurately focused. The technique of drawing free-body diagrams is covered in Chapter 3 where they are first used.

In applying the laws of statics, numerical values of the quantities may be used directly in proceeding toward the solution, or algebraic symbols may be used to represent the quantities involved and the answer left as a formula. With numerical substitution the magnitude of each quantity expressed in its particular units is evident at each stage of the calculation. This approach offers advantage when the practical significance of the magnitude of each term is important. The symbolic solution, however, has several advantages over the numerical solution. First, the abbreviation achieved by the use of symbols aids in focusing attention on the connection between the physical situation and its related mathematical description. Second, a symbolic solution permits a dimensional check to be made at every step, whereas dimensional homogeneity may be lost when numerical values only are used. Third, a symbolic solution may be used repeatedly for obtaining answers to the same problem when different sets and sizes of units are used. Facility with both forms of solution is essential, and ample practice with each should be sought in the problem work.

The student will find that solutions to the problems of statics may be obtained in one of three ways. First, a direct mathematical solution by hand calculation may be utilized where answers appear either as algebraic symbols or as numerical results. The large majority of problems come under this category. Second, certain problems are readily handled by graphical solutions. Third, the modern digital computer is of particular advantage where a large number of equations or repeated data are involved in numerical form. The student who has ready access to digital computation facilities may wish to solve a selected few of his problems by this means. In order to reduce computation time in the problem work the data for most of the problems are given in simple numbers. The choice of the most expedient method of solution is an important aspect of the experience to be gained from the problem work.

2 FORCE SYSTEMS

8 Introduction. In this chapter and in the chapters that follow the properties and effects of various kinds of forces as they act on engineering structures and mechanisms will be examined. The experience gained through this examination will prove to be of fundamental use throughout the study of mechanics and in the study of other subjects such as stress analysis, design of machine elements, and fluid flow. The foundation for a basic understanding of not only statics but also of the entire subject of mechanics is laid in this chapter, and the student is urged to master this material thoroughly.

9 Force. Before dealing with a group or *system* of forces it is necessary to examine the properties of a single force in some detail. A force has been defined as the action of one body on another. It is found that force is a vector quantity, since its effect depends on the direction as well as on the magnitude of the action and since forces may be combined according to the parallelogram law of vector combination. The action of the cable tension P on the bracket in Fig. 5a is represented in Fig. 5b by the force vector of magnitude P. The effect of this action on the bracket will depend on P, the angle θ, and the location of the point of application A. Changing any one of these three specifications will alter the effect on the bracket, as could be detected, for instance, by the force in one of the bolts which secure the bracket to the base or the internal stress and strain in the material of the bracket at any point. It is seen, therefore, that the complete specification of the action of a force requires a knowledge of its *magnitude, direction,* and *point of application.*

Force is applied either by direct mechanical contact or by remote action. Gravitational and electric and magnetic forces are applied by remote action. All other actual forces are applied through direct physical contact.

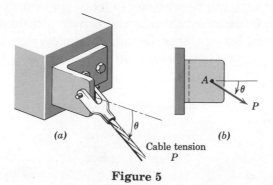

(a)   Cable tension P *(b)*

Figure 5

13

The action of a force on a body can be separated into two effects, *external* and *internal*. For the bracket of Fig. 5 the effects of *P* external to the bracket are the reactions or forces (not shown) exerted on the bracket by the foundation and bolts in consequence of the action of *P*. Forces external to a body are then of two kinds, *applied* forces and *reactive* forces. The effects of *P* internal to the bracket are the resulting internal stresses and strains distributed throughout the material of the bracket. The relation between internal forces and internal strains involves the material properties of the body and is studied in the subjects of strength of materials, elasticity, and plasticity.

In dealing with the mechanics of rigid bodies, where concern is given only to the net *external* effects of forces, experience shows that it is not necessary to restrict the action of an applied force to a given point. Hence the force *P* acting on the rigid plate in Fig. 6 may be applied at *A* or at *B* or at any other point on its action line, and the net external effects of *P* on the bracket will not change. The external effects are the force exerted on the plate by the bearing support at *O* and the force exerted on the plate by the roller support at *C*. This conclusion is described by the *principle of transmissibility,* which states that a force may be applied at any point on its given line of action without altering the resultant effects of the force *external* to the *rigid* body on which it acts. When the resultant external effects only of a force are to be investigated, the force may be treated as a *sliding* vector, and it is necessary and sufficient to specify the *magnitude, direction,* and *line of action* of the force. Since this book deals essentially with the mechanics of rigid bodies, almost all forces will be treated as sliding vectors with respect to the rigid body on which they act.

Forces may be either concentrated or distributed. Actually every contact force is applied over a finite area and is therefore a distributed force. When the dimensions of the area are negligible compared with the other dimensions of the body, the force may be considered concentrated at a point. Force may be distributed over an area, as in the case of mechanical contact, or it may be distributed over a volume when gravity or magnetic force is acting. The "weight" of a body is the force of gravity distributed over its volume and may be taken as a concentrated force acting through the center of gravity. The position of the center of gravity is usually obvious from considerations of symmetry. If the position is not clear, then a separate calculation, explained in Chapter 5, will be necessary to locate the center of gravity.

A force may be measured either by comparison with other known forces, using a mechanical balance, or by the calibrated movement of an elastic element. All such comparisons or calibrations have as their basis a primary

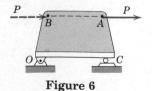

Figure 6

standard. The standard pound for the United States is legally defined as 0.4535924277 times the international kilogram and is that force required to support this portion of the standard kilogram in a vacuum and under the standard conditions of sea level and a latitude of 45 deg.*

The characteristic of a force expressed by Newton's third law must be carefully observed. The action of a force is always accompanied by an equal and opposite reaction. It is essential to fix clearly in mind which force of the pair is being considered. The answer is always clear when the body in question is *isolated* and the force exerted *on* that body (not *by* the body) is represented. It is very easy to make a careless mistake and consider the wrong force of the pair unless careful distinction between every action and reaction is made.

Two forces $\mathbf{F}_1$ and $\mathbf{F}_2$ that are concurrent may be added by the parallelogram law in their common plane to obtain their sum or *resultant* $\mathbf{R}$ as shown in Fig. 7a. If the two concurrent forces lie in the same plane but are applied at two different points as in Fig. 7b, by the principle of transmissibility they may be moved along their lines of action and their vector sum $\mathbf{R}$ completed at the point of concurrency. The resultant $\mathbf{R}$ may replace $\mathbf{F}_1$ and $\mathbf{F}_2$ without altering the external effects on the body upon which they act. The triangle law may also be used to obtain $\mathbf{R}$, but it will require moving the line of action of one of the forces as shown in Fig. 7c. In Fig. 7d the same two forces are added, and although the correct magnitude and direction of $\mathbf{R}$ are preserved, the correct line of action is lost, since $\mathbf{R}$ obtained in this way does not pass through A. This type of combination should be avoided. Mathematically the sum of the two forces may be written by the vector equation

$$\mathbf{R} = \mathbf{F}_1 + \mathbf{F}_2$$

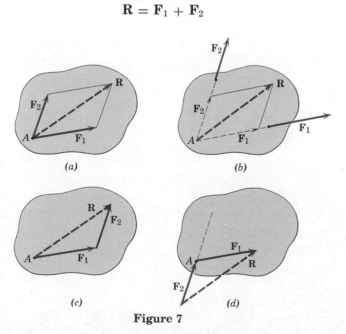

Figure 7

---

* More precisely, the "standard conditions" refer to any location at which the measured acceleration due to gravity is 32.1740 ft/sec².

In addition to the need for combining forces to obtain their resultant, there is often the need to replace a force by its *components* which act in two specified directions. Thus the force **R** in Fig. 7a may be replaced by or *resolved* into two components **F₁** and **F₂** with these specified directions merely by completing the parallelogram as shown to obtain the magnitudes of **F₁** and **F₂**.

A special case of addition is presented when the two forces **F₁** and **F₂** are parallel, Fig. 8. They may be combined by first adding two equal, opposite, and collinear forces **F** and −**F** of convenient magnitude which taken together produce no external effect on the body. Adding **F₁** and **F** to produce **R₁** and combining with the sum **R₂** of **F₂** and −**F** yield the resultant **R** correct in magnitude, direction, and line of action. The procedure here is also useful in obtaining a graphical combination of two forces that are almost parallel and hence have a remote point of concurrency.

In accordance with Art. 3 of Chapter 1, a force vector **F**, which acts at a point O, Fig. 9, may be *resolved* into *rectangular components* F_x, F_y, F_z, where

$$F_x = F \cos \theta_x \qquad F = \sqrt{F_x^2 + F_y^2 + F_z^2}$$
$$F_y = F \cos \theta_y \qquad \mathbf{F} = \mathbf{i}F_x + \mathbf{j}F_y + \mathbf{k}F_z \qquad (4)$$
$$F_z = F \cos \theta_z \qquad \mathbf{F} = F(\mathbf{i} \cos \theta_x + \mathbf{j} \cos \theta_y + \mathbf{k} \cos \theta_z)$$

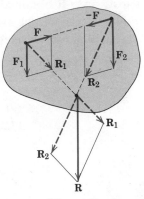

Figure 8

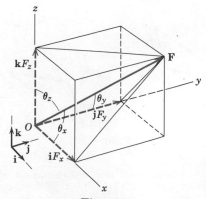

Figure 9

The unit vectors **i**, **j**, **k** are in the x-, y-, and z-directions, respectively, and the choice of orientation of the coordinate system is quite arbitrary, with convenience being the prime consideration. For two-dimensional representation with, say, the z-component absent, the resolution gives

$$F_x = F \cos \theta_x \qquad F_y = F \sin \theta_x \qquad \tan \theta_x = \frac{F_y}{F_x}$$

To eliminate ambiguity between the representation of a force and the representation of its components it is desirable to show the components in dotted lines and the resultant as a full line (or vice versa). With this understanding it will always be clear that a single force is being represented and not several separate forces.

Rectangular components of a force **F** (or other vector) may be written alternatively with the aid of the vector operation known as the *dot* or *scalar product* (see Arts. B1, B2, and B3 of Appendix B for an introduction to vector analysis). By definition, the dot product of two vectors **P** and **Q**, Fig. 10a, is the product of their magnitudes times the cosine of the angle α between them. This product may be viewed either as the projection (component) $P \cos \alpha$ of **P** in the direction of **Q** multiplied by **Q** or as the projection (component) $Q \cos \alpha$ of **Q** in the direction of **P** multiplied by **P**. In either case the dot product of the two vectors is a scalar quantity and is written as **P · Q** $= PQ \cos \alpha$. Thus the component $F_x = F \cos \theta_x$ of the force **F** in Fig. 9, for instance, may be written as $F_x = $ **F · i** where **i** is the unit vector in the x-direction. In more general terms, if **s** is a unit vector in a specified direction, the component of **F** in the **s**-direction, Fig. 10b, has the magnitude $F_s = $ **F · s**. If it is desired to write the component vector in the **s**-direction as a vector quantity, then its scalar magnitude, expressed by **F · s**, must be multiplied by the unit vector **s** to give $\mathbf{F}_s = (\mathbf{F \cdot s})\mathbf{s}$, which may be written merely as $\mathbf{F}_s = \mathbf{F \cdot ss}$.

If **s** has the direction cosines α, β, γ and **F** has the direction cosines l, m, n with respect to reference axes x-y-z, then the component of **F** in the **s**-direction becomes

$$F_s = \mathbf{F \cdot s} = F(\mathbf{i}l + \mathbf{j}m + \mathbf{k}n) \cdot (\mathbf{i}\alpha + \mathbf{j}\beta + \mathbf{k}\gamma)$$
$$= F(l\alpha + m\beta + n\gamma)$$

since

$$\mathbf{i \cdot i} = \mathbf{j \cdot j} = \mathbf{k \cdot k} = 1 \quad \text{and} \quad \mathbf{i \cdot j} = \mathbf{j \cdot i} = \mathbf{i \cdot k} = \mathbf{k \cdot i} = \mathbf{j \cdot k} = \mathbf{k \cdot j} = 0$$

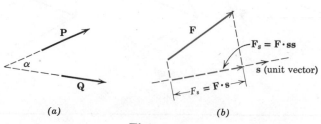

(a)   (b)

Figure 10

Sample Problems

2/1 The 100-lb force F is applied to the fixed bracket as shown. Determine the rectangular components of F in (1) the x- and y-directions and (2) in the x'- and y'-directions. Also (3) find the components of F in the x'- and y-directions.

Solution. Part (1). The x- and y-components of F are shown in the b-part of the figure and are

$$F_x = F\cos\theta_x = 100\cos 20° = 94.0 \text{ lb}$$

$$F_y = F\cos\theta_y = 100\cos 70° = 34.2 \text{ lb} \qquad\qquad Ans.$$

Part (2). The x'- and y'-components of F are the projections on these axes as shown in the c-part of the figure and are

$$F_{x'} = F\cos\theta_{x'} = 100\cos 50° = 64.3 \text{ lb}$$

$$F_{y'} = F\cos\theta_{y'} = 100\cos 40° = 76.6 \text{ lb} \qquad\qquad Ans.$$

Part (3). The components of F in the x'- and y-directions are nonrectangular and are obtained by completing the parallelogram as shown in the d-part of the figure. The components may be calculated by the law of sines which gives

$$\frac{F_{x'}}{\sin 70°} = \frac{F}{\sin 60°} \qquad F_{x'} = \frac{0.940}{0.866}100 = 108.5 \text{ lb}$$

$$\frac{F_y}{\sin 50°} = \frac{F}{\sin 60°} \qquad F_y = \frac{0.766}{0.866}100 = 88.5 \text{ lb} \qquad Ans.$$

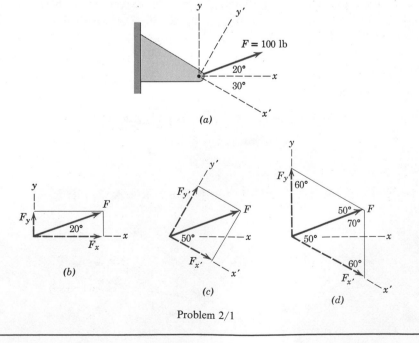

(a)

(b)

(c)

(d)

Problem 2/1

2/2 A force $F = 100$ lb is applied at the origin O of the axes x-y-z as shown. The line of action of **F** passes through a point A whose coordinates are 3 in., 4 in., and 5 in. Determine (a) the x-, y-, and z-components of **F**, (b) the projection of **F** on the x-y plane, and (c) the component F_s of **F** in the direction of the line O-s which passes through point B as shown.

Solution. Part (a). The direction cosines of **F** are

$$l = \frac{3}{7.071} = 0.424 \qquad m = \frac{4}{7.071} = 0.566 \qquad n = \frac{5}{7.071} = 0.707$$

where the diagonal to point A is $\sqrt{3^2 + 4^2 + 5^2} = \sqrt{50} = 7.071$ in. The components become

$$F_x = Fl = 100(0.424) = 42.4 \text{ lb}$$
$$F_y = Fm = 100(0.566) = 56.6 \text{ lb}$$
$$F_z = Fn = 100(0.707) = 70.7 \text{ lb} \qquad\qquad Ans.$$

Part (b). The cosine of the angle θ_{xy} between **F** and the x-y plane is

$$\cos \theta_{xy} = \frac{\sqrt{3^2 + 4^2}}{7.071} = 0.707$$

so that $F_{xy} = F \cos \theta_{xy} = 100(0.707) = 70.7$ lb. $\qquad\qquad Ans.$

Part (c). The direction cosines of a unit vector **s** along O-s are

$$\alpha = \beta = \frac{6}{\sqrt{6^2 + 6^2 + 2^2}} = 0.688 \qquad \gamma = \frac{2}{\sqrt{6^2 + 6^2 + 2^2}} = 0.229$$

Thus the component of **F** along O-s becomes

$$F_s = \mathbf{F} \cdot \mathbf{s} = 100(0.424\mathbf{i} + 0.566\mathbf{j} + 0.707\mathbf{k}) \cdot (0.688\mathbf{i} + 0.688\mathbf{j} + 0.229\mathbf{k})$$
$$= 100[(0.424)(0.688) + (0.566)(0.688) + (0.707)(0.229)]$$
$$= 84.3 \text{ lb} \qquad\qquad Ans.$$

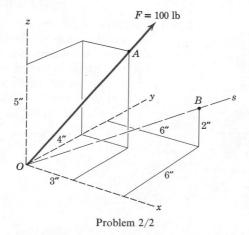

Problem 2/2

Problems

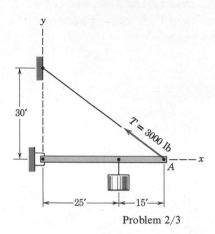

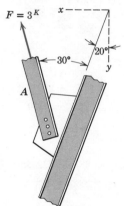

Problem 2/3

2/3 Calculate the *x*- and *y*-components of the cable tension $T = 3000$ lb shown acting on the beam at *A*.   *Ans.* $T_x = -2400$ lb, $T_y = 1800$ lb

Problem 2/4

2/4 The structural member *A* is acted upon by a tensile load of 3 kips (1 kip = 1000 lb). Determine the *x*- and *y*-components of this force.
      Ans. $F_x = 0.521$ kips, $F_y = -2.95$ kips

Problem 2/5

2/5 The contact force between the cam follower and the smooth circular cam is normal to the surface of the cam and is limited in magnitude to *F* for $\theta = \pi/2$. For this position write the expression for the component F' of the force in the direction of the center line of the follower. This component would be required for the design of the proper spring.

2/6 The hydraulic cylinder exerts a force of 4000 lb in the direction of its shaft against the load that it is hoisting. Determine the components F_n and F_t normal and tangent to AB for the position $\theta = 30$ deg.

Ans. $F_n = 2252$ lb, $F_t = 3306$ lb

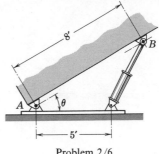

Problem 2/6

2/7 At what angle θ must F_1 be applied in order to have the combined effect of F_1 and F_2 be equal to that of a 200-lb force? *Ans. $\theta = 75°31'$*

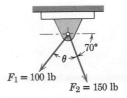

Problem 2/7

2/8 To find the forces exerted on the pin connections at A and C, the 4000-lb force may be resolved into two components, one along AB and the other along BC. Determine these components.

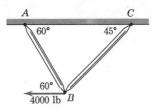

Problem 2/8

2/9 The force $P = 100$ lb is directed along the diagonal AB of the 8-in. square. Determine the magnitude P' of the component of P in the direction of OC with the positive sense taken from O to C. *Ans. $P' = -31.6$ lb*

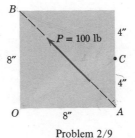

Problem 2/9

2/10 The rigid member ABC is supported by the pin A and the hinged link D and is subjected to a force F at C. Can it be concluded from the principle of transmissibility that the reaction on the pin at A would be the same if F were applied either at D or at E rather than at C?

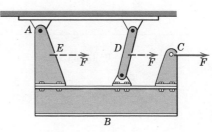

Problem 2/10

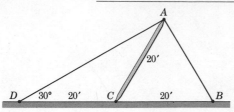

Problem 2/11

2/11 The cable from A to B is stretched until its tension T is 3000 lb. Resolve this tension acting on joint A into components T_n and T_t normal to and along strut AC.

2/12 Resolve the 3000-lb tension acting in cable AB of Prob. 2/11 into components in the directions of AC and AD.

Ans. $T_{AD} = 5196$ lb, $T_{AC} = 6000$ lb

2/13 A 1000-lb force which makes an angle of 45 deg with the horizontal x-axis is to be replaced by two forces, a horizontal force F and a second force of 800-lb magnitude. Determine F.

Ans. $F = 333$ lb or 1081 lb

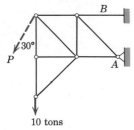

10 tons

Problem 2/14

2/14 The resultant of the 10-ton load and the accompanying tension T in member B passes through point A and results in a certain force on the pin which supports the truss at this point. If the 10-ton load is replaced by a force P applied along the dotted line shown, determine the magnitude of P which will result in the same effect on the pin at A as when the 10-ton force was applied. Specify the corresponding increment ΔT in the tension of member B. All interior angles of the truss are either 45 or 90 deg.

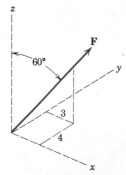

Problem 2/15

2/15 If the y-component of the force $\mathbf{F}$ is 4000 lb, write $\mathbf{F}$ as a vector.

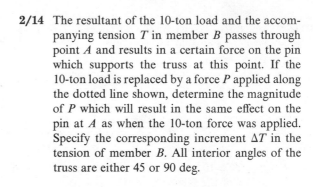

2/16 The cable AB exerts a tension $T = 1500$ lb on the fixed bracket at A. Write the expression for this force as a vector $\mathbf{T}$.

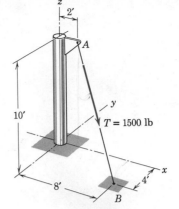

Problem 2/16

2/17 The line of action of a force $\mathbf{F}$ makes an angle of 60 deg with the x-axis and 70 deg with the y-axis. If the z-component of $\mathbf{F}$ is 50 lb, write the vector expression for $\mathbf{F}$.

Ans. $\mathbf{F} = 62.8(0.5\mathbf{i} + 0.342\mathbf{j} + 0.796\mathbf{k})$ lb

2/18 The tension in the supporting cable AB is 1000 lb. Express this tension as a force vector $\mathbf{T}$ acting on BC.

Ans. $\mathbf{T} = \dfrac{1000}{\sqrt{389}} (8\mathbf{i} - 15\mathbf{j} + 10\mathbf{k})$ lb

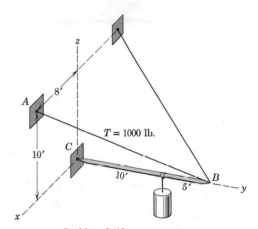

Problem 2/18

2/19 Three points in the x-y plane have coordinates expressed in feet as follows: $A(2, 3)$, $B(4, 6)$, $C(6, 5)$. A force $F = 100$ lb is applied at A and is directed toward B. Determine the vector expression for the component $\mathbf{F}_t$ of the force in the direction of AC.

2/20 Three points have x-y-z coordinates expressed in feet as follows: $A(4, 4, 5)$, $B(-2, -4, 3)$, $C(3, -6, -2)$. A force $F = 100$ lb is applied at A and is directed toward B. Determine the vector expression for the component $\mathbf{F}_t$ of the force in the direction of AC.

Ans. $\mathbf{F}_t = 80.1(-0.082\mathbf{i} - 0.816\mathbf{j} - 0.571\mathbf{k})$ lb

2/21 Calculate the magnitude of the projection F_{CD} of the 100-lb force on the face diagonal CD of the cube.

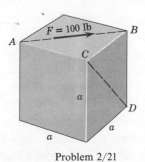

Problem 2/21

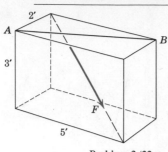

Problem 2/22

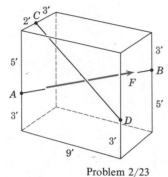

Problem 2/23

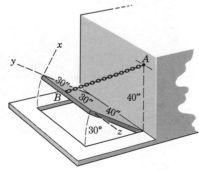

Problem 2/24

◀ **2/22** Calculate the magnitude F_{AB} of the projection of the force $F = 100$ lb on the diagonal AB of the upper face of the rectangular parallelepiped.
Ans. $F_{AB} = 63.3$ lb

◀ **2/23** Determine an expression for the magnitude F_{CD} of the component of force **F** along the line CD.
Ans. $F_{CD} = 0.555F$

◀ **2/24** The chain AB secures the hinged access door in the open position shown. If the tension T in the chain has a z-component of 600 lb, determine the x-component of T.
Ans. $T_x = 693$ lb

10 Moment. In addition to the tendency to move a body in the direction of its application, a force also tends to rotate the body about any axis which does not intersect the line of action of the force and which is not parallel to it. This tendency is known as the *moment* **M** of the force about the given axis. The moment of a force is also frequently referred to as *torque*.

Figure 11*a* shows a perfectly general body with a force **R** applied to it at point A. Axis O-O is any line in the body which does not intersect the line of action of **R**. The force **R** may be resolved into two components, one **P** parallel to O-O and the other **F** which lies in a plane perpendicular to O-O. Clearly the parallel component **P** has no tendency to rotate the body about O-O. The component **F**, on the other hand, lies in a plane a perpendicular to the axis O-O and will have a tendency to rotate the body about O-O. The magnitude of this tendency is proportional both to the magnitude of $\mathbf{F}$ and to the perpendicular *moment arm d*. The scalar magnitude of the moment, then, is

$$M = Fd \tag{5}$$

The sense of the moment depends on the direction in which $\mathbf{F}$ tends to rotate the body. The right-hand rule, Fig. 11*b*, is used to identify this sense, and the moment of $\mathbf{F}$ about *O-O* may be represented as a vector pointing in the direction of the thumb with the fingers curled in the direction of the tendency to rotate. The moment $\mathbf{M}$ obeys all the rules of vector combination and may be considered a sliding vector with a line of action coinciding with the moment axis. The units of moment are lb-in. or lb-ft and are generally written in this order to distinguish a moment from an energy term whose units are written in.-lb or ft-lb.

When dealing with forces all of which act in a given plane it is customary to speak of the moment about a point. Actually the moment with respect to an axis normal to the plane and passing through the point is implied. Thus the moment of force $\mathbf{F}$ about point *O* in Fig. 12 has the magnitude $M_O = Fd$ and is counterclockwise. Vector representation of moments for coplanar forces is unnecessary, since the vectors are either out from the paper (counterclockwise) or into the paper (clockwise). Since the addition of parallel free vectors may be accomplished with *scalar* algebra, the moment directions may be accounted for by using a plus sign (+) for counterclockwise moments and a minus sign (−) for clockwise moments, or vice versa. It is necessary only to be consistent within a given problem in using either sign convention.

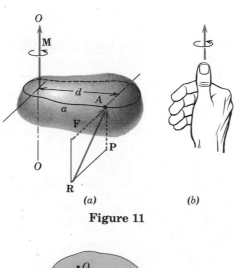

(a)   (b)

Figure 11

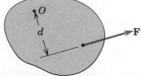

Figure 12

One of the most important principles of mechanics is *Varignon's theorem,* or the *principle of moments,* which for coplanar forces states that the moment of a force about any point is equal to the sum of the moments of the components of the force about the same point. To prove this statement consider a force R and two equivalent components P and Q acting at point A, Fig. 13. Point O is selected arbitrarily as the moment center. Construct the line AO and project the three vectors onto the normal to this line. Also construct the moment arms p, q, r of the three forces to point O and designate the angles of the vectors to the line AO by α, β, γ as shown in the figure. Since the parallelogram whose sides are P and Q requires that $\overline{ac} = \overline{bd}$, it is evident that

$$\overline{ad} = \overline{ab} + \overline{bd} = \overline{ab} + \overline{ac}$$

or

$$R \sin \gamma = P \sin \alpha + Q \sin \beta$$

Multiplying by the distance $\overline{AO}$ and substituting the values of p, q, r give

$$Rr = Pp + Qq$$

which proves that the moment of a force about any point equals the sum of the moments of its two components about the same point. Varignon's theorem need not be restricted to the case of only two components but applies equally well to three or more, since it is always possible by direct combination to reduce the number of components to two for which the theorem was proved. The theorem may also be applied to the moments of other fixed or sliding vectors.

A somewhat more general formulation of the concept of moment which is particularly useful in the analysis of three-dimensional force systems will now be developed. Consider a force $\mathbf{F}$ with an established line of action, Fig. 14*a*, and any point O not on this line. Point O and the line of $\mathbf{F}$ establish a plane *a*. The moment $\mathbf{M}_O$ of $\mathbf{F}$ about an axis through O normal to the plane has the magnitude $M_O = Fd$, where d is the perpendicular distance from O to the line of $\mathbf{F}$. This moment is also referred to as the moment of $\mathbf{F}$ about the *point* O. The vector $\mathbf{M}_O$ is normal to the plane and along the axis through O. Both the magnitude and the direction of $\mathbf{M}_O$ may be described by the vector operation known as the *cross* or *vector product* (see Art. B4 of Appendix B). A vector $\mathbf{r}$ is introduced that extends from O to *any* point on the line of action of $\mathbf{F}$. By definition, the cross product of $\mathbf{r}$ and $\mathbf{F}$ is written

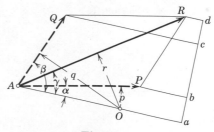

Figure 13

$\mathbf{r} \times \mathbf{F}$ and has the magnitude $(r \sin \alpha)F$, which is the same as Fd, the magnitude of $\mathbf{M}_O$. The correct direction and sense of the moment are established by the right-hand rule, described previously in this article. Thus, with $\mathbf{r}$ and $\mathbf{F}$ treated as free vectors, Fig. 14*b*, the thumb points in the direction of $\mathbf{M}_O$ if the fingers of the right hand curl in the direction of rotation from $\mathbf{r}$ to $\mathbf{F}$. Therefore the moment of $\mathbf{F}$ about the axis through O may be written as

▶
$$\mathbf{M}_O = \mathbf{r} \times \mathbf{F} \tag{6}$$

The order $\mathbf{r} \times \mathbf{F}$ of the vectors *must* be maintained, since $\mathbf{F} \times \mathbf{r}$ would produce a vector with a sense opposite to that of $\mathbf{M}_O$ or $\mathbf{F} \times \mathbf{r} = -\mathbf{M}_O$.

The cross product expression for $\mathbf{M}_O$ may be expanded into determinant form (see Eqs. B12 and B12*a* in Appendix B), which gives

▶
$$\mathbf{M} = \begin{vmatrix} \mathbf{i} & \mathbf{j} & \mathbf{k} \\ r_x & r_y & r_z \\ F_x & F_y & F_z \end{vmatrix} \tag{7}$$

The symmetry and order of the terms should be carefully noted.

The moment $\mathbf{M}_\lambda$ of $\mathbf{F}$ about *any* axis λ through O in Fig. 14*a* may now be written. If $\mathbf{n}$ is a unit vector in the λ-direction, then by using the dot product expression for the component of a vector as described in Art. 9, the component of $\mathbf{M}_O$ in the direction of λ is merely $\mathbf{M}_O \cdot \mathbf{n}$, which is the scalar magnitude of the moment $\mathbf{M}_\lambda$ of $\mathbf{F}$ about λ. To obtain the vector expression for the moment of $\mathbf{F}$ about λ, the magnitude must be multiplied by the unit directional vector $\mathbf{n}$ to give

$$\mathbf{M}_\lambda = (\mathbf{r} \times \mathbf{F} \cdot \mathbf{n})\mathbf{n} \tag{8}$$

where $\mathbf{r} \times \mathbf{F}$ replaces $\mathbf{M}_O$. The expression $\mathbf{r} \times \mathbf{F} \cdot \mathbf{n}$, known as the *triple scalar product* (see Art. B5, Appendix B), need not be written $(\mathbf{r} \times \mathbf{F}) \cdot \mathbf{n}$, since the association $\mathbf{r} \times (\mathbf{F} \cdot \mathbf{n})$ would have no meaning because a cross product cannot be formed by a vector and a scalar. From Eq. B14 in Appendix B the

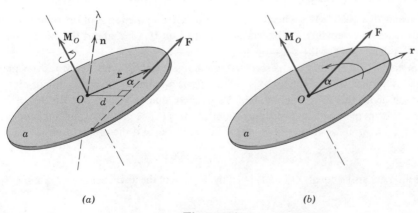

(a)   (b)

Figure 14

triple scalar product may be written in determinant form, so that Eq. 8 may be expressed alternatively as

$$\mathbf{M}_\lambda = \begin{vmatrix} r_x & r_y & r_z \\ F_x & F_y & F_z \\ \alpha & \beta & \gamma \end{vmatrix} (\mathbf{i}\alpha + \mathbf{j}\beta + \mathbf{k}\gamma) \tag{8a}$$

where α, β, γ are the direction cosines of the unit vector **n**.

Consider now two forces $\mathbf{F}_1$ and $\mathbf{F}_2$ concurrent at point A, Fig. 15. The position vector of A from any other point O is **r**. The two vector moments about the point O due to the two forces may be added to obtain

$$\mathbf{M}_1 + \mathbf{M}_2 = \mathbf{r} \times \mathbf{F}_1 + \mathbf{r} \times \mathbf{F}_2$$

and their vector sum is

$$\mathbf{M} = \mathbf{r} \times (\mathbf{F}_1 + \mathbf{F}_2)$$

This expression is the three-dimensional statement of Varignon's theorem, which says that the sum **M** of the moments about any point O of two forces concurrent at a different point equals the moment about the same point O of their sum $\mathbf{F}_1 + \mathbf{F}_2$. Although only two forces are shown, the theorem is clearly applicable to any number of concurrent forces.

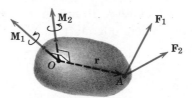

Figure 15

Sample Problem

2/25 A tension $T = 1000$ lb is applied to the cable attached to the top A of the rigid mast and secured to the ground at B. Determine the moment M_z of T about the z-axis passing through the base O of the mast.

Solution (a). The force T is resolved into components T_z and T_{xy} in the x-y plane, which is normal to the moment axis z. Since T_z is parallel to the z-axis, it can exert no moment about this axis. The moment M_z is, then, due only to T_{xy} and is $M_z = T_{xy} d$ where d is the perpendicular distance from T_{xy} to O. The cosine of the angle between T and T_{xy} is $\sqrt{50^2 + 40^2}/\sqrt{50^2 + 40^2 + 30^2} = 0.906$, and therefore

$$T_{xy} = 1000(0.906) = 906 \text{ lb}$$

The moment arm d equals $\overline{OA}$ multiplied by the sine of the angle between T_{xy} and OA, or

$$d = 50 \frac{40}{\sqrt{40^2 + 50^2}} = 31.2 \text{ ft}$$

Hence the moment of T about the z-axis is

$$M_z = 906(31.2) = 28{,}300 \text{ lb-ft} \qquad\qquad Ans.$$

and is clockwise when viewed in the x-y plane.

Solution (b). The moment is also easily calculated by resolving T_{xy} into its components T_x and T_y. It is clear that T_y exerts no moment about the z-axis since it passes through it, so that the required moment is due to T_x alone. The direction cosine of T with respect to the x-axis is $40/\sqrt{30^2 + 40^2 + 50^2} = 0.566$ so that $T_x = 1000(0.566) = 566$ lb. Thus

$$M_z = 566(50) = 28{,}300 \text{ lb-ft} \qquad\qquad Ans.$$

Solution (c). The required moment may be obtained by vector methods from the moment $\mathbf{M}_O$ of T about point O. The vector $\mathbf{M}_O$ is normal to the plane defined by $\mathbf{T}$ and point O as shown in the right-hand figure. In using Eq. 6 to find $\mathbf{M}_O$, the vector $\mathbf{r}$ is any vector from point O to the line of action of $\mathbf{T}$. The simplest choice is the vector from O to A, which is written as $\mathbf{r} = 50\mathbf{j}$ ft. The vector expression for $\mathbf{T}$ requires its direction cosines, which are $40/\overline{AB} = 0.566$, $-50/\overline{AB} = -0.707$, and $30/\overline{AB} = 0.424$. Therefore

$$\mathbf{T} = 1000(0.566\mathbf{i} - 0.707\mathbf{j} + 0.424\mathbf{k}) \text{ lb}$$

From Eq. 6,

$$\mathbf{M}_O = 50\mathbf{j} \times 1000(0.566\mathbf{i} - 0.707\mathbf{j} + 0.424\mathbf{k})$$

$$= 50{,}000(-0.566\mathbf{k} + 0.424\mathbf{i}) \text{ lb-ft}$$

The magnitude M_z of the desired moment is the component of $\mathbf{M}_O$ in the z-direction or $M_z = \mathbf{M}_O \cdot \mathbf{k}$. Therefore

$$M_z = 50{,}000(-0.566\mathbf{k} + 0.424\mathbf{i}) \cdot \mathbf{k} = -28{,}300 \text{ lb-ft} \qquad\qquad Ans.$$

The minus sign indicates that the vector $\mathbf{M}_z$ is in the negative z-direction. Expressed as a vector, the moment is

$$\mathbf{M}_z = -28{,}300\mathbf{k} \text{ lb-ft}$$

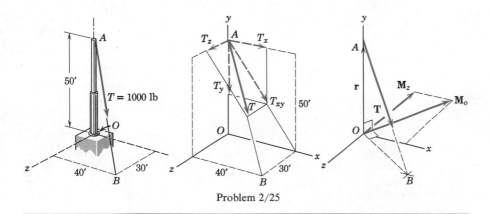

Problem 2/25

Problems

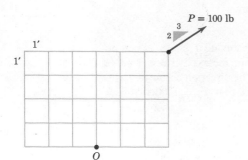

Problem 2/26

2/26 Calculate the moment M_O of the 100-lb force P about point O in two ways, using the principle of transmissibility to eliminate, first, the moment of the horizontal component of P and second, the moment of the vertical component of P. The plate upon which P acts is divided into 1-ft squares.

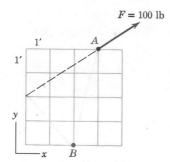

Problem 2/27

2/27 The rectangular plate is made up of 1-ft squares as shown. A force $F = 100$ lb is applied at point A in the direction shown. Calculate the moment M_B of F about point B. *Ans.* $M_B = 277$ lb-ft clockwise

2/28 Determine the vector expression for the moment $\mathbf{M}_B$ of the 100-lb force in Prob. 2/27 about point B.

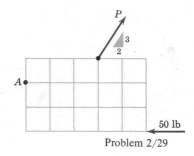

Problem 2/29

2/29 If the combined moment about point A of the 50-lb force and the force P is zero, determine both graphically and algebraically the magnitude of P. The plate upon which the forces act is divided into squares.

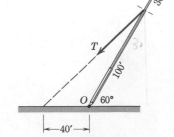

Problem 2/30

2/30 In raising the flagpole from the position shown, the tension T in the cable must supply a moment about O of 24,000 lb-ft. Determine T.
Ans. $T = 865$ lb

2/31 A force of 40 lb is applied to the end of the wrench to tighten a flange bolt which holds the wheel to the axle. Determine the moment M produced by this force about the center O of the wheel for the position of the wrench shown.

Ans. $M = 626$ lb-in.

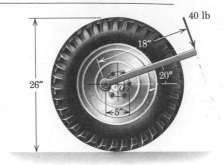

Problem 2/31

2/32 In the slider-crank mechanism shown, the connecting rod AB of length l supports a variable compressive force C. Derive an expression for the moment of C about the crank axis O in terms of C, r, l, and the variable angle θ.

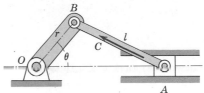

Problem 2/32

2/33 In picking up a load from position A a cable tension T of 4200 lb is developed. Calculate the moment that T produces about each of the coordinate axes through the base of the construction crane.

2/34 Use the vector cross-product expression to determine the moment $\mathbf{M}_O$ of the 4200-lb tension T in Prob. 2/33 about the base O at the origin of coordinates.

Ans. $\mathbf{M}_O = -249{,}000\mathbf{i} + 62{,}300\mathbf{j} - 37{,}400\mathbf{k}$ lb-ft

2/35 If the crane of Prob. 2/33 picks up a load at B rather than at A and develops an initial 4200-lb tension T in its cable, determine the moment $\mathbf{M}_O$ of this force about the origin O.

Ans. $\mathbf{M}_O = -184{,}900\mathbf{i} + 61{,}600\mathbf{j} - 37{,}000\mathbf{k}$ lb-ft

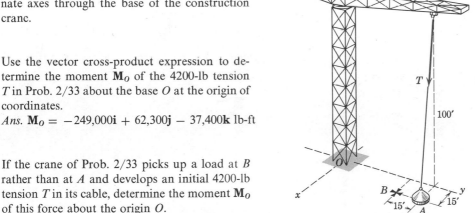

Problem 2/33

2/36 Compute the moment M_O of the 250-lb force about the axis O-O.

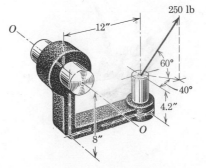

Problem 2/36

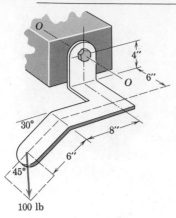

100 lb

Problem 2/37

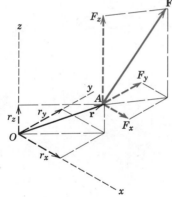

Problem 2/40

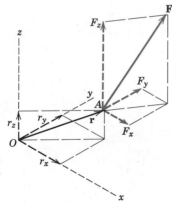

Problem 2/41

2/37 The bolt resists the torque (moment) about its axis $O\text{-}O$ produced by the 100-lb force acting on the bent bracket. Determine this torque M_O.

Ans. $M_O = 37.9$ lb-in.

2/38 A 50-lb force with direction cosines proportional to 2, 6, 9 passes through a point P whose x-, y-, z-coordinates in inches are 3, 2, -5. Evaluate the moment $\mathbf{M}$ of the force about a point whose coordinates in inches are 2, 2, -3.

2/39 A 100-lb force passes through the two points A and B in the sense from A to B. The x-, y-, z-coordinates of A and B expressed in inches are -3, -1, 4 and 3, 4, 5, respectively. Calculate the moment $\mathbf{M}$ of the force about point C whose coordinates in inches are 2, -2, 1.

Ans. $\mathbf{M} = \dfrac{100}{\sqrt{62}} \left(-14\mathbf{i} + 23\mathbf{j} - 31\mathbf{k} \right)$ lb-in.

2/40 The force $\mathbf{F}$ is applied to a body at point A which is located by the vector $\mathbf{r}$ from the origin O of a rectangular coordinate system. Expand the expression $\mathbf{r} \times \mathbf{F}$ for the moment of $\mathbf{F}$ about O and identify each term by observing the corresponding moment of each of the force components about the coordinate axes through O.

2/41 Determine the magnitude M of the moment of the 100-lb force about the diagonal axis $O\text{-}O$ of the rectangular block. *Ans.* $M = 145.6$ lb-ft

2/42 Write the vector expression for the moment $\mathbf{M}$ of the 100-lb force about the axis O_1-O_2 in the direction from O_1 to O_2.

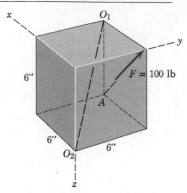

Problem 2/42

◄2/43 The force $\mathbf{F}$ exerts a certain moment $\mathbf{M}$ about the axis O-O. Determine which of the expressions cited correctly describes $\mathbf{M}$. The vector $\mathbf{n}$ is a unit vector along O-O, and the box is rectangular.

$$\mathbf{r}_1 \times \mathbf{F} \cdot \mathbf{n} \qquad\qquad (\mathbf{r}_1 \times \mathbf{F} \cdot \mathbf{n})\mathbf{n}$$

$$(-\mathbf{r}_4 \times \mathbf{F} \cdot \mathbf{n})\mathbf{n} \qquad [\mathbf{F} \times (\mathbf{r}_2 + \mathbf{r}_3) \cdot \mathbf{n}]\mathbf{n}$$

$$|\mathbf{r}_5|F\mathbf{n} \cos \beta \cos \alpha$$

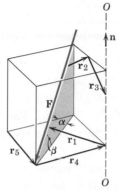

Problem 2/43

◄2/44 A rectangular steel plate 4.33 ft long and 2.69 ft wide is supported at its corner C in the horizontal x-y plane and at its corner D on the pedestal shown so that its edge CD makes an angle of 19.9 deg with the horizontal plane. The plate is then tilted backward so that its plane makes an angle of 52 deg with the vertical plane through CD. A tension T of 1000 lb is required in the cable AB to support the plate. If the cable makes an angle of 16.45 deg with the plane of the plate in the angular position shown, determine the moment M of the tension T about the lower edge CD.   *Ans.* $M = 762$ lb-ft

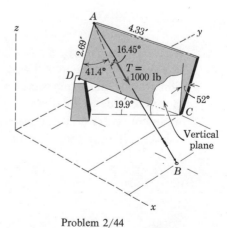

Problem 2/44

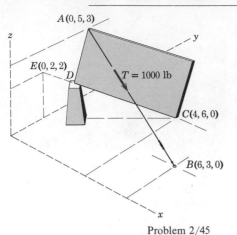

Problem 2/45

◀ **2/45** The identical plate of Prob. 2/44 is repeated here, but the coordinates of the intersections of CD and AB with the x-y and y-z planes, expressed in feet, are specified rather than the various angles as in Prob. 2/45. Determine the moment of the 1000-lb cable tension about CD. Discuss the advantages and limitations of the respective solutions for the two problems.

11 Couple. The moment produced by two equal and opposite and noncollinear forces is known as a *couple*. A couple has certain unique properties and has important applications in mechanics.

Consider the action of two equal and opposite forces $\mathbf{F}$ and $-\mathbf{F}$ a distance d apart, Fig. 16a. These two forces cannot be combined into a single force, since their sum in every direction is zero. Their effect is entirely to produce a tendency of rotation. The combined moment of the two forces about an axis normal to their plane and passing through any point in their plane such as O is the *couple* $\mathbf{M}$. It has a magnitude

$$M = F(a + d) - Fa$$

or

$$M = Fd$$

and is in the counterclockwise direction when viewed from above for the case illustrated. This expression for the magnitude of the couple does not contain any reference to the dimension a which locates the forces with respect to the moment center O. It follows that a couple has the same value for *all* moment centers. It may therefore be represented by a *free* vector $\mathbf{M}$, as shown in

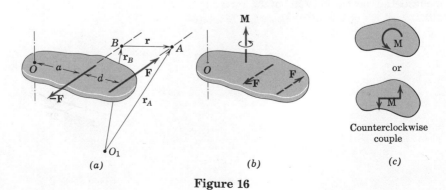

Figure 16

Fig. 16*b*, where the direction of **M** is normal to the plane of the couple and the sense of the vector is established by the right-hand convention.

The free-vector property of the couple may be demonstrated alternatively in a somewhat more general manner by combining the vector moments of the two forces about any reference point such as O_1 in Fig. 16*a*. Points *A* and *B* with position vectors $\mathbf{r}_A$ and $\mathbf{r}_B$ are any two points on the respective lines of action of **F** and $-\mathbf{F}$. With this notation the combined moment of the two forces about O_1, using Eq. 6, becomes

$$\mathbf{M} = \mathbf{r}_A \times \mathbf{F} + \mathbf{r}_B \times (-\mathbf{F})$$
$$= (\mathbf{r}_A - \mathbf{r}_B) \times \mathbf{F}$$
$$= \mathbf{r} \times \mathbf{F}$$

Since *d* is the projection of **r** along the normal to **F**, it is observed that the magnitude of this expression is $M = Fd$, which is the magnitude of the couple. It is also noted that the direction of **M** is normal to the plane of **r** and **F** as previously described. Inasmuch as $\mathbf{r} \times \mathbf{F}$ contains no reference to point O_1, it follows that **M** is the same for all reference points and, hence, may be treated as a free vector. A couple is unchanged as long as the magnitude and direction of its vector remain constant. Consequently a given couple will not be altered by changing the values of *F* and *d* as long as their product remains the same. Likewise a couple is not affected by allowing the forces to act in any one of parallel planes. Figure 17 shows four different configurations of the same couple **M**. In each of the four cases the couple is described by the identical free vector that represents the identical tendencies to twist the bodies in the direction shown.

For couples due to forces all of which act in the same or parallel planes, the couple vectors will be perpendicular to the plane or planes. In this case it is more convenient to represent such a couple by either of the conventions shown in Fig. 16*c*, where the counterclockwise couple may be taken as positive and a clockwise couple negative or vice versa.

Couples that act in nonparallel planes may be added by the ordinary rules of vector combination. Thus in Fig. 18*a* the couple $\mathbf{M}_1$ due to forces $\mathbf{F}_1$ and the couple $\mathbf{M}_2$ due to forces $\mathbf{F}_2$, acting in the two different planes shown, may be replaced by their vector sum **M** shown in Fig. 18*b*. This combination may be seen by forming the couple **M** from the forces **F**, which represent the vector sum of $\mathbf{F}_1$ and $\mathbf{F}_2$.

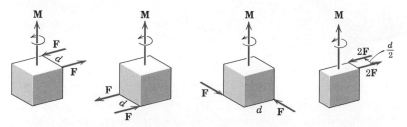

Figure 17

The effect of a force acting on a body has been described in terms of the tendency to push or pull the body in the direction of the force and to rotate the body about any axis which does not intersect the line of the force. The representation of this dual effect is often facilitated by replacing the given force by an equal parallel force and a couple to compensate for the change in the moment of the force. This resolution of a force into a force and a couple is illustrated in Fig. 19, where the given force **F** acting at point A is replaced by the same force shifted to some point O and the counterclockwise couple $M = Fd$. The transfer is seen from the middle figure, where the equal and opposite forces **F** and $-\mathbf{F}$ are added at point O without introducing any net external effects on the body. It is now seen that the original force at A and the equal and opposite one at O constitute the couple $M = Fd$, which is counterclockwise for the sample chosen, as shown in the right-hand part of the figure. Thus the original force at A has been replaced by the same force acting at a different point O and a couple without altering the external effects of the original force on the body. It follows also that a given couple and a force which lies in the plane of the couple (normal to the couple vector) may be combined to produce a single force. In vector notation the moment of the couple may be represented by the expression $\mathbf{M} = \mathbf{r} \times \mathbf{F}$, where **r** is any vector from point O to a point on the line of action of **F**, as described in Fig. 14a. The resolution of a force into an equivalent force and couple is a step that finds repeated application in mechanics and should be thoroughly mastered.

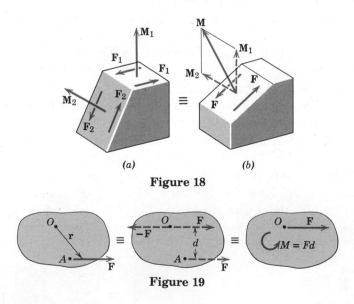

Figure 18

Figure 19

Sample Problems

2/46 Determine the magnitude and direction of the couple **M** that will replace the two given couples and still produce the same external effect on the block. Specify the two forces **F** and $-\mathbf{F}$, applied in two of the faces of the block parallel to the y-z plane, that may replace the four given forces.

Solution. The couple due to the 30-lb forces has the magnitude $M_1 = 30(6) = 180$ lb-in. The direction of M_1 is normal to the plane defined by the two forces, and the sense, shown in the middle figure, is established by the right-hand convention. The couple due to the 25-lb forces has the magnitude $M_2 = 25(10) = 250$ lb-in. with the direction and sense shown in the same figure. The two couple vectors combine to give the components

$$M_y = 180 \sin 60° = 155.9 \text{ lb-in.}$$

$$M_z = -250 + 180 \cos 60° = -160 \text{ lb-in.}$$

Thus

$$M = \sqrt{(155.9)^2 + (-160)^2} = 223 \text{ lb-in.} \qquad Ans.$$

with

$$\theta = \tan^{-1} \frac{155.9}{160} = \tan^{-1} 0.974 = 44°15' \qquad Ans.$$

The forces **F** lie in a plane normal to the couple **M**, and their moment arm as seen from the right-hand figure is 10 in. Thus each force has the magnitude

$$F = \tfrac{223}{10} = 22.3 \text{ lb} \qquad Ans.$$

and the direction $\theta = 44°15'$.

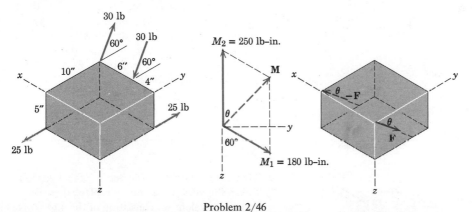

Problem 2/46

2/47 A force of 40 lb is applied at A to the handle of the control lever that is attached to the fixed shaft OB. In determining the effect of the force on the shaft at a cross section such as at O, the force may be replaced by an equivalent force at O and a couple. Describe this couple as a vector **M**.

Solution. Moving the 40-lb force through a distance $d = \sqrt{5^2 + 8^2} = 9.43$ in. to a parallel position through O requires the addition of a couple **M** whose magnitude is

$$M = Fd = 40(9.43) = 377 \text{ lb-in.} \qquad Ans.$$

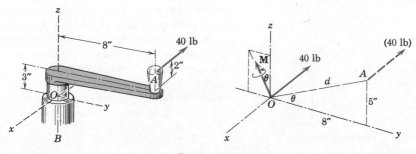

Problem 2/47

The couple vector is perpendicular to the plane in which the force is shifted, and its sense is that of the moment of the given force about *O*. The direction of **M** in the *y-z* plane is given by

$$\theta = \tan^{-1} \tfrac{5}{8} = 32°0' \qquad Ans.$$

Alternatively the couple may be expressed in vector notation as $\mathbf{M} = \mathbf{r} \times \mathbf{F}$ where $\mathbf{r} = \overrightarrow{OA} = 8\mathbf{j} + 5\mathbf{k}$ in. and $\mathbf{F} = -40\mathbf{i}$ lb. Thus

$$\mathbf{M} = (8\mathbf{j} + 5\mathbf{k}) \times (-40\mathbf{i})$$
$$= -200\mathbf{j} + 320\mathbf{k} \text{ lb-in.} \qquad Ans.$$

from which the magnitude and direction of **M** may be written.

Problems

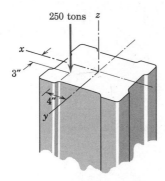

Problem 2/48

2/48 A reinforced concrete column with the cross section shown supports a vertical compressive load of 250 tons. The load is eccentric with respect to the center line of the column by the amount shown. Replace the load by one along the center line and a corresponding couple, which tends to bend the column. Specify the *x*- and *y*-components of this couple.

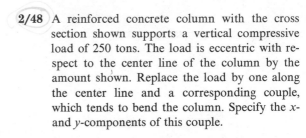

Problem 2/49

2/49 The angle plate is subjected to the two 50-lb forces shown. It is desired to replace these forces by an equivalent set consisting of the 40-lb force applied at *A* and a second force applied at *B*. Determine the *y*-coordinate of *B*.

Ans. y = 17.32 in.

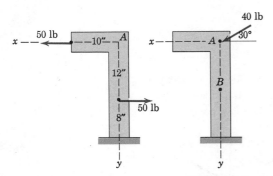

Problem 2/50

2/50 The simple truss supports a load of 4 tons. The vertical wall exerts a horizontal force against the supporting roller at *A*, and the hinged connection at *B* exerts the additional force on the truss required to maintain equilibrium. The 4-ton load and the vertical component of the reaction at *B* constitute a couple that is equal and opposite to the couple due to the two horizontal forces. Calculate the magnitude *B* of the force acting on the pin connection at *B*.

2/51 The rear wheel of an accelerating car is acted upon by a friction force F of 600 lb and a torque on the axle which amounts to a couple M. If the force and the couple can be replaced by an equivalent force which acts through a point 0.5 in. directly above the center of the wheel, find M. *Ans.* $M = 9300$ lb-in.

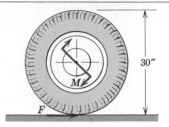

Problem 2/51

2/52 The control lever is subjected to a clockwise couple of 60 lb-ft exerted by its shaft at A and is to be designed to operate with a 50-lb pull as shown. If the resultant of the couple and the force passes through A, determine the proper dimension x of the lever.

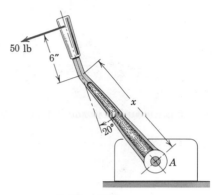

Problem 2/52

2/53 Replace the couple and force shown by a single force F applied at a point D. Locate D by determining the distance b. *Ans.* $b = 8.87$ in.

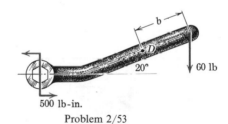

Problem 2/53

2/54 The directions of rotation of the input shaft A and output shaft B of the 10:1 worm-gear reducer are indicated by the curved arrows. An input torque (couple) of 60 lb-ft is applied to shaft A in the direction of rotation. The output shaft B supplies a torque of 240 lb-ft to the machine that it drives (not shown). The shaft of the driven machine exerts an equal and opposite reacting torque on the output shaft of the reducer. Determine the resultant $\mathbf{M}$ of the two couples that act on the reducer unit and calculate the direction cosine of $\mathbf{M}$ with respect to the x-axis.

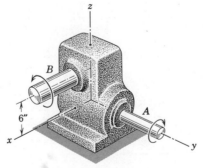

Problem 2/54

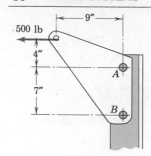

Problem 2/55

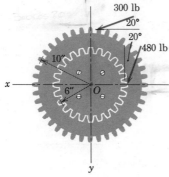

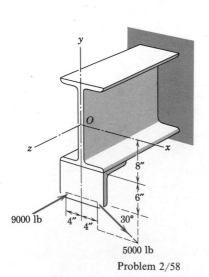

Problem 2/56

2/55 The bracket is fastened to the girder by means of the two rivets A and B. For equilibrium of the bracket the resultant of the forces exerted by the rivets on the bracket must be equal and opposite to and collinear with the 500-lb applied force. Find the force supported by each rivet by replacing the applied load by a force along a horizontal center line midway between the rivets and a couple. Then distribute this resulting system appropriately to the rivets.

2/56 The figure represents two integral gears subjected to the tooth-contact forces shown. Replace the two forces by an equivalent single force R at the rotation axis O and a corresponding couple M. If the gears start from rest under the action of the tooth loads shown, what direction would rotation take place?

> *Ans.* $R = 711$ lb, $\theta_x = 51°9'$
> $M = 112.8$ lb-in. counterclockwise

2/57 Force $\mathbf{F}_1 = 200(14\mathbf{i} - 2\mathbf{j} + 5\mathbf{k})$ lb is applied to a body at a point A whose coordinates in feet are $(2, 2, -3)$. Force $\mathbf{F}_2 = 200(2\mathbf{i} - 6\mathbf{j} - 9\mathbf{k})$ lb is applied to the same body at a point B whose coordinates in feet are $(4, -3, -1)$. Determine the force $\mathbf{F}_3$ which, when applied at point B, will result in the three forces constituting a couple $\mathbf{M}$. Express $\mathbf{M}$ as a vector.

> *Ans.* $\mathbf{F}_3 = 800(-4\mathbf{i} + 2\mathbf{j} + \mathbf{k})$ lb
> $\mathbf{M} = 600(7\mathbf{i} - 6\mathbf{j} - 22\mathbf{k})$ lb-ft

2/58 A right-angle bracket is welded to the flange of the I-beam to support the 9000-lb force, applied parallel to the axis of the beam, and the 5000-lb force, applied in the end plane of the beam. In the analysis of the capacity of the beam to withstand the applied loads, it is convenient to replace them by an equivalent force at the center O of the beam section and a corresponding couple $\mathbf{M}$. Determine the components of this couple.

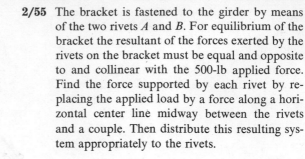

Problem 2/58

2/59 Replace the two forces that act on the 3-ft cube by an equivalent single force **F** at *A* and a couple **M**.

Ans. $\mathbf{F} = -1.72\mathbf{j} + 28.28\mathbf{k}$ lb
$\mathbf{M} = 5.15\mathbf{i} - 90\mathbf{k}$ lb-ft

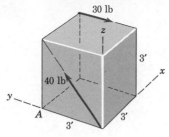

Problem 2/59

2/60 If the resultant of the four forces is a couple **M**, find *F*, θ, and **M**.

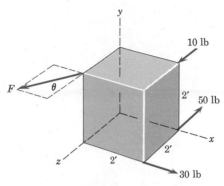

Problem 2/60

2/61 A force-and-couple system consists of the force **F** applied at point *A* and a couple **M**. Show that this system may be combined into a single force provided that $F_x M_x + F_y M_y + F_z M_z = 0$.

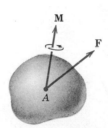

Problem 2/61

2/62 Use the procedure for the resolution or transformation of a force into a force and a couple and replace the two forces shown acting on the pipe wrenches by a single force *F* acting at a point *P* in the *x-y* plane. Find the coordinates of *P*.

Ans. $F = 20$ lb, $x = 9$ in., $y = 40$ in.

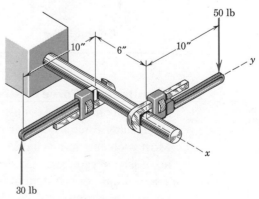

Problem 2/62

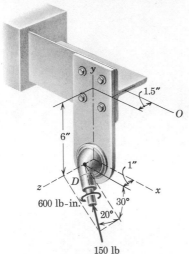

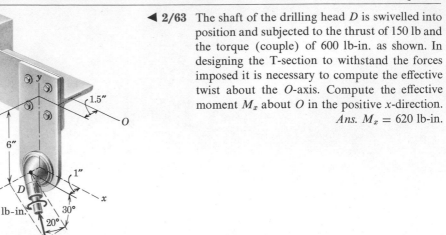

◀ **2/63** The shaft of the drilling head D is swivelled into position and subjected to the thrust of 150 lb and the torque (couple) of 600 lb-in. as shown. In designing the T-section to withstand the forces imposed it is necessary to compute the effective twist about the O-axis. Compute the effective moment M_x about O in the positive x-direction.

Ans. $M_x = 620$ lb-in.

Problem 2/63

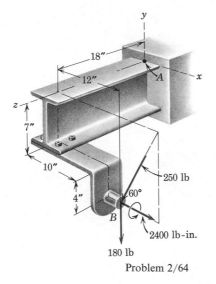

◀ **2/64** A torque (couple) of 2400 lb-in. with the rotational sense shown is applied to the pin B of the rigid bracket along with the two forces indicated. If the two forces were applied at A rather than B, calculate the resultant couple **M** (including the given couple) applied at B which would fully compensate for the shift of the forces insofar as the response of the bracket as a rigid body is concerned.

Ans. $\mathbf{M} = 9537\mathbf{i} - 2250\mathbf{j} - 6133\mathbf{k}$ lb-in.

Problem 2/64

---

12 **Resultants of Force Systems.** In the previous three articles the properties of force, moment, and couple were developed. With the aid of this description the resultant action of a group or *system* of forces may now be described. Most problems in mechanics deal with a system of forces, and it is generally necessary to reduce the system to its simplest form in describing its action. The resultant of a system of forces is the simplest force combination that can replace the original forces without altering the external effect of the system on the rigid body to which the forces may be applied. The equilibrium of a body is the condition where the resultant of all forces that act on it is zero. When the resultant of all forces on a body is not zero, the acceleration of the body is described by equating the force resultant to the product of the

mass and acceleration of the body. Thus the determination of resultants is basic to both statics and dynamics.

It was shown in Art. 11 with the aid of Fig. 19 that a force F may be replaced by the same force, moved to a parallel position through an arbitrary point O, and a couple Fd, where d is the moment arm from O to the original position of F. Or in vector notation, $\mathbf{M} = \mathbf{r} \times \mathbf{F}$, where $\mathbf{r}$ is a vector from O to any point on the line of action of the force $\mathbf{F}$. For a general system of forces $\mathbf{F}_1$, $\mathbf{F}_2$, $\mathbf{F}_3$, $\cdots$ in space, it follows that each of them may be moved to a parallel position so as to act through the same point O provided that a couple is added for each of the forces so transferred. Thus for the system illustrated schematically in Fig. 20a, the forces may all be considered to be acting through the arbitrary point O with the addition of the corresponding couples, Fig. 20b. The concurrent forces may then be added vectorially to produce a resultant force $\mathbf{R}$, and the couples may also be added to produce a resultant couple $\mathbf{M}$, Fig. 20c. The general force system, then, is reduced to

$$\mathbf{R} = \mathbf{F}_1 + \mathbf{F}_2 + \mathbf{F}_3 + \cdots = \Sigma\mathbf{F}$$
$$\mathbf{M} = \mathbf{M}_1 + \mathbf{M}_2 + \mathbf{M}_3 + \cdots = \Sigma(\mathbf{r} \times \mathbf{F})$$

(9)

The couple vectors are shown through point O, but since they are free vectors, they may be represented in any parallel positions. The magnitudes of the resultants and their components are

$$R_x = \Sigma F_x \qquad R_y = \Sigma F_y \qquad R_z = \Sigma F_z$$
$$R = \sqrt{(\Sigma F_x)^2 + (\Sigma F_y)^2 + (\Sigma F_z)^2}$$
$$M_x = \Sigma(\mathbf{r} \times \mathbf{F})_x \qquad M_y = \Sigma(\mathbf{r} \times \mathbf{F})_y \qquad M_z = \Sigma(\mathbf{r} \times \mathbf{F})_z$$
$$M = \sqrt{M_x{}^2 + M_y{}^2 + M_z{}^2}$$

(10)

The point O that is selected as the point of concurrency for the forces is arbitrary, and the magnitude and direction of $\mathbf{M}$ will depend on the particular point O selected. The magnitude and direction of $\mathbf{R}$, however, are the same no matter which point is selected. In general any system of forces may be replaced by its resultant force $\mathbf{R}$ and the resultant couple $\mathbf{M}$. In dynamics the mass center is usually selected as the reference point, and the change in the linear motion of the body is determined by the resultant force and the

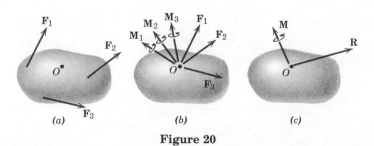

Figure 20

change in the angular motion of the body is determined by the resultant couple.

With the establishment of the resultants **R** and **M** for any general force system, the determination of the resultants for several special force systems will now be described.

Coplanar Force Systems. The most common type of force system occurs when the forces all act in a single plane, say the *x*-*y* plane, as illustrated by the system of three forces $\mathbf{F}_1$, $\mathbf{F}_2$, and $\mathbf{F}_3$ in Fig. 21*a*. The resultant force **R** is obtained in magnitude and direction by forming the *force polygon* in the *b*-part of the figure where the forces are added head-to-tail in any sequence to obtain

$$R_x = \Sigma F_x \qquad R_y = \Sigma F_y \qquad R = \sqrt{(\Sigma F_x)^2 + (\Sigma F_y)^2}$$

$$\theta = \tan^{-1}\frac{\Sigma F_y}{\Sigma F_x} \tag{11}$$

Graphically the correct line of action of **R** may be obtained by preserving the correct lines of action of the forces and adding them by the parallelogram law as indicated in the *a*-part of the figure where the sum $\mathbf{R}_1$ of $\mathbf{F}_2$ and $\mathbf{F}_3$ is added to $\mathbf{F}_1$ to obtain **R**. In this process the principle of transmissibility has been used. A well-known graphical construction known as a *funicular polygon* may also be used to establish the resultant, and the student is referred to treat-

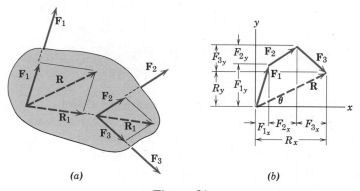

(a)   (b)

Figure 21

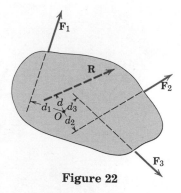

Figure 22

ments on elementary structures for a description of the technique. Algebraically, the resultant force may be located by using Varignon's moment principle with the selection of some convenient point O as a moment center, Fig. 22. Thus the unknown arm d is computed from

$$Rd = \Sigma Fd = F_1 d_1 - F_2 d_2 + F_3 d_3$$

If R is zero, then the resultant of the system is either a couple or zero. The three forces in Fig. 23, for instance, have a zero resultant force but have a resultant clockwise couple $M = F_3 d$. The resultant force of a coplanar system may be applied through any point not on its unique line of action by adding the corresponding couple.

Parallel Force System. The resultant of a parallel force system, Fig. 24, has a magnitude that is clearly equal to the scalar sum of the individual forces of the system. The position of the line of action of the resultant is found from Varignon's theorem, since the moment of the resultant about any axis must equal the sum of the moments of its components about the same axis. Thus for the case illustrated,

$$R = F_1 + F_2 + F_3$$

$$\bar{x}R = F_1 x_1 + F_2 x_2 + F_3 x_3 \qquad \bar{y}R = F_1 y_1 + F_2 y_2 + F_3 y_3$$

or in general,

$$R = \Sigma F \qquad \bar{x} = \frac{\Sigma (Fx)}{R} \qquad \bar{y} = \frac{\Sigma (Fy)}{R} \tag{12}$$

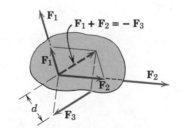

Figure 23

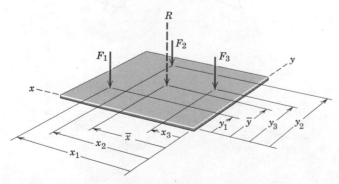

Figure 24

Concurrent Force System. When the lines of action of all forces in a system are concurrent, the resultant **R** through that point is given by Eqs. 10, with **M** clearly equal to zero.

Wrench Resultant. When the resultant couple vector **M** is parallel to the resultant force, Fig. 25, the resultant is said to be a *wrench*. A wrench is positive if the couple and force vectors point in the same direction and negative if they point in opposite directions. A common example of a positive wrench is found with the application of a screw driver where a thrust and a twist are both exerted on the screw in the direction of its axis.

Any general force system may be represented by a wrench applied along a unique line of action. This reduction is illustrated in Fig. 26, where the *a*-part of the figure represents for the general force system the resultant force **R** acting at some point O and the corresponding resultant couple **M**. Although **M** is a free vector, for convenience it is represented through O'. In the *b*-part of the figure, **M** is resolved into components $\mathbf{M}_1$ along the direction of **R** and $\mathbf{M}_2$ normal to **R**. In the *c*-part of the figure the couple $\mathbf{M}_2$ is replaced by its equivalent of two forces **R** and $-\mathbf{R}$ separated a distance $d = M_2/R$ with $-\mathbf{R}$ applied at O to cancel the original **R**. This step leaves the resultant **R**, which acts along a new and unique line of action, and the parallel couple $\mathbf{M}_1$, which is a free vector, as shown in the *d*-part of the figure. Thus the resultants of the original general force system have been transformed into a

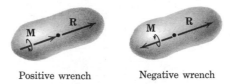

Positive wrench   Negative wrench

Figure 25

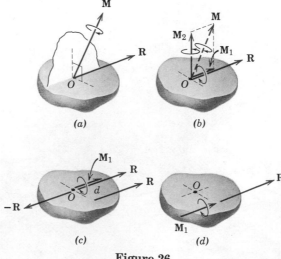

(a)   (b)

(c)   (d)

Figure 26

wrench (positive in this illustration) with its unique axis defined by the new position of **R**. It is seen from Fig. 26 that the axis of the wrench resultant lies in the plane through O normal to the plane defined by **R** and **M**. The wrench is the simplest form in which the resultant of a general force system may be expressed. This form of the resultant, however, has limited application, since it is usually more convenient to use as the reference point some point O such as the mass center of the body or other convenient origin of coordinates not on the unique axis of the wrench.

Sample Problems

2/65   Determine the resultant of the four forces and one couple that act on the plate shown.

    Solution. Point O is selected arbitrarily as a convenient origin of coordinates and moment center. The components R_x and R_y, the resultant R, and the angle θ made by R with the x-axis become

$$[R_x = \Sigma F_x] \qquad\qquad R_x = 40 + 80 \cos 30° - 60 \cos 45° = 66.9 \text{ lb}$$

$$[R_y = \Sigma F_y] \qquad\qquad R_y = 50 + 80 \sin 30° + 60 \sin 45° = 132.4 \text{ lb}$$

$$[R = \sqrt{R_x^2 + R_y^2}] \qquad R = \sqrt{(66.9)^2 + (132.4)^2} = 148.3 \text{ lb} \qquad\qquad\qquad Ans.$$

$$\left[\theta = \tan^{-1}\frac{R_y}{R_x}\right] \qquad \theta = \tan^{-1}\frac{132.4}{66.9} = 63°12' \qquad\qquad\qquad Ans.$$

    Although the couple has no influence on the magnitude and direction of R, it does influence the moment of the resultant, which will now be determined. The position of the line of action of R is found from the principle of moments (Varignon's theorem). With O as the moment center, with d as the moment arm of R, and with the counterclockwise sense chosen arbitrarily as positive, this principle requires

$$[Rd = \Sigma M_O] \qquad 148.3d = 140 - 50(5) + 60 \cos 45°(4) - 60 \sin 45°(7)$$

$$d = -1.60 \text{ ft}$$

The negative sign indicates that the moment of the resultant is acting in a clockwise rather than counterclockwise sense about O. Hence the resultant may be applied at any point on a line, making an angle of $63°12'$ with the x-axis and tangent to a circle of 1.60-ft radius about O as shown in the right-hand figure. The clockwise moment of R requires that the line of action of R be tangent at point A and not at point B, as would have been the case if the moment had been acting in a counterclockwise sense.

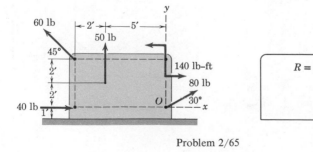

Problem 2/65

It is noted that the choice of point O as a moment center eliminated any moments due to the two forces that pass through O. The careful selection of a convenient moment center that eliminates as many terms as possible from the moment equations is an important simplification in mechanics calculations.

The given force system may also be combined graphically using the parallelogram law, the principle of transmissibility, and the procedure for the transformation of a couple and a force into a single force.

2/66   Replace the two forces and the negative wrench by a single force $\mathbf{R}$ applied at A and the corresponding couple $\mathbf{M}$.

Solution. The resultant force has the components

$[R_x = \Sigma F_x]$ $R_x = 50 \sin 40° + 70 \sin 60° = 92.8$ lb

$[R_y = \Sigma F_y]$ $R_y = 60 + 50 \cos 40° \cos 45° = 87.1$ lb

$[R_z = \Sigma F_z]$ $R_z = 70 \cos 60° + 50 \cos 40° \sin 45° = 62.1$ lb

Thus $\mathbf{R} = 92.8\mathbf{i} + 87.1\mathbf{j} + 62.1\mathbf{k}$ lb

and $R = \sqrt{(92.8)^2 + (87.1)^2 + (62.1)^2} = 141.6$ lb   *Ans.*

The couple to be added as a result of moving the 50-lb force is

$[\mathbf{M} = \mathbf{r} \times \mathbf{F}]$ $\mathbf{M}_{50} = (8\mathbf{i} + 12\mathbf{j} + 5\mathbf{k}) \times 50(\mathbf{i} \sin 40° + \mathbf{j} \cos 40° \cos 45°$
$+ \mathbf{k} \cos 40° \sin 45°)$

where $\mathbf{r}$ is the vector from A to B.

The term-by-term, or determinant, expansion gives

$$\mathbf{M}_{50} = 189.5\mathbf{i} - 55.9\mathbf{j} - 169.0\mathbf{k} \text{ lb-in.}$$

The moment of the 60-lb force about A is written by inspection of its x- and z-components, which give

$$\mathbf{M}_{60} = 360\mathbf{i} + 240\mathbf{k} \text{ lb-in.}$$

The moment of the 70-lb force about A is easily obtained from the moments of the x- and z-components of the force. The result becomes

$$\mathbf{M}_{70} = 105\mathbf{i} - 713.7\mathbf{j} - 181.9\mathbf{k} \text{ lb-in.}$$

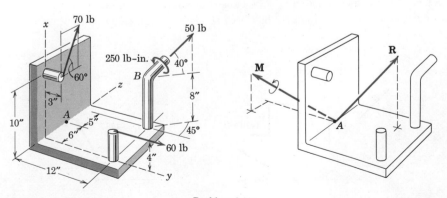

Problem 2/66

Also, the couple of the wrench may be written

$$\mathbf{M}' = 250(-\mathbf{i}\sin 40° - \mathbf{j}\cos 40°\cos 45° - \mathbf{k}\cos 40°\sin 45°)$$

$$= -160.7\mathbf{i} - 135.4\mathbf{j} - 135.4\mathbf{k}\ \text{lb-in.}$$

Therefore, the resultant couple upon adding together the **i**-, **j**-, and **k**-terms is

$$\mathbf{M} = 493.8\mathbf{i} - 905.0\mathbf{j} - 246.3\mathbf{k}\ \text{lb-in.}$$

and $\qquad M = \sqrt{(493.8)^2 + (905.0)^2 + (246.3)^2} = 1060\ \text{lb-in.}$ *Ans.*

Problems

2/67 The mast supports the two horizontal cable tensions shown. How far above the base of the mast would the resultant of these two forces act?

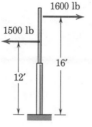

Problem 2/67

2/68 Determine the force R that could replace the four forces that act on the cantilever beam and not change the reaction on the end of the beam at the supporting weld at A. Locate R by finding its distance b to the left of A.

$\qquad$ *Ans.* $R = 300$ lb down, $\quad b = 11$ ft

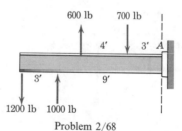

Problem 2/68

2/69 The two forces are applied normal to the plane of the plate. Calculate the coordinates of the point in this plane through which their resultant acts.

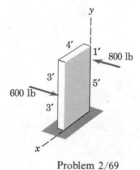

Problem 2/69

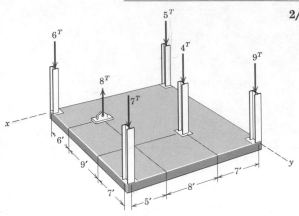

Problem 2/70

2/70 The concrete slab supports the six vertical loads shown. Determine the resultant of these forces and the *x*- and *y*-coordinates of a point through which it acts.

Ans. x = 7.30 ft, y = 15.83 ft

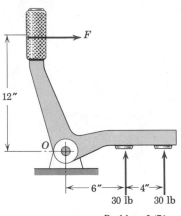

Problem 2/71

2/71 Determine the magnitude *F* of the force applied to the handle which will make the resultant of the three forces pass through *O*.

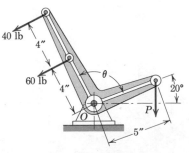

Problem 2/72

2/72 In the equilibrium position shown the resultant of the three forces acting on the bell crank passes through the bearing *O*. Determine the vertical force *P*. Does the result depend on *θ*?

Ans. P = 119.3 lb

2/73 The eyebolt supports the four forces shown. If the x-component of their resultant is 85 lb and the magnitude of the resultant is 100 lb, determine F and θ. The angle θ is limited to positive values less than 90 deg.

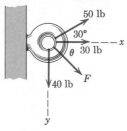

Problem 2/73

2/74 The fixed plate is subjected to the three 100-lb forces. Replace these forces by an equivalent system composed of a single force F at A and a couple M.

> *Ans.* $F = 241$ lb, $\theta_x = 45$ deg
> $M = 95$ lb-in. counterclockwise

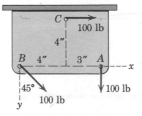

Problem 2/74

2/75 The gear reducer shown is subjected to the two couples, its 50-lb weight, and a vertical force at each of the mountings A and B. If the resultant of this system of two couples and three forces is zero, determine forces A and B.

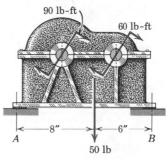

Problem 2/75

2/76 Determine the resultant $\mathbf{R}$ of the three forces and two couples shown. Find the coordinate x of the point on the x-axis through which $\mathbf{R}$ passes.

> *Ans.* $\mathbf{R} = -150\mathbf{i} - 200\mathbf{j}$ lb, $x = 2.90$ in.

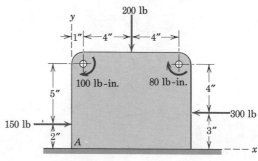

Problem 2/76

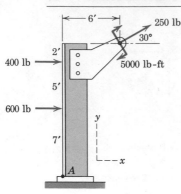

Problem 2/77

2/77 Represent the resultant of the applied loads and couple shown acting on the vertical support by a force R acting at A and a couple M.

Ans. $R = 1223$ lb

$M = 16{,}280$ lb-ft clockwise

2/78 Show that the most general force system may be represented by two nonintersecting forces.

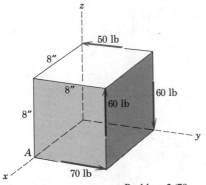

Problem 2/79

2/79 The four forces are acting along the edges of the 8-in. cube as shown. Represent the resultant of these forces by a force $\mathbf{R}$ through point A and a couple $\mathbf{M}$.

Ans. $\mathbf{R} = 20\mathbf{j}$ lb, $\mathbf{M} = 80(5\mathbf{i} - 6\mathbf{j} + 5\mathbf{k})$ lb-in.

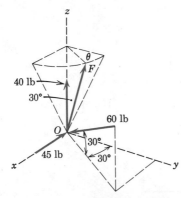

Problem 2/80

2/80 The four forces are concurrent at the origin O of coordinates. If the x-component of their resultant $\mathbf{R}$ is -50 lb and the z-component is 100 lb, determine F, θ, and R.

Ans. $F = 103.9$ lb, $\theta = 23°49'$, $R = 111.8$ lb

2/81 Replace the two forces of equal magnitude F by a single force $\mathbf{R}$ passing through point A and a couple $\mathbf{M}$.

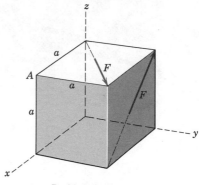

Problem 2/81

2/82 The combined action of the two forces on the base at O may be obtained by establishing their resultant through O. Determine $\mathbf{R}$ and the accompanying couple $\mathbf{M}$.

> *Ans.* $\mathbf{R} = 354\mathbf{i} - 954\mathbf{k}$ lb
> $\mathbf{M} = -739\mathbf{i} + 8196\mathbf{j} + 1061\mathbf{k}$ lb-ft

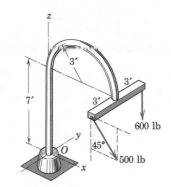

Problem 2/82

2/83 The 40-lb motor is mounted on the bracket and its shaft resists the 30-lb thrust and 200-lb-in. couple applied to it. Determine the resultant of the force system shown in terms of a force $\mathbf{R}$ at A and a couple $\mathbf{M}$.

> *Ans.* $\mathbf{R} = -30\mathbf{i} - 40\mathbf{k}$ lb
> $\mathbf{M} = -120\mathbf{i} + 90\mathbf{j} + 240\mathbf{k}$ lb-in.

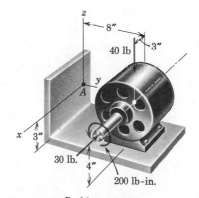

Problem 2/83

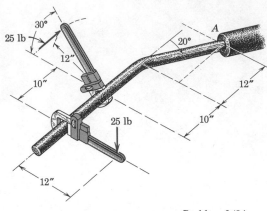

25 lb

Problem 2/84

◄ **2/84** Two pipe wrenches are applied to the bent pipe as shown to screw it into the fitting at *A*. If a single pipe wrench applied to the pipe at *A* can accomplish the same job, what force *F* must be exerted on its handle at a moment arm of 12 in.?

Ans. $F = 55.1$ lb

◄ **2/85** The resultant of a general force system may be expressed as a wrench along a unique line of action. For the force system of Prob. 2/83 determine the coordinates of the point *P* which is the intersection of the line of action of the wrench with the *x-y* plane. (*Hint:* The direction cosines of the resultant couple of the wrench must be plus or minus the direction cosines of the resultant force.)

Ans. $x = 2.25$ in., $y = 4.80$ in.

3 EQUILIBRIUM

13 Introduction. The subject of statics deals primarily with the description of the conditions of force that are both necessary and sufficient to maintain the equilibrium state of engineering structures. This chapter on equilibrium, therefore, constitutes the most central portion of statics and should be mastered thoroughly. It makes continuous use of the concepts and quantities developed in Chapter 2 and further provides a comprehensive introduction to the approach used in the solutions to countless problems in mechanics and in other engineering areas as well. This approach is basic to the successful mastery of statics, and the student is urged to read and study the following articles with special attention and effort.

14 Mechanical System Isolation. In Art. 9 the basic characteristics of force were described with primary attention focused on the vector properties of force. It was noted that forces are applied both by direct physical contact and by remote action and that forces may be either internal or external to the body under consideration. It was further observed that the application of external forces is accompanied by reactive forces and that both applied and reactive forces may be either concentrated or distributed. Additionally the principle of transmissibility was introduced which permits the treatment of force as a sliding vector insofar as its external effects on a rigid body are concerned. These characteristics of force will now be used in developing the analytical model of an isolated mechanical system to which the equations of equilibrium will then be applied.

A mechanical system is defined as a body or group of bodies that can be isolated from all other bodies. Such a system may be a single body or a combination of connected bodies. The bodies may be rigid or nonrigid. The system may also be a defined fluid mass, liquid or gas, or the system may be a combination of fluids and solids. In statics, attention is directed primarily to a description of the forces that act on rigid bodies at rest, although consideration is also given to the statics of fluids. Once a decision is reached about which body or combination of bodies is to be analyzed, then this body or combination considered as a single body is *isolated* from all surrounding bodies. This isolation is accomplished by means of the *free-body diagram,* which is a diagrammatic representation of the isolated body or combination of bodies showing all forces applied to it by other bodies that are considered to be removed. Only after such a diagram has been carefully drawn should the various force calculations be carried out.

Before an attempt is made to draw free-body diagrams, the mechanical characteristics of force application need further description. Figure 27 shows the common types of force application on mechanical systems. In each example the force exerted *on* the body to be isolated *by* the body to be removed is indicated. Newton's third law, which notes the existence of an equal and opposite reaction to every action, must be carefully observed.

In example 1 the action of a flexible cable, belt, rope, or chain on the body to which it is attached is depicted. Because of its flexibility a flexible cable is unable to offer any resistance to bending, shear, or compression and therefore exerts a tension force in a direction tangent to the cable at its point of attachment. The force exerted *by* the cable *on* the body to which it is attached is always *away* from the body. When the tension T is large compared with the weight of the cable, the cable may be assumed to form a straight line. When the cable weight is not negligible compared with its tension, the sag of the cable becomes important, and the tension in the cable changes direction and magnitude along its length. At its attachment it exerts a force tangent to itself.

When the smooth surfaces of two bodies are in contact, as in example 2, the force exerted by one on the other is *normal* to the tangency of the surfaces and is compressive. Although no actual surfaces are completely smooth, this assumption is justified in a practical sense in many instances.

When mating surfaces of contacting bodies are rough, example 3, the force of contact may not necessarily be normal to the tangent to the surfaces but may be resolved into a *tangential* or frictional component F and a *normal component N*.

Example 4 illustrates a number of forms of mechanical support which effectively eliminate tangential friction forces, and here the net reaction is normal to the supporting surface.

Example 5 shows the action of a smooth guide on the body it supports. Resistance parallel to the guide is absent.

Example 6 illustrates the action of a pin connection. Such a connection is able to support force in any direction normal to the axis of the pin. This action is normally represented in terms of two rectangular components. If the joint is free to turn about the pin, only the force R can be supported. If the joint is not free to turn, a resisting couple M may also be supported.

Example 7 shows the resultants of the rather complex distribution of force over the cross section of a slender bar or beam at a built-in or fixed support.

The three-dimensional counterpart of the pin connection is the ball-and-socket joint, example 8, which is capable of supporting all three components of a force. If the ball-and-socket is free to pivot, only the force $\mathbf{R}$ can be supported. If the joint is not free to pivot, as would be the case if it were welded at the point, then a couple $\mathbf{M}$ with all three components could also be supported.

One of the most common forces is that due to gravitational attraction, example 9. This force affects all elements of a body and is, therefore, distributed throughout it. The resultant of the gravitational forces on all ele-

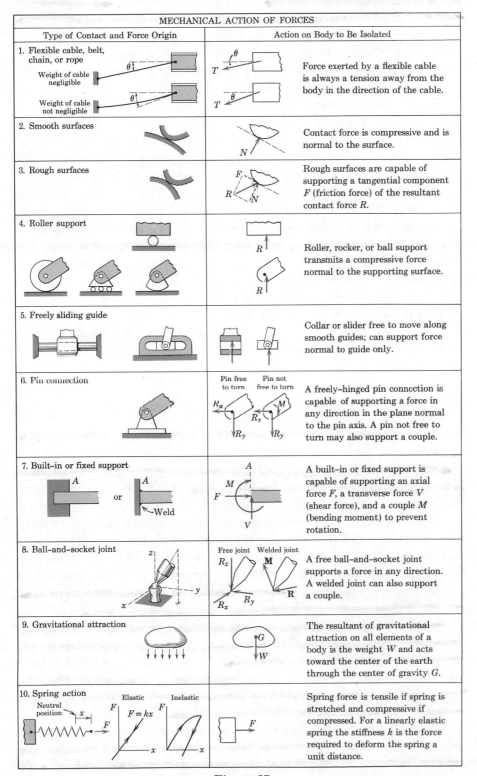

MECHANICAL ACTION OF FORCES	
Type of Contact and Force Origin	Action on Body to Be Isolated
1. Flexible cable, belt, chain, or rope Weight of cable negligible Weight of cable not negligible	Force exerted by a flexible cable is always a tension away from the body in the direction of the cable.
2. Smooth surfaces	Contact force is compressive and is normal to the surface.
3. Rough surfaces	Rough surfaces are capable of supporting a tangential component F (friction force) of the resultant contact force R.
4. Roller support	Roller, rocker, or ball support transmits a compressive force normal to the supporting surface.
5. Freely sliding guide	Collar or slider free to move along smooth guides; can support force normal to guide only.
6. Pin connection	A freely-hinged pin connection is capable of supporting a force in any direction in the plane normal to the pin axis. A pin not free to turn may also support a couple.
7. Built-in or fixed support	A built-in or fixed support is capable of supporting an axial force F, a transverse force V (shear force), and a couple M (bending moment) to prevent rotation.
8. Ball-and-socket joint	A free ball-and-socket joint supports a force in any direction. A welded joint can also support a couple.
9. Gravitational attraction	The resultant of gravitational attraction on all elements of a body is the weight W and acts toward the center of the earth through the center of gravity G.
10. Spring action	Spring force is tensile if spring is stretched and compressive if compressed. For a linearly elastic spring the stiffness k is the force required to deform the spring a unit distance.

Figure 27

ments is the *weight W* of the body, which passes through the center of gravity *G* and is directed toward the center of the earth for earthbound structures. The position of *G* is usually obvious from the geometry of the body, particularly where conditions of symmetry exist. When the position is not readily apparent, the location of *G* must be calculated or determined by experiment. Similar remarks apply to the remote action of magnetic and electric forces. These forces of remote action have the same overall effect on a rigid body as forces of equal magnitude and direction applied by direct external contact.

Example 10 illustrates the action of a linear elastic spring and of a nonlinear spring. The force exerted by a linear spring, tension or compression, is given by $F = kx$, where k is the stiffness of the spring and x is its deformation measured from the neutral or undeformed position. The linearity of the force-deformation relation describes equal force for equal deformation during loading or unloading of the spring. For the nonlinear spring the force for a given deformation is not the same for loading and unloading conditions.

The student is urged to study these ten conditions and to identify them in the problem work so that the correct free-body diagrams may be drawn. The representations in Fig. 27 are *not* free-body diagrams but are merely elements in the construction of free-body diagrams.

The full procedure for drawing a free-body diagram which accomplishes the isolation of the body or system under consideration will now be described. The following steps are involved.

Step 1. A clear decision is made which body or combination of bodies is to be isolated. The body chosen will involve one or more of the desired unknown quantities.

Step 2. The body or combination chosen is next isolated by a diagram that represents its complete external boundary. When the problem is three-dimensional, a single pictorial sketch of the exterior boundary may be drawn, or the alternative representation of two or three orthogonal views of the body may be preferable. The diagram of the external boundary should represent a *closed surface* in space which defines the isolation of the body from *all* other contacting or attracting bodies, which are considered removed. This step is often the most crucial of all. The student should always be certain that he has *completely isolated* the body before proceeding with the next step.

Step 3. All forces that act *on* the isolated body as applied *by* the removed contacting and attracting bodies are next represented in their proper positions on the diagram of the isolated body. A systematic traverse of the entire boundary will disclose all such forces. Weights, where appreciable, should be included. Known forces should be represented by vector arrows with their proper magnitude, direction, and sense indicated. Unknown forces should be represented by vector arrows with the unknown magnitude or direction indicated by symbol. If the sense of the vector is also unknown, it may be arbitrarily assumed. The calculations will reveal a positive quantity if the correct sense was assumed and a negative magnitude if the incorrect sense was assumed. It is necessary to be *consistent* with the assigned characteristics of unknown forces throughout all of the calculations.

Step 4. The choice of coordinate axes should be indicated directly on the diagram. Pertinent geometric dimensions may also be represented for convenience. Note, however, that the free-body diagram serves the purpose of focusing accurate attention on the action of the external forces, and therefore the diagram should not be cluttered with excessive extraneous information. Force arrows should be clearly distinguished from any other arrows which may appear so that confusion will not result.

When the foregoing four steps are completed, a correct free-body diagram will result, and the way will be clear for a straightforward and successful application of the principles of mechanics, both in statics and in dynamics.

Many students will be tempted to omit from the free-body diagram certain forces that may not appear at first glance to be needed in the calculations. To yield to this temptation is to invite serious error. It is only through *complete* isolation and a systematic representation of *all* external forces that a reliable accounting of the effects of all applied and reactive forces can be made. Very often a force that at first glance may not appear to influence a desired result does indeed have an influence. Hence the only safe procedure is to make certain that all forces whose magnitudes are not negligible appear on the free-body diagram.

The free-body diagram has been explained in some detail because of its great importance in mechanics. The free-body method ensures an accurate definition of a mechanical system and focuses attention on the exact meaning and application of the force laws of statics and dynamics. Indeed the free-body method is so important that the student is strongly urged to reread this section several times in conjunction with his study of the sample free-body diagrams shown in Fig. 28 and the sample problems which appear at the end of the next article.

Figure 28 gives four examples of mechanisms and structures together with their correct free-body diagrams. Dimensions and magnitudes are omitted for simplicity. In each case the entire system is treated as a single body, so that the internal forces are not shown. The characteristics of the various types of contact forces illustrated in Fig. 27 are included in the four examples as they apply.

In example 1 the truss is composed of structural elements that, taken all together, constitute a rigid framework. Thus the entire truss may be removed from its supporting foundation and treated as a single rigid body. In addition to the applied external load P, the free-body diagram must disclose the reactions on the truss at A and B. The rocker at B can support a vertical force only, and this force is transmitted to the structure at B (example 4 of Fig. 27). The pin connection at A (example 6 of Fig. 27) is capable of supplying both a horizontal and a vertical component of force to the truss. In this relatively simple example it is clear that the vertical component A_y must be directed down to prevent the truss from rotating clockwise about B. Also, the horizontal component A_x will be to the left to keep the truss from moving to the right under the influence of the horizontal component of P. If the total weight of the truss members is appreciable compared with P and the forces at A and B, then the weights of the members would have to be included on the free-body diagram as external forces.

In example 2 the boom AC is in equilibrium under the action of noncoplanar forces and is, therefore, a three-dimensional problem. Two orthogonal views of the isolated boom are shown, which are related by their common z-dimension. The tension T_B, applied by the cable to the boom at the attachment point B, is in the direction of the cable (example 1 of Fig. 27) and is disclosed in each of the two views. In the x-z view the vector arrow is the component or projection of T_B in that plane, and similarly the y-z component is shown in the y-z view. The z-component of T_B is common to both views. The tension T_C is similarly represented. The ball-and-socket joint at A (example 8 of Fig. 27) is capable of exerting all three components of supporting force on the boom as shown. The proper sense of R_z is clearly to the right to balance the action of the z-components of T_B and T_C. The positive sense for R_x cannot be determined by inspection and is arbitrarily assigned. The positive sense for R_y may also be assigned arbitrarily, although in this particular case it can be observed that R_y must be directed up in the y-z view to prevent the boom from rotating counterclockwise about point B. The free-body diagrams are completed by adding the weights w and W. A single three-dimensional free-body diagram of the boom could be drawn in place of the pair of two-dimensional diagrams if preferred.

In example 3 the weight W is shown acting through the center of gravity of the beam, which is assumed known (example 9 of Fig. 27). The force exerted by the corner A on the beam is normal to the smooth surface of the beam (example 2 of Fig. 27). If the contacting surfaces at the corner were not smooth, a tangential frictional component of force could be developed. In addition to the applied force P and couple M, there is the pin connection at B, which exerts both an x- and a y-component of force on the beam. The positive senses of these components are assigned arbitrarily.

In example 4 the free-body diagram of the entire isolated mechanism discloses three unknown quantities for equilibrium with the given loads W and P. Any one of many internal configurations for securing the cable leading from W would be possible without affecting the external response of the mechanism as a whole, and this fact is brought out by the free-body diagram.

The positive senses of R_x in example 2, B_x and B_y in example 3, and B_y in example 4 are assumed on the free-body diagrams, and the correctness of the assumptions would be proved or disproved according to whether the algebraic signs of the terms were plus or minus when the calculations were carried out in the actual problems.

The isolation of the mechanical system under consideration will be recognized as a crucial step in the formulation of the mathematical model. The student is again urged to devote special attention to this step. Before direct use is made of the free-body diagram in the application of the principles of force equilibrium in the next article, some initial practice with the drawing of free-body diagrams is helpful. To this end the problems that follow are designed to provide this practice for selected simple and direct examples.

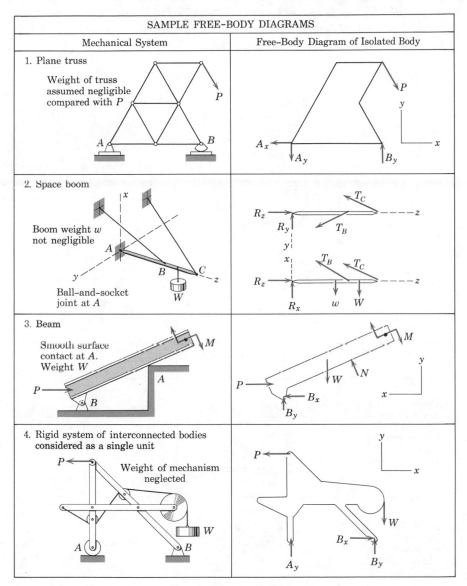

SAMPLE FREE-BODY DIAGRAMS

Mechanical System	Free-Body Diagram of Isolated Body
1. Plane truss — Weight of truss assumed negligible compared with P	
2. Space boom — Boom weight w not negligible. Ball-and-socket joint at A	
3. Beam — Smooth surface contact at A. Weight W	
4. Rigid system of interconnected bodies considered as a single unit — Weight of mechanism neglected	

Figure 28

Problems

3/1 In each of the five following examples, the body to be isolated is shown in the left-hand diagram, and an *incomplete* free-body diagram (FBD) of the isolated body is shown on the right. Add whatever forces are necessary in each case to form a complete free-body diagram. The weights of the bodies are negligible unless otherwise indicated. Dimensions and numerical values are omitted for simplicity.

	Body	Incomplete *FBD*
1. Bell crank supporting load W	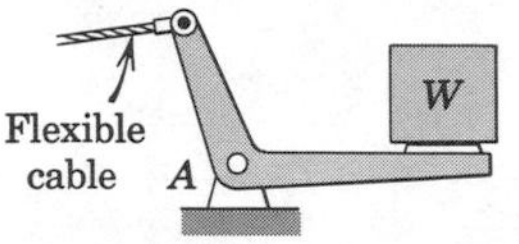	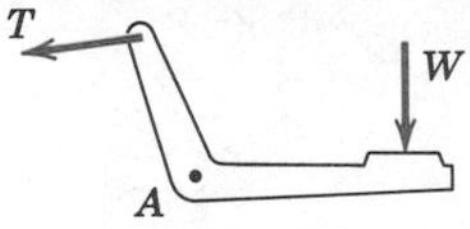
2. Control lever applying torque to shaft at O	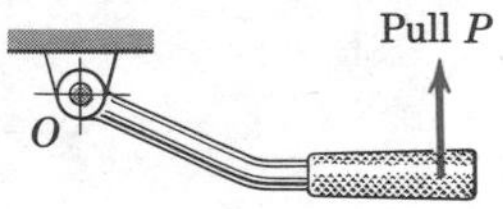	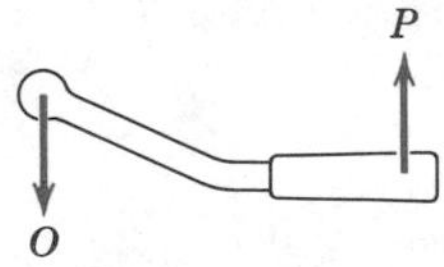
3. Mast of weight W loaded and supported by the two guying cables from the horizontal plane and ball–and–socket support at A	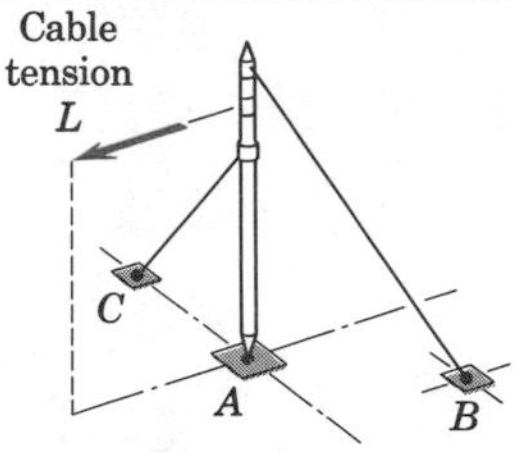	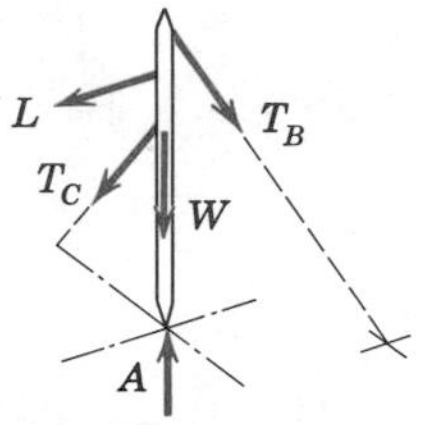
4. Uniform crate of weight W leaning against smooth vertical wall and supported on a rough horizontal surface	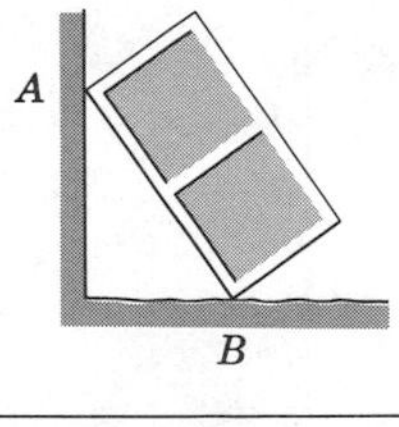	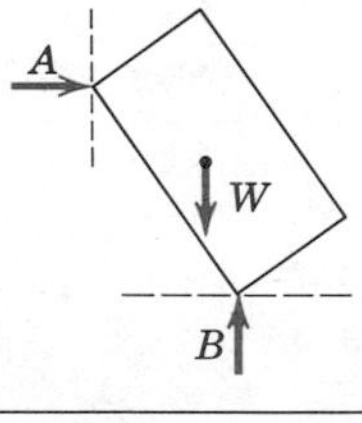
5. Loaded bracket supported by pin connection at A and fixed pin in smooth slot at B	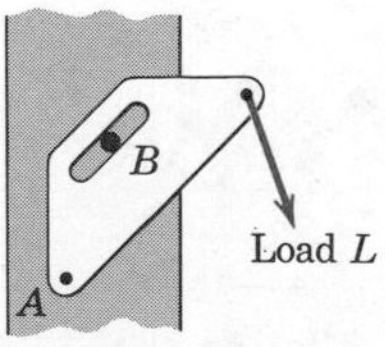	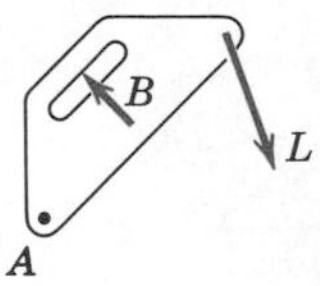

Problem 3/1

3/2 In each of the five following examples, the body to be isolated is shown in the left-hand diagram, and either a *wrong* or an *incomplete* free-body diagram (FBD) is shown on the right. Make whatever changes or additions are necessary in each case to form a correct and complete free-body diagram. The weights of the bodies are negligible unless otherwise indicated. Dimensions and numerical values are omitted for simplicity.

	Body	Wrong or Incomplete *FBD*
1. Lawn roller of weight W being pushed up incline θ	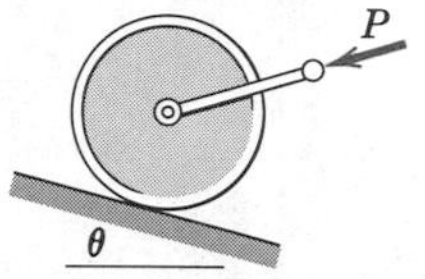	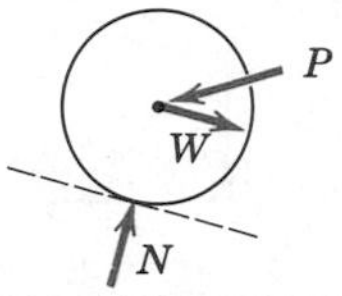
2. Pry bar lifting load W having smooth horizontal surface. Bar rests on horizontal rough surface	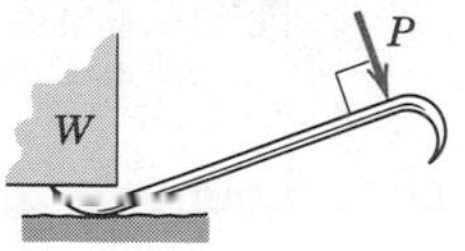	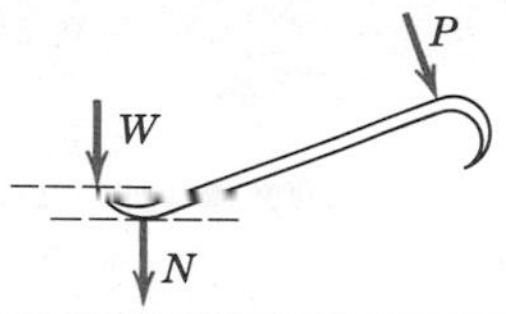
3. Uniform pole of weight W being hoisted into position by winch. Horizontal supporting surface notched to prevent slipping of pole	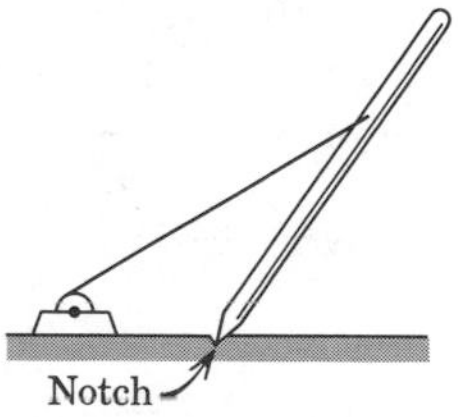	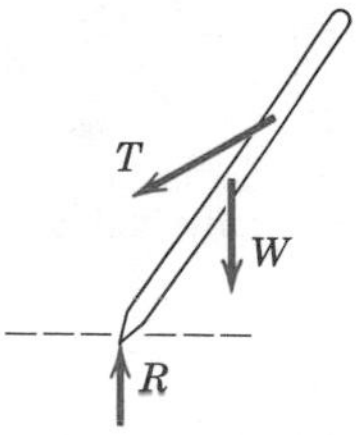
4. Supporting angle bracket for frame. Pin joints	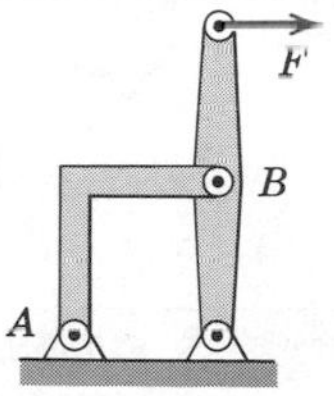	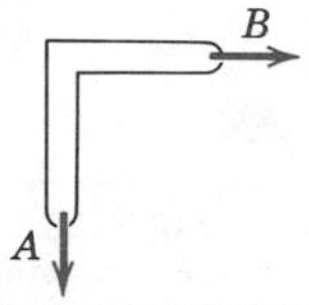
5. Bent rod welded to support at A and subjected to two forces and couple	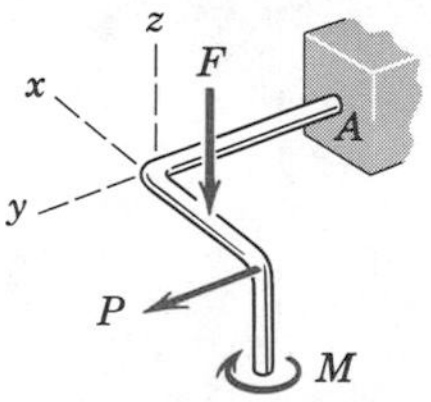	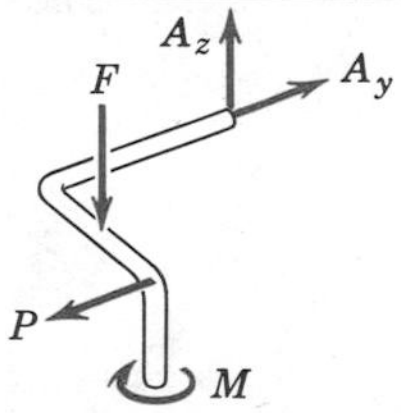

Problem 3/2

3/3 Draw a complete and correct free-body diagram of each of the bodies designated in the statements. The weights of the bodies are significant only if stated. All forces, known and unknown, should be labeled. (*Note:* The sense of some reaction components cannot always be determined without numerical calculation.)

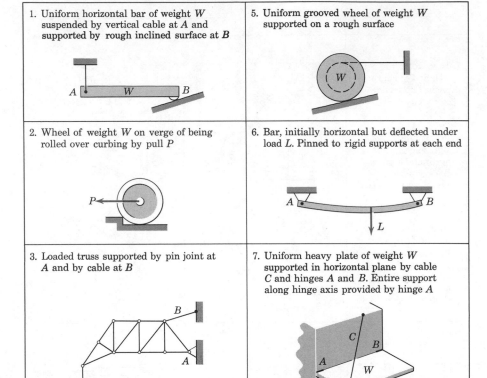

1. Uniform horizontal bar of weight W suspended by vertical cable at A and supported by rough inclined surface at B

5. Uniform grooved wheel of weight W supported on a rough surface

2. Wheel of weight W on verge of being rolled over curbing by pull P

6. Bar, initially horizontal but deflected under load L. Pinned to rigid supports at each end

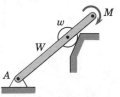

3. Loaded truss supported by pin joint at A and by cable at B

7. Uniform heavy plate of weight W supported in horizontal plane by cable C and hinges A and B. Entire support along hinge axis provided by hinge A

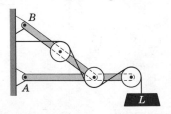

4. Uniform bar of weight W and roller of weight w taken together. Subjected to couple M and supported as shown

8. Entire frame, pulleys, and contacting cable to be isolated as a single unit

Problem 3/3

15 Equilibrium Conditions. The concept of equilibrium is derived from the condition in which the forces that act on a body are in balance. Stated another way, equilibrium is the condition for which the resultant of all forces acting on a given body is zero. In Art. 12 it was shown that the most general force system that can act on a body can be expressed in terms of a resultant force **R** and a resultant couple **M**. Hence in order for the forces to be in complete balance and the body to be in equilibrium, the relations **R** = **0** and **M** = **0** must be satisfied. Physically these vector equations mean that for a body in equilibrium there is as much force acting on it in one direction as in the opposite direction and that there is as much twist or moment applied to it about any axis in one sense as in the opposite sense. Since each of the terms in the two relations represents, respectively, a vector sum of all external forces that act on the body in question and a vector sum of all corresponding couples, it follows that the requirements for the equilibrium of any body may be written as

▶ $$\Sigma\mathbf{F} = \mathbf{0} \qquad \Sigma\mathbf{M} = \mathbf{0} \tag{13}$$

These basic vector equations also express the fact that the space polygon of forces and the space polygon of corresponding couple vectors must both close.

Equations 13 are the necessary and sufficient conditions for complete equilibrium. They are necessary conditions because, if not satisfied, there can be no force or moment balance. They are sufficient conditions since once satisfied there can be no unbalance, and equilibrium is assured.

The equations relating force and acceleration for rigid-body motion are developed in *Dynamics* from Newton's second law of motion. These equations show that when the resultant of a force system is expressed by a single force **R** acting through the center of mass, and a corresponding couple **M**, the linear acceleration of the center of mass of the body is proportional to **R**, and the time rate of change of angular momentum about the center of mass is proportional to **M**. Hence the first of Eqs. 13 applies not only to bodies at rest, but also to any body whose mass center moves with a constant velocity (no acceleration). Likewise the second of Eqs. 13 describes not only a body at rest but also one rotating with a constant angular velocity. Although the subject of statics, as its name implies, concerns bodies which are static, i.e., which have no motion or change in motion, the case of motion with constant linear velocity of the mass center and constant angular velocity must be included when using the statical equations of equilibrium.

For a body to be in complete equilibrium it is necessary for both of Eqs. 13 to hold. These equations, however, represent two independent conditions, and either may hold without the other. If **R** = **0** and **M** ≠ **0**, the mass center of the body either is at rest or is moving with a constant velocity, and the body is undergoing a change in its angular momentum. Under these conditions it may be said that the body is in equilibrium only insofar as its linear motion is concerned. On the other hand if **R** ≠ **0** and **M** = **0**, the mass center of the body has a linear acceleration but there is no change in angular momentum. Such a body may be considered to be in rotational equilibrium. If the linear acceleration of a body is in the x-direction, then the body may be considered to be in equilibrium in the y- and z-directions, since it has no acceleration in these directions. The term equilibrium, however, is most commonly used to describe a body which is completely at rest.

The two vector expressions of Eqs. 13 are equivalent to the six scalar equations

$$
\begin{aligned}
\Sigma F_x = 0 \qquad \Sigma M_x = 0 \\
\Sigma F_y = 0 \qquad \Sigma M_y = 0 \\
\Sigma F_z = 0 \qquad \Sigma M_z = 0
\end{aligned}
\tag{14}
$$

which merely express the fact that complete equilibrium requires a zero force sum in all three directions and a zero moment sum about any three mutually perpendicular axes. Equations 14 are independent. Any one of them may hold irrespective of the others. Reliable application of the equations of equilibrium is ensured by the free-body diagram from which the necessary summations of forces and moments are obtained.

Applications of Eqs. 13 and 14 fall naturally into a number of categories that are easily recognized. These categories of force systems acting on bodies in equilibrium are summarized in Fig. 29 and explained further as follows:

Case 1, equilibrium under collinear forces, clearly requires only the one force equation in the direction of the forces (x-direction), since all other equations are automatically satisfied.

Case 2, equilibrium of forces that lie in a plane (x-y plane) and are concurrent at a point O, requires the two force equations only, since the moment sum about O, i.e. about a z-axis through O, is necessarily zero.

Case 3, equilibrium of forces in space that are concurrent at a point O, requires all three force equations but no moment equations since their moment about any axis through O and about every other axis is zero.

Case 4, equilibrium of forces in space that are concurrent with a line, requires all equations except the moment equation about that line, which is automatically satisfied.

Case 5, equilibrium of parallel forces in a plane, requires the one force equation in the direction of the forces (x-direction) and one moment equation about an axis (z-axis) normal to the plane of the forces.

Case 6, equilibrium of parallel forces in space, requires only the one force equation in the direction of the forces (x-direction) but two moment equations about axes (y and z) that are normal to the direction of the forces.

Case 7, equilibrium of a general system of forces in a plane (x-y), requires the two force equations in the plane and one moment equation about an axis (z-axis) normal to the plane.

Case 8, equilibrium of a general system of forces in space, requires all three force equations and all three moment equations.

CATEGORIES OF FORCE SYSTEMS IN EQUILIBRIUM			
Category	Dimensions	Figure	Independent Equations
1. Collinear	1 D	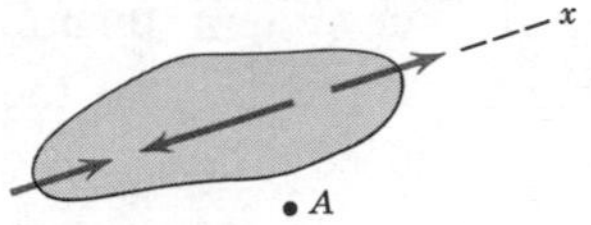	$\Sigma F_x = 0$
2. Concurrent at a point	2 D	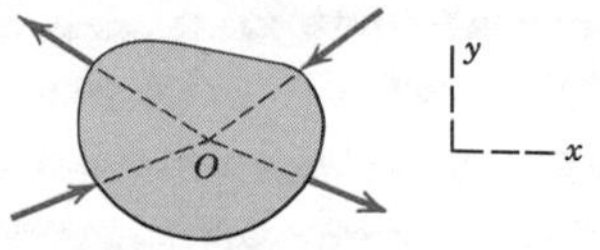	$\Sigma F_x = 0$ $\Sigma F_y = 0$
3. Concurrent at a point	3 D	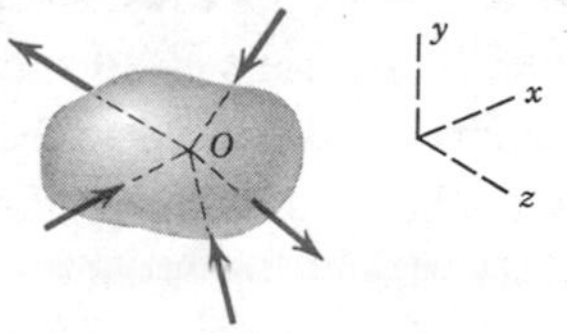	$\Sigma F_x = 0$ $\Sigma F_y = 0$ $\Sigma F_z = 0$
4. Concurrent with a line	3 D	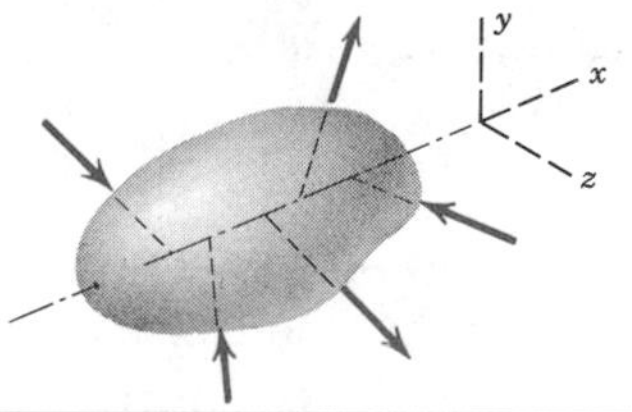	$\Sigma F_x = 0 \qquad \Sigma M_y = 0$ $\Sigma F_y = 0 \qquad \Sigma M_z = 0$ $\Sigma F_z = 0$
5. Parallel	2 D	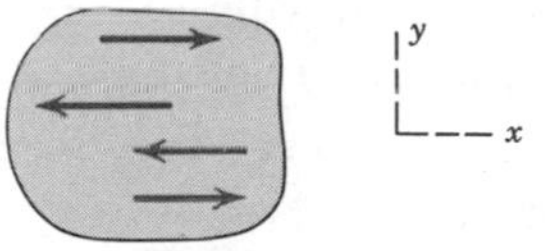	$\Sigma F_x = 0 \qquad \Sigma M_z = 0$
6. Parallel	3 D	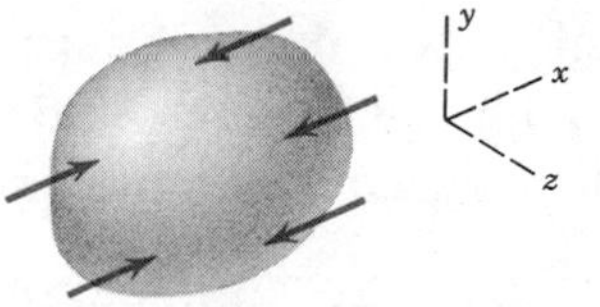	$\Sigma F_x = 0 \qquad \Sigma M_y = 0$ $\qquad\qquad\quad \Sigma M_z = 0$
7. General	2 D	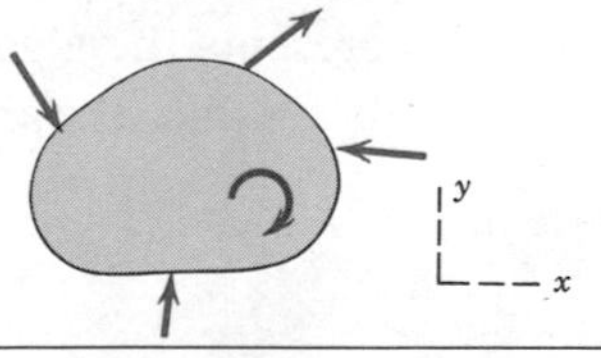	$\Sigma F_x = 0 \qquad \Sigma M_z = 0$ $\Sigma F_y = 0$
8. General	3 D	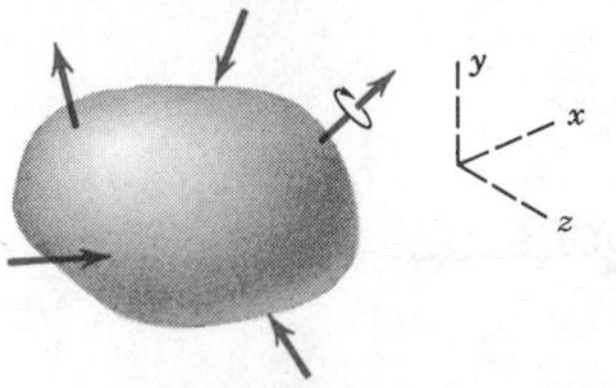	$\Sigma F_x = 0 \qquad \Sigma M_x = 0$ $\Sigma F_y = 0 \qquad \Sigma M_y = 0$ $\Sigma F_z = 0 \qquad \Sigma M_z = 0$

Figure 29

In each of the cases cited, other than the most general case 8, those of the six equilibrium equations that are absent are automatically satisfied by virtue of the nature of the particular force system. It should be unnecessary to commit the results shown in Fig. 29 to memory, as they are easily deduced from an understanding of the six scalar equations of equilibrium and their applicability as illustrated in Fig. 29.

There are two equilibrium situations that occur frequently to which the student should be alerted. The first situation is the equilibrium of a body under the action of two forces only. Two examples are shown in Fig. 30*a*, and it is clear that for such a *two-force member* the forces must be equal, opposite, and collinear. The shape of the member should not obscure this simple requirement. In the illustrations cited the weights of the members are considered negligible compared with the applied forces.

The second situation is the equilibrium of a body under the action of three forces, Fig. 30*b*. Since the closed space triangle of three forces must lie in a plane, the problem is necessarily two-dimensional. Furthermore, the lines of action of the three forces must be concurrent. If they were not concurrent, then one of the forces would exert a resultant moment about the point of concurrency of the other two, which would violate the requirement of zero moment about every point. The only exception occurs when the three forces are parallel. In this case the point of concurrency can be considered to be at infinity. The principle of the concurrency of three forces in equilibrium is of considerable use in carrying out a graphical solution of the force equations. In this connection a body in equilibrium under the action of more than three forces may often be reduced to a *three-force member* by a combination of two or more of the known forces.

By virtue of the frequency with which the equilibrium analysis of coplanar forces occurs, it is helpful to extend the discussion of this category beyond that given with case 7 of Fig. 29. For the equilibrium of a coplanar force system in, say, the *x-y* plane the equilibrium requirements may be written as

$$\Sigma F_x = 0 \qquad \Sigma F_y = 0 \qquad \Sigma M_O = 0 \qquad (15)$$

where ΣM_O replaces ΣM_z and represents the algebraic sum of the moments of all forces acting on the body about an axis parallel to the *z*-direction and

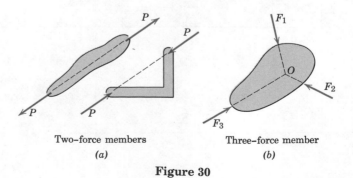

Two–force members   Three–force member

(a)   (b)

Figure 30

passing through any point O on the body or off the body but in the x-y plane. More frequently this summation is referred to as the sum of the moments about point O.

There are two additional ways of expressing the necessary conditions for the equilibrium of forces in two dimensions. For the body shown in Fig. 31*a*, if $\Sigma M_A = 0$, then the resultant R, if it still exists, cannot be a couple but must be a force R passing through A. If now the equation $\Sigma F_x = 0$ holds, where the x-direction is perfectly arbitrary, it follows from Fig. 31*b* that the resultant force R, if it still exists, not only must pass through A, but also must be perpendicular to the x-direction as shown. Now, if $\Sigma M_B = 0$, where B is any point such that the line AB is not perpendicular to the x-direction, it is clear that R must be zero, and hence the body is in equilibrium. Therefore an alternative set of equilibrium equations is

$$\Sigma F_x = 0 \qquad \Sigma M_A = 0 \qquad \Sigma M_B = 0$$

where the two points A and B must not lie on a line perpendicular to the x-direction.

A third formulation of the conditions of equilibrium may be made for a coplanar force system. Again, if $\Sigma M_A = 0$ for any body such as shown in Fig. 31*c*, the resultant, if it exists, must be a force R through A. In addition if $\Sigma M_B = 0$, the resultant, if one still exists, must pass through B as shown in Fig. 31*d*. Such a force cannot exist, however, if $\Sigma M_C = 0$, where C is not collinear with A and B. Hence the equations of equilibrium may be written

$$\Sigma M_A = 0 \qquad \Sigma M_B = 0 \qquad \Sigma M_C = 0$$

where A, B, and C are any three points not on the same straight line.

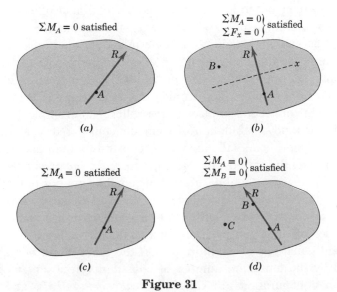

Figure 31

A brief discussion of alternative sets of equilibrium equations for other classes of force systems is presented at the end of the article in the section marked for greater difficulty.

Whereas the equilibrium equations developed in this article are both necessary and sufficient conditions to establish the equilibrium of a body, they do not necessarily provide sufficient information to calculate all of the unknown forces that may act on a body in equilibrium. The question of sufficiency lies in the characteristics of the constraints to possible movement of the body provided by its supports. By constraint is meant the prevention of movement. In example 4 of Fig. 27 the roller, ball, and rocker provide constraint normal to the surface of contact but none tangent to the surface. Hence a tangential force cannot be supported. For the collar and slider of example 5 constraint is possible only normal to the guide. In example 6 the fixed-pin connection provides constraint in both directions but offers no resistance to rotation about the pin unless the pin is not free to turn. The fixed support of example 7, however, offers constraint to rotation as well as constraint to lateral movement. In the free ball-and-socket joint of example 8, constraint to motion in all three directions is provided, although no resistance to rotation about the center of the ball exists. For the welded connection, on the other hand, constraint to rotation also exists.

If the rocker that supports the truss of example 1 in Fig. 28 were replaced by a pin joint, as at A, there would be one additional constraint beyond that required to support an equilibrium configuration without collapse. The three conditions of equilibrium, Eqs. 15, would not be sufficient to determine all four unknowns, since A_x and B_x could not be separated. These two components of force would be dependent on the deformation of the members of the truss as influenced by their corresponding stiffness properties. The horizontal reactions would also be dependent on any initial deformation required to fit the dimensions of the structure to those of the foundation between A and B. Again referring to Fig. 28, if the connection at A of example 2 were welded, two additional unknown moments at A about the x- and y-axes would be generated that could not be determined by the available equations of equilibrium without knowing the internal stiffness characteristics of the boom. Similar examples of more constraint than is necessary to ensure a stable equilibrium configuration would be a welded pin at B in example 3 and a fixed-pin connection in place of the roller A in example 4.

A body, or rigid combination of elements considered as a single body, that possesses more external supports or constraints than are necessary to maintain an equilibrium position is called *statically indeterminate.* Supports that can be removed without destroying the equilibrium position of the body are said to be *redundant.* The number of redundant supporting elements present corresponds to the degree of statical indeterminacy and equals the total number of unknown external forces minus the number of available independent equations of equilibrium. On the other hand, bodies that are supported by the minimum number of constraints necessary to ensure an equilibrium configuration are called *statically determinate,* and for such bodies the equilibrium equations are sufficient to determine the unknown external forces.

The problems on equilibrium included in this article and throughout *Statics* are generally restricted to statically determinate bodies where the constraints are just sufficient to ensure a stable position and where the unknown supporting forces can be completely determined by the available independent equations of equilibrium. The student is alerted at this point by this brief discussion, however, to the fact that he should be aware of the nature of the constraints before he attempts to solve an equilibrium problem. A body will be recognized as statically indeterminate when there are more unknown external reactions than there are available independent equilibrium equations for the force system involved. It is always well to count the number of unknown forces on a given body and to be certain that an equal number of independent equations may be written; otherwise effort may be wasted in attempting an impossible solution with the aid of the equilibrium equations only.

The question of the adequacy of constraint is developed in more detail in Art. 16. A full development of the analysis of statically indeterminate bodies or structures is, however, beyond the purpose of this book on statics and, therefore, only occasional reference is made to such problems.

In applying the equilibrium equations one of the most useful observations is an expeditious choice of reference axes, particularly the moment axis. Generally, the best choice of a moment axis is one through which as many unknown forces pass as possible. Simultaneous solutions of equilibrium equations are frequently necessary but can be minimized by a careful choice of axis.

The sample problems at the end of the article illustrate the application of free-body diagrams and the equations of equilibrium to typical statics problems. The student is urged to study these examples in detail before proceeding with the problem work. It is strongly recommended that each major application of a principle of mechanics be preceded by a symbolic statement of the principle or governing equation which is involved. In the sample problems these statements are set forth in brackets to the left of the calculations and serve as a reminder of the justification for each major step. Also, the recommendations set forth in Art. 7 will be valuable to the student particularly at this stage as he begins to form his habits of approach to the solution of engineering problems.

▼   ▼   ▼   ▼   ▼

Alternative equilibrium equations were developed for the general two-dimensional force system of case 7 in Fig. 29. Alternative equilibrium equations may also be written for the other force systems listed in Fig. 29, and several examples will be cited here without a full discussion. The reader may wish to formulate the procedures and proof for himself in more detail.

For the collinear forces of case 1, it should be clear that a moment equation about any point A not on the line of action of the forces can be used in place of the single force equation.

For the plane concurrent forces in case 2, moment equations about two different points that are not collinear with the point of concurrency O are alternative expressions to the two force equations.

For the equilibrium of concurrent forces in space, case 3, moment equations may be written about any three nonparallel axes chosen so that a line from the concurrency point O will not intersect all three axes.

For the equilibrium of a space force system concurrent with a line, case 4, the equations $\Sigma F_y = 0$ and $\Sigma F_z = 0$ of the set of five equations cited may be replaced by two moment equations about axes parallel to the x-axis, $\Sigma M_{A_x} = 0$ and $\Sigma M_{B_x} = 0$, and passing through two points A and B that are not on a line that intersects the x-axis.

For the equilibrium of parallel forces in a plane, case 5, moment equations about two points that do not lie on a line parallel to the direction of the forces can replace the force and moment equations.

In case 6 for the equilibrium of parallel forces in space, an alternative set of independent equations consists of two moment equations about axes parallel to the y-axis and passing through two points A and B that do not lie in a plane parallel to the x-y plane, and a moment equation about the z-axis, or $\Sigma M_{A_y} = 0$, $\Sigma M_{B_y} = 0$, and $\Sigma M_z = 0$.

In case 8 for the general force system in space, alternative sets of equations are somewhat more complex than those in the previous categories. One such set will be outlined with the aid of Fig. 32, which shows three noncollinear points A, B, and C located in, off, or on the body to which the force system is applied. For simplicity let the origin of coordinates be at A with the x-axis along AB and with the x-y plane containing C. If a resultant did exist, it could be represented in terms of a resultant force $\mathbf{R}$ at A and a corresponding couple $\mathbf{M}_A$. If the equations $\Sigma M_{A_x} = 0$, $\Sigma M_{A_y} = 0$, and $\Sigma M_{A_z} = 0$ hold, they would eliminate the possible existence of $\mathbf{M}_A$ but would not restrict $\mathbf{R}$ at this point. Adding the requirements $\Sigma M_{B_{y'}} = 0$ and $\Sigma M_{B_{z'}} = 0$ would, however, eliminate all possibilities for the existence of $\mathbf{R}$ except for the one position along AB for its line of action. The equation $\Sigma M_{C_{z''}} = 0$

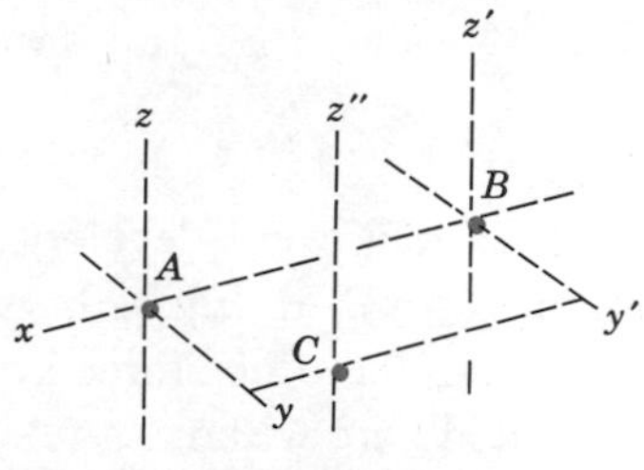

Figure 32

would eliminate this possibility and ensure that $\mathbf{R} = \mathbf{0}$. Thus the alternative set of equations would be

$$\Sigma M_{A_x} = 0 \qquad \Sigma M_{A_y} = 0 \qquad \Sigma M_{A_z} = 0$$
$$\Sigma M_{B_{y'}} = 0 \qquad \Sigma M_{B_{z'}} = 0 \qquad \Sigma M_{C_{z''}} = 0$$

It should also be mentioned that the equations of equilibrium hold for a nonorthogonal system of coordinates. Any three coordinate axes that are nonparallel may be used for the force and moment summations. For geometric simplicity, however, it is generally more convenient to use an orthogonal coordinate system.

Sample Problems

3/4 Determine the tension T in the supporting cable and the force on the pin at A for the jib crane shown. The beam AB is a standard 18-in. I-beam weighing 65 lb per foot of length.

Algebraic Solution. It is clear that the system is symmetrical about the vertical plane O-O as seen from the end view, so the problem may be analyzed as the equilibrium of a coplanar force system. The free-body diagram of the beam is shown in the *a*-part of the figure with the pin reaction at A separated in terms of its two rectangular components. The weight of the beam is $65(15) = 975$ lb and acts through its center. In applying the moment equation about A it is simpler to consider the moments of the x- and y-components of T than it is to compute the perpendicular distance from T to A. Hence

$$[\Sigma M_A = 0] \qquad (T\cos 25°)\tfrac{9}{12} + (T\sin 25°)(15 - \tfrac{4}{12})$$
$$- 2000(15 - 4.5 - \tfrac{4}{12}) - 975(7.5 - \tfrac{4}{12}) = 0$$

from which

$$T = 3980 \text{ lb} \qquad\qquad Ans.$$

Equating the sum of forces in the x- and y-directions to zero gives

$[\Sigma F_x = 0] \qquad A_x - 3980 \cos 25° = 0, \qquad A_x = 3610$ lb

$[\Sigma F_y = 0] \qquad A_y + 3980 \sin 25° - 975 - 2000 = 0, \qquad A_y = 1293$ lb

$[A = \sqrt{A_x^2 + A_y^2}] \qquad A = \sqrt{(3610)^2 + (1293)^2}, \qquad A = 3830$ lb $\qquad Ans.$

Graphical Solution. The principle that three forces in equilibrium must be concurrent is utilized for a graphical solution by combining the two known vertical forces of 975 lb and 2000 lb into a single 2975-lb force located as shown on the modified free-body diagram of the beam in the *b*-part of the figure. The position of this resultant load may be determined graphically or algebraically. The intersection of the 2975-lb force with the line of action of the unknown tension T defines the point of concurrency O through which the pin reaction A must pass. The unknown magnitudes of T and A may now be found by constructing the closed equilibrium polygon of forces. After the known vertical load is laid off to a convenient scale, as shown in the lower part of the figure, a line representing the given direction of the tension T is drawn through the tip of the 2975-lb vector. Likewise a line representing the direction of the pin reaction A, determined from the concurrency established with the free-body diagram, is drawn through the tail of the 2975-lb vector. The intersection of the lines representing vectors T and A establishes the magnitudes of both T and A which are necessary to make the vector sum of the forces equal to zero. These magnitudes may be scaled directly from the diagram. The *x*- and *y*-components of *A* may be constructed on the force polygon if desired.

The student should recognize that the polygon is begun with the known forces and that the order of addition of the remaining two vectors is immaterial.

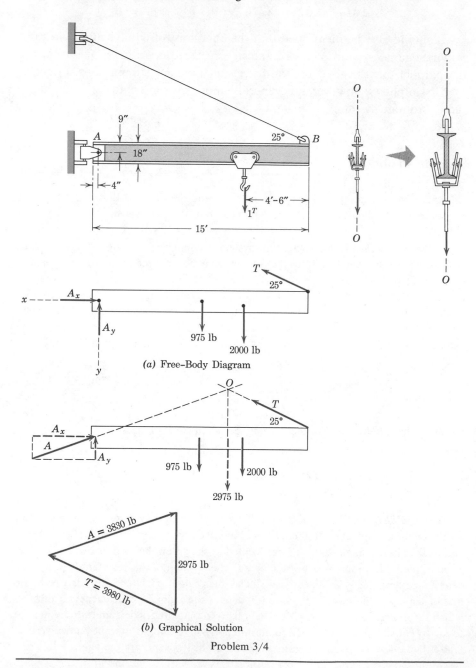

(a) Free–Body Diagram

(b) Graphical Solution

Problem 3/4

3/5 A 50-lb force is applied to the handle of the hoist in the direction shown. The bearing *A* supports the thrust (force in the direction of the shaft axis) while bearing *B* supports only radial load (load normal to the shaft axis). Determine the weight *W* which can be supported and the total radial force exerted on the shaft by each bearing.

Solution. The system is clearly three-dimensional with no lines or planes of symmetry, and therefore the problem must be analyzed as a general space system of forces. A scalar solution is used here to illustrate this approach, although a solution using vector notation would be equally satisfactory (see Sample Prob. 3/6). The free-body diagram of the shaft, lever, and drum considered a single body could be shown by a space view if desired but is represented here by its three orthogonal projections. The 50-lb applied force is resolved into its three components, and each of the three views shows two of these components. The correct directions of A_x and B_x may be seen by inspection by observing that the line of action of the resultant of the two 17.7-lb forces passes between A and B. The correct sense of the forces A_y and B_y cannot be determined until the magnitudes of the moments are obtained, so they may be arbitrarily assigned. The x-y projection of the bearing forces is shown in terms of the sums of the unknown x- and y-components. The addition of A_z and W completes the free-body diagrams. It should be noted that the three views represent three two-dimensional problems related by the corresponding components of the forces.

From the x-y projection

$$[\Sigma M_O = 0] \qquad\qquad 4W - 10(43.3) = 0, \qquad W = 108.3 \text{ lb} \qquad\qquad Ans.$$

From the x-z projection

$$[\Sigma M_A = 0] \qquad\qquad 6B_x + 7(17.7) - 10(17.7) = 0, \qquad B_x = 8.85 \text{ lb}$$

$$[\Sigma F_x = 0] \qquad\qquad A_x + 8.85 - 17.7 = 0, \qquad A_x = 8.85 \text{ lb}$$

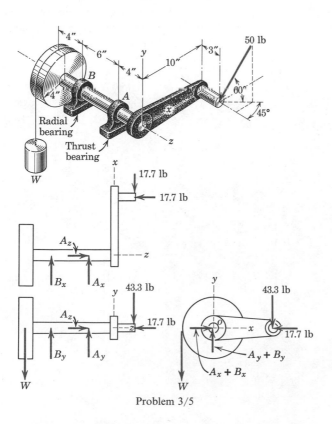

Problem 3/5

The *y-z* view gives

$[\Sigma M_A = 0]$ $\qquad$ $6B_y + 7(43.3) - 10(108.3) = 0$, $\qquad B_y = 130$ lb

$[\Sigma F_y = 0]$ $\qquad$ $A_y + 130 - 43.3 - 108.3 = 0$, $\qquad A_y = 21.6$ lb

$[\Sigma F_z = 0]$ $\qquad$ $A_z = 17.7$ lb

The total radial forces on the bearings become

$[A_r = \sqrt{A_x^2 + A_y^2}]$ $\qquad$ $A_r = \sqrt{(8.85)^2 + (21.6)^2} = 23.3$ lb $\qquad$ *Ans.*

$[B = \sqrt{B_z^2 + B_y^2}]$ $\qquad$ $B = \sqrt{(8.85)^2 + (130)^2} = 130.2$ lb $\qquad$ *Ans.*

3/6 The welded tubular frame is secured to the horizontal *x-y* plane by a ball-and-socket joint at *A* and receives support from the loose-fitting ring at *B*. Under the action of the 500-lb load, rotation about the line *AB* is prevented by the cable *CD*, and the frame is stable in the position shown. Neglect the weight of the frame compared with the applied load and determine the tension *T* in the cable, the reactions at the ring, and the reaction components at *A*.

Solution. The system is clearly three-dimensional with no lines or planes of symmetry, and therefore the problem must be analyzed as a general space system of forces. A solution using vector methods is employed here to illustrate this more general approach, in contrast to the scalar approach used with Sample Prob. 3/5. The free-body diagram is drawn, where the ring reaction is shown in terms of its two components. All unknowns except **T** may be eliminated by a moment sum about the line *AB*. The direction of *AB* is specified by the unit vector $\mathbf{n} = \frac{1}{5}(3\mathbf{j} + 4\mathbf{k})$. The moment of **T** about *AB* is the component in the direction of *AB* of the vector moment about the point *A*. Thus the moment of **T** about *AB* is given by $\mathbf{r}_1 \times \mathbf{T} \cdot \mathbf{n}$. Similarly the moment of the applied load **F** about *AB* is $\mathbf{r}_2 \times \mathbf{F} \cdot \mathbf{n}$. The vector expressions for **T**, **F**, $\mathbf{r}_1$, and $\mathbf{r}_2$ are

$$\mathbf{T} = \frac{T}{\sqrt{185}}(4\mathbf{i} + 5\mathbf{j} - 12\mathbf{k}), \qquad \mathbf{F} = 500\mathbf{j} \text{ lb}$$

$$\mathbf{r}_1 = -2\mathbf{i} + 5\mathbf{j} \text{ ft}, \qquad \mathbf{r}_2 = 5\mathbf{i} + 12\mathbf{k} \text{ ft}$$

where $\overline{CD} = \sqrt{185}$ ft.

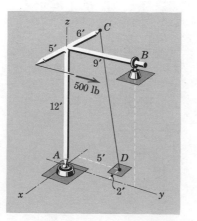
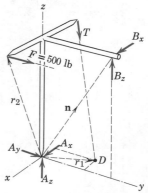

Problem 3/6

The moment equation now becomes

$$[\Sigma M_{AB} = 0] \qquad (-2\mathbf{i} + 5\mathbf{j}) \times \frac{T}{\sqrt{185}} (4\mathbf{i} + 5\mathbf{j} - 12\mathbf{k}) \cdot \tfrac{1}{5}(3\mathbf{j} + 4\mathbf{k})$$

$$+ (5\mathbf{i} + 12\mathbf{k}) \times (500\mathbf{j}) \cdot \tfrac{1}{5}(3\mathbf{j} + 4\mathbf{k}) = 0$$

Completion of the vector operations gives

$$-\frac{192T}{\sqrt{185}} + 10{,}000 = 0, \qquad T = 708 \text{ lb} \qquad\qquad Ans.$$

The components of **T** are, therefore,

$$T_x = 208 \text{ lb} \qquad T_y = 260 \text{ lb} \qquad T_z = -625 \text{ lb}$$

The remaining unknowns are easily found by moment and force summations as follows.

$[\Sigma M_z = 0]$	$500(5) - 9B_x - 260(6) = 0$	$B_x = 104.4$ lb	*Ans.*
$[\Sigma M_x = 0]$	$9B_z - 500(12) - 260(12) = 0$	$B_z = 1013$ lb	*Ans.*
$[\Sigma F_x = 0]$	$A_x + 104.4 + 208 = 0$	$A_x = -312$ lb	*Ans.*
$[\Sigma F_y = 0]$	$A_y + 500 + 260 = 0$	$A_y = -760$ lb	*Ans.*
$[\Sigma F_z = 0]$	$A_z + 1013 - 625 = 0$	$A_z = -388$ lb	*Ans.*

The negative signs with the A-components indicate that they are in the opposite direction to the chosen positive directions shown.

In setting up the expression for the moment of **T** about A, the vector from A to C could have been used instead of $\mathbf{r}_1$. The advantage of using vector notation in this problem is the freedom to take moments directly about any axis. In this problem this freedom permits the choice of an axis that eliminates five of the unknowns.

Problems

3/7 The homogeneous smooth ball weighs 50 lb and rests on the 30-deg incline at A and bears against the vertical surface at B. Calculate the contact forces at A and B.

Ans. $A = 57.7$ lb, $B = 28.9$ lb

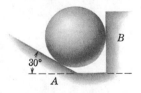

Problem 3/7

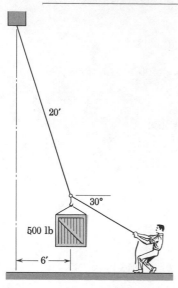

Problem 3/8

3/8 What pull *P* on the rope must the man exert in order to suspend the 500-lb crate in the deflected position shown?

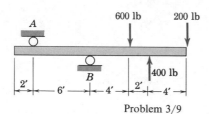

Problem 3/9

3/9 The uniform 18-ft beam weighs 500 lb and is loaded in the vertical plane by the parallel forces shown. Calculate the reactions at the roller supports *A* and *B*.

Ans. $A = 417$ lb, $B = 1317$ lb

3/10 If the screw *B* of the wood clamp is tightened so that the two blocks are under a compression of 100 lb, determine the force in screw *A*.

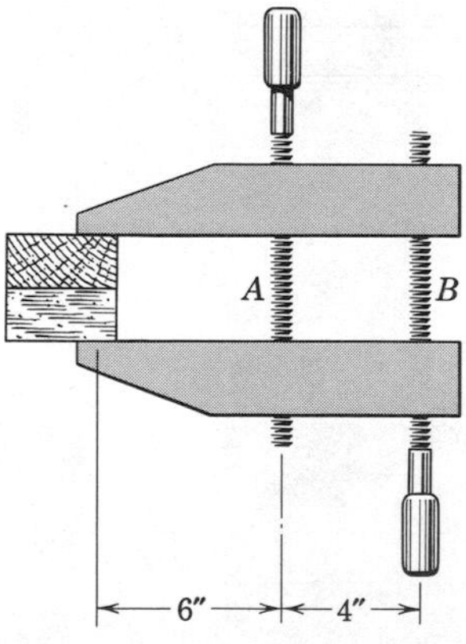

Problem 3/10

3/11 A small postal scale consists of the plate suspended at A from a link which acts as the pointer. The scale tilts according to the weight hung from the hook at B. In the unloaded position shown, AB is horizontal. The scale weighs 4 oz exclusive of pointer, and its center of gravity is $2\frac{1}{4}$ in. from A. The weight of the hook is negligible. Determine the angle θ through which the scale tilts when a 6-oz weight is attached to the hook.

Ans. $\theta = 53°8'$

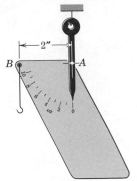

Problem 3/11

3/12 Determine the tension T in the indicated cable for the pulley combination that supports the weight W

Problem 3/12

3/13 Determine the force P that the 180-lb worker must exert on the rope in order to support himself in the bosun's chair. What force R does the man exert on the seat of the chair?

Ans. $P = 36$ lb, $R = 144$ lb

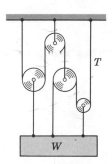

Problem 3/13

3/14 As a check on the balance of the aircraft each of the three wheels is run onto a scale, and the force readings are $A = 5060$ lb, $B = 5130$ lb, $C = 790$ lb. Calculate the x-y coordinates of the center of gravity of the plane.

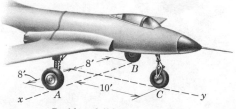

Problem 3/14

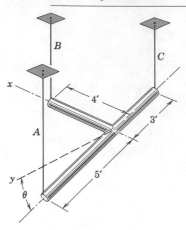

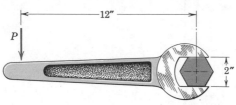

Problem 3/15

3/15 The two rods are welded together at right angles to one another and supported at their ends by vertical wires A, B, and C with the longer rod making an angle θ with the horizontal x-y plane. If the rod material weighs 30 lb per foot of length, calculate the tension in each wire.

Ans. $T_A = 142$ lb, $T_B = 60$ lb, $T_C = 158$ lb

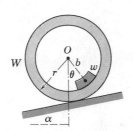

Problem 3/16

3/16 The jaws of the open-ended wrench are smooth case-hardened steel and, for the given thickness of the wrench, can withstand up to 1500 lb of concentrated force against the corner of a hexagonal hardened bolt without surface damage. Calculate the maximum force P at a lever arm of 12 in. that can safely be applied to the wrench. Assume that there is a very slight clearance between the head of the bolt and the jaws of the wrench.   *Ans.* $P = 138$ lb

Problem 3/17

3/17 A uniform ring of weight W and radius r carries an eccentric weight w at a radius b and is placed on an incline that makes an angle α with the horizontal. If the surfaces are rough enough to prevent any slipping, write the expression for θ which defines the equilibrium position.

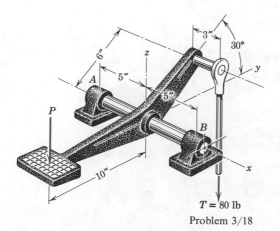

$T = 80$ lb

Problem 3/18

3/18 A vertical force P on the foot pedal of the bell crank is required to produce a tension T of 80 lb in the vertical control rod. Determine the corresponding bearing reactions at A and B.

Ans. $A = 36.8$ lb, $B = 84.8$ lb

3/19 If the weight of the boom is negligible compared with the load *L*, find the force *F* on the ball joint at *A* and show that *F* is constant for all values of *θ*. Determine the limiting value of *T* as *θ* approaches 90 deg.

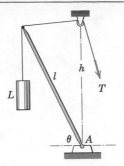

Problem 3/19

3/20 A jet airplane weighing 8 tons is flying horizontally at a constant speed of 600 mi/hr under a thrust of 4000 lb from its turbojet engines. If the pilot increases the fuel rate to give a thrust of 5000 lb and noses the plane upward to maintain a constant 600 mi/hr air speed, determine the angle *θ* made by the new line of flight with the horizontal. Note that the air resistance in the line of flight at the particular altitude involved is a function only of air speed.

3/21 A smooth homogeneous sphere of weight *W* rests in the inclined V-groove and is prevented from rolling by contact against the smooth vertical surface *A* which is normal to the *x-y* plane of symmetry. Determine the expression for the force of contact *R* between the sphere and each side of the groove. *Ans. R = 2W/3*

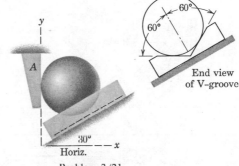

Problem 3/21

3/22 The uniform 30-ft pole weighs 150 lb and is supported by its smooth ends against the vertical walls and by the tension *T* in the vertical cable. Compute the reactions at *A* and *B*.
Ans. A = B = 33.3 lb

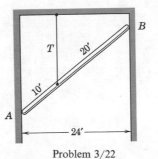

Problem 3/22

3/23 The roller-band device consists of two rollers, each of radius *r*, encircled by a flexible band of negligible thickness and subjected to the two tensions *T*. Write the expression for the contact force *R* between the band and the flat supporting surfaces at *A* and *B*. The action is in the horizontal plane, so that the weights of the rollers and band are not involved.

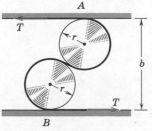

Problem 3/23

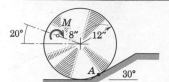

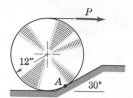

Problem 3/24

3/24 Calculate the value of the couple M required to roll the 80-lb wheel up the incline. Also determine the contact force R at A. The surface of the incline is sufficiently rough to prevent slipping.
Ans. $M = 480$ lb-in., $R = 80$ lb

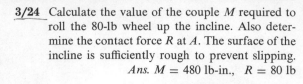

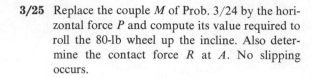

Problem 3/25

3/25 Replace the couple M of Prob. 3/24 by the horizontal force P and compute its value required to roll the 80-lb wheel up the incline. Also determine the contact force R at A. No slipping occurs.

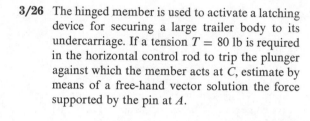

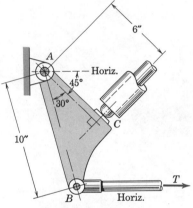

Problem 3/26

3/26 The hinged member is used to activate a latching device for securing a large trailer body to its undercarriage. If a tension $T = 80$ lb is required in the horizontal control rod to trip the plunger against which the member acts at C, estimate by means of a free-hand vector solution the force supported by the pin at A.

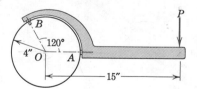

Problem 3/27

3/27 The hook wrench or pin spanner is used to turn shafts and collars. If a moment of 60 lb-ft is required to turn the 8-in. diameter collar about its center O under the action of the applied force P, determine the contact force R on the smooth surface at A. Engagement of the pin at B may be considered to occur at the periphery of the collar.
Ans. $R = 236$ lb

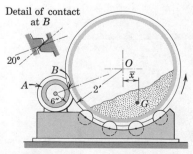

Problem 3/28

3/28 A large symmetrical drum for drying sand is operated by the geared motor drive shown. If the sand weighs 1500 lb and an average gear-tooth force of 580 lb is supplied by the motor pinion A to the drum gear normal to the contacting surfaces at B, calculate the average offset $\bar{x}$ of the center of gravity G of the sand from the vertical center line. Neglect all friction in the supporting rollers. *Ans.* $\bar{x} = 0.727$ ft

3/29 The elevating structure for a rocket test stand and the rocket within it have a combined weight of 1,400,000 lb with center of gravity at G. For the position in which $x = 20$ ft determine the equilibrium value of the force on the hinge axis A. Solve graphically.

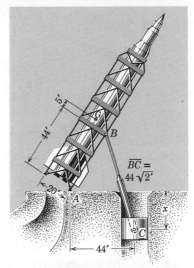

Problem 3/29

3/30 The light bracket ABC is freely hinged at A and is constrained by the fixed pin in the smooth slot at B. Calculate the force R supported by the pin at A if the bracket is subjected to the clockwise couple of 60 lb-ft at C.

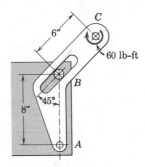

Problem 3/30

3/31 The 60-lb-ft couple of Prob. 3/30 is replaced by the horizontal force of 100 lb applied to the slotted bracket as shown. Calculate the force supported by the pin at A. The surfaces of the slot are smooth.   *Ans.* $A = 162$ lb

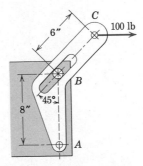

Problem 3/31

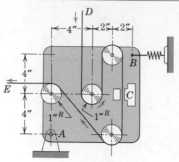

Problem 3/32

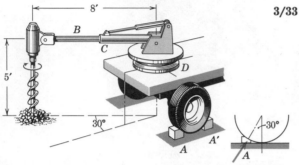

Problem 3/33

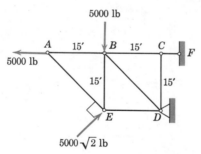

Problem 3/34

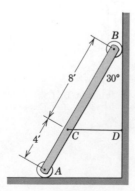

Problem 3/35

3/32 Magnetic tape under a tension of 2.00 lb at D passes around the guide pulleys and through the erasing head at C at constant speed. As a result of a small amount of friction in the bearings of the pulleys, the tape at E is under a tension of 2.20 lb. Determine the tension T in the supporting spring at B. The plate is horizontal and is mounted on a precision needle bearing at A.

Ans. $T = 2.13$ lb

3/33 The power unit of the post-hole digger supplies a torque of 4000 lb-in. to the auger. The arm B is free to slide in the supporting sleeve C but is not free to rotate about the horizontal axis of C. If the unit is free to swivel about the vertical axis of the mount D, determine the force exerted against the right rear wheel by the block A (or A') which prevents the unbraked truck from rolling.

3/34 The rigid truss $ABCDE$ supports the applied loads shown. Neglect the weight of the truss and determine the reaction at D and the force in the link CF.

3/35 The uniform bar with end rollers weighs 100 lb and is held in equilibrium in the position shown by the horizontal cord CD. Determine the reaction at B by writing only one equation of equilibrium. Find the tension T in CD by writing only one equilibrium equation, without involving the reaction at B. Obtain the reaction at A by inspection.

Ans. $B = 43.3$ lb, $T = 43.3$ lb, $A = 100$ lb

3/36 The uniform concrete slab shown in edge view has a weight of 50,000 lb and is being hoisted slowly into a vertical position by the tension P in the hoisting cable. For the position where $\theta = 60$ deg calculate the tension T in the horizontal anchor cable by using only one equation of equilibrium.

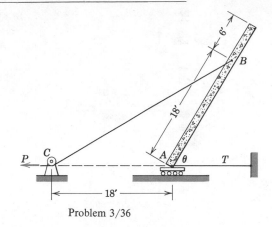

Problem 3/36

3/37 The truss weighs 16,000 lb and supports the two loads shown. If the reaction at the roller support at B is 50,000 lb, determine the horizontal distance a from AB to the center of gravity of the truss. Also find the total force supported by the pin connection at A.

> *Ans.* $a = 21.5$ ft, $A = 60,900$ lb

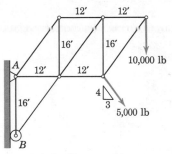

Problem 3/37

3/38 The weight of the rigid truss $ABCDE$ is small compared with the 12-ton load which it supports. Compute the force in the horizontal link at B and the force supported by the pin joint at A.

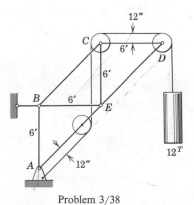

Problem 3/38

3/39 The member OBC and sheave at C together weigh 500 lb, with a combined center of gravity at G. Calculate the force supported by the pin connection at O when the 300-lb load is applied. The collar at A can provide support in the horizontal direction only. *Ans.* $O = 1345$ lb

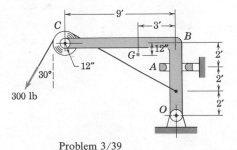

Problem 3/39

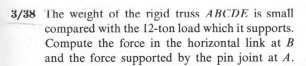

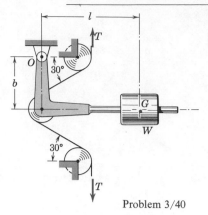

Problem 3/40

3/40 Determine the dimension *l* if the weight *W* maintains a specified tension *T* in the belt for the position shown. Neglect the weight of the arm and central pulley compared with *W*. Also find the force *R* supported by the pin at *O*.

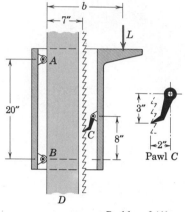

Problem 3/41

3/41 The device shown in section can support the load *L* at various heights by resetting the pawl *C* in another tooth at the desired height on the fixed vertical column *D*. Determine the distance *b* at which the load should be positioned in order for the two rollers *A* and *B* to support equal forces. The weight of the device is negligible compared with *L*.   *Ans. b* = 10.33 in.

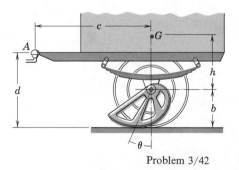

Problem 3/42

3/42 A built-in trailer jack consists of two hinged segments of spiral shape that can be lowered into position under the axle, one under each wheel. The trailer is then pulled forward and rolls on the spiral segment of increasing radial dimension. Friction with the ground prevents slipping at the contact point. For a spiral of constant angle θ between the radial line and the tangent to the curve, determine the expression for the horizontal pull *P* on the hitch at *A* required to jack up the trailer on both segments. The trailer has a total weight *W* with center of gravity at *G*.

$$Ans. \ P = \frac{Wc}{c \ \text{ctn} \ \theta + d - b}$$

3/43 The I-beam and symmetrical end brackets together weigh 1500 lb and are supported as shown. Determine the couple M which would be required to reduce the reaction at B to one-half the value which would exist with no applied couple. The contact surfaces at B are smooth.

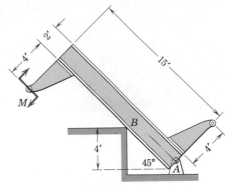

Problem 3/43

3/44 When the 60-lb force is applied to the tire wrench as shown, the wheel is prevented from turning because of the friction force between it and the ground. By using only one equation of equilibrium determine the horizontal component O_x of the reaction exerted on the wheel by the fixed bearing at O. Analyze the wrench and wheel as a single body.　　　　*Ans.* $O_x = 102$ lb

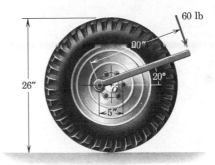

Problem 3/44

3/45 The supporting slippers, which hold the missile to the launcher, slide in T-slots in the guiding rail. In a static test of a missile the slipper at A is clamped securely to the rail whereas the slipper at B is not clamped. If the missile has a weight of 3200 lb with center of gravity at G and if the static thrust is 4400 lb, determine the force supported by the pin that connects the missile to the slipper at B.

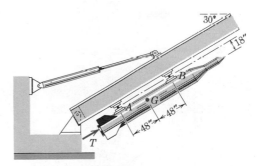

Problem 3/45

3/46 An experimental boat is equipped with four hydrofoils, two on each side as shown. The boat has a total weight W with center of gravity at G. Thrust is provided with the air screw. The ratio of lift to drag for each foil is n. Lift is the vertical force supported by each foil, and drag is the horizontal resistance to motion through the water. For a given thrust T of the propeller write the expression for the drag D on each of the two forward foils.

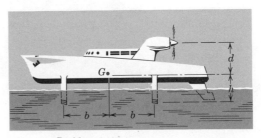

Problem 3/46

$$Ans.\ D = \frac{1}{4n}\left(W + T\frac{d+h}{b}\right)$$

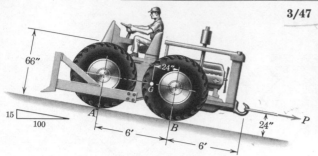

Problem 3/47

Problem 3/48

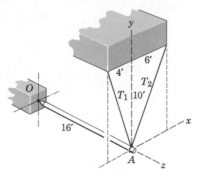

Problem 3/49

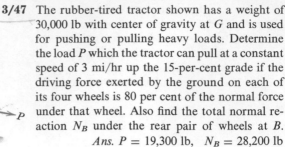

3/47 The rubber-tired tractor shown has a weight of 30,000 lb with center of gravity at G and is used for pushing or pulling heavy loads. Determine the load P which the tractor can pull at a constant speed of 3 mi/hr up the 15-per-cent grade if the driving force exerted by the ground on each of its four wheels is 80 per cent of the normal force under that wheel. Also find the total normal reaction N_B under the rear pair of wheels at B.
Ans. $P = 19{,}300$ lb, $N_B = 28{,}200$ lb

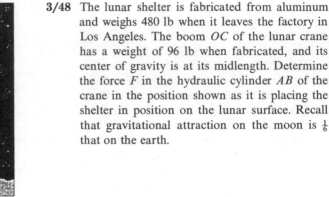

3/48 The lunar shelter is fabricated from aluminum and weighs 480 lb when it leaves the factory in Los Angeles. The boom OC of the lunar crane has a weight of 96 lb when fabricated, and its center of gravity is at its midlength. Determine the force F in the hydraulic cylinder AB of the crane in the position shown as it is placing the shelter in position on the lunar surface. Recall that gravitational attraction on the moon is $\frac{1}{6}$ that on the earth.

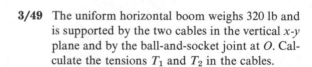

3/49 The uniform horizontal boom weighs 320 lb and is supported by the two cables in the vertical x-y plane and by the ball-and-socket joint at O. Calculate the tensions T_1 and T_2 in the cables.

3/50 The uniform horizontal boom weighs 320 lb and is supported by the two cables anchored at B and C and by the ball-and-socket joint at O. Calculate the tension T in cable AC.

Ans. $T = 178.5$ lb

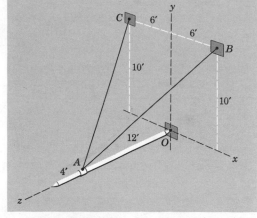

Problem 3/50

3/51 A flexible shaft operates in a rigid tube bent into the shape shown and supported at A and B. The support at A is free to slide in the y-direction but can exert a restraining force against the tube in the z-direction. The hole in the support at B is slightly larger than the tube, so that contact occurs at the ends C and D. Determine the contact forces at C, D, and A while the shaft is running at constant speed under the action of a 200-lb-in. input torque. One revolution at the input end results in one revolution at the output end, and with negligible friction the output torque will equal the input torque. Neglect the weight of the shaft and tube.

Ans. $A = 20$ lb, $C = 104.4$ lb, $D = 111.8$ lb

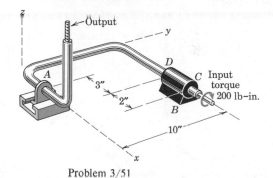

Problem 3/51

3/52 The bar bender consists of the hand-operated lever pivoted about O and carrying the two bending rollers at A and O which are free to turn about their centers. The $\frac{1}{2}$-in. bar is held in clamps at C. If a 60-lb force is required on the handle for the position shown, calculate the corresponding forces that act on the lever at A and O. For the configuration specified, the value of θ is $48°11'$.

Ans. $A = 523$ lb, $O = 467$ lb

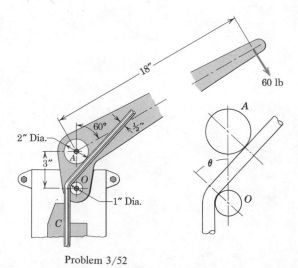

Problem 3/52

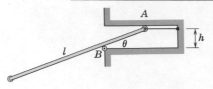
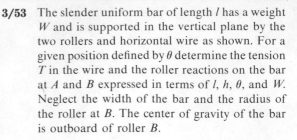

Problem 3/53

3/53 The slender uniform bar of length l has a weight W and is supported in the vertical plane by the two rollers and horizontal wire as shown. For a given position defined by θ determine the tension T in the wire and the roller reactions on the bar at A and B expressed in terms of l, h, θ, and W. Neglect the width of the bar and the radius of the roller at B. The center of gravity of the bar is outboard of roller B.

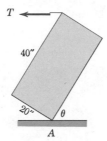

Problem 3/54

3/54 The uniform rectangular block rests on its corner edge A and is supported by the horizontal tension T in the cord attached to the diagonally opposite corner. If the tangential (horizontal) component of the contact force at A cannot exceed 20 per cent of the normal component of this force before slipping occurs, calculate the minimum value of θ that can be maintained for equilibrium as the cord is gradually relaxed but still kept horizontal.

Ans. $\theta_{min} = 41°38'$

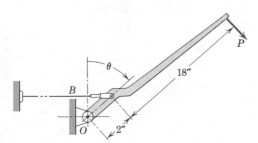

Problem 3/55

3/55 The lever is used to stretch a long wire B within its elastic limit. The overall stiffness of the entire length of wire is $k = EA/l$ where E is the elastic modulus of the wire material, A is the cross-sectional area of the wire, and l is its total unstretched length. For large values of l the wire may be assumed to remain horizontal during the motion of the lever. If 100 ft of No. 10 gage steel wire (diameter 0.102 in. and cross-sectional area 8.17 (10^{-3}) in.^{2}) is to be stretched, determine the maximum value of the force P and the corresponding value of θ if the stretch begins at $\theta = 0$. Compute the total shear force Q supported by the pin at O for this maximum condition.

Ans. $P_{max} = 20.4$ lb at $\theta = \pi/4$, $Q = 275$ lb

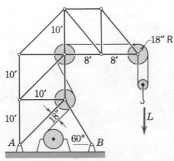

Problem 3/56

3/56 If the crane load L is increased by 4 tons, compute the corresponding increment ΔA in the force supported by the pin at A.

Ans. $\Delta A = 5.66$ tons

3/57 The two uniform rectangular plates each weighing 800 lb are hinged about their common edge and supported by a central cable and four symmetrical corner cables as shown. Calculate the tension T_0 in the 3-ft central cable and the tension T in each of the corner cables.

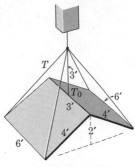

Problem 3/57

3/58 If the anchorage B for the cable AB of Prob. 3/50 were placed 4 ft from the vertical y-z plane rather than 6 ft with the other conditions remaining unchanged, calculate the tension T in cable AC.

Ans. $T = 142.8$ lb

3/59 Determine the force in each member of the tripod. Each of the three members is secured at its ends by ball-and-socket connections and is capable of supporting tension or compression. The weights of the members may be neglected.

Ans. $A = 204.1$ lb tension
$B = 86.1$ lb compression
$C = 126.9$ lb compression

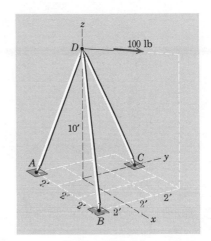

Problem 3/59

3/60 The access door in the horizontal deck is a uniform square plate weighing 180 lb and is hinged at its corners A and B. The hydraulic cylinder CD opens and closes the door. Determine the force supported by each hinge for the 60-deg position shown.

Ans. $A = 57.0$ lb, $B = 70.0$ lb

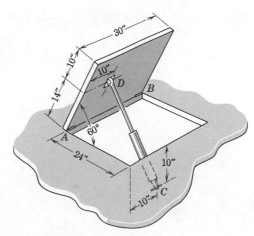

Problem 3/60

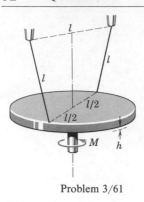

Problem 3/61

3/61 The uniform circular disk of radius $l/2$ is suspended by two wires of length l from two points in a horizontal plane a distance l apart. The disk and small central shaft have a combined weight W. Determine the height h which the disk will rise to the equilibrium position established by the couple M applied to the shaft. What is the value of M as h approaches l?

$$Ans. \ h = l\left(1 - \sqrt{1 - \left(\frac{2M}{Wl}\right)^2}\right)$$

$$\text{For } h \to l, \qquad M = \frac{Wl}{2}$$

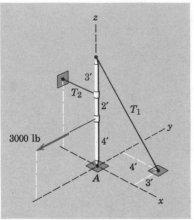

Problem 3/62

3/62 If the weight of the mast is negligible compared with the applied 3000-lb load, determine the two cable tensions T_1 and T_2 and the force A acting at the ball joint at A.

$$Ans. \ T_1 = 4580 \text{ lb}$$
$$T_2 = 2667 \text{ lb}$$
$$A = 4420 \text{ lb}$$

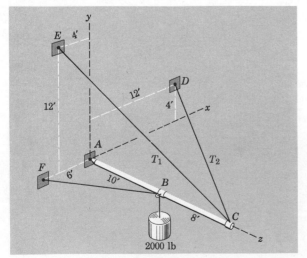

Problem 3/63

3/63 The 18-ft steel boom weighs 600 lb with center of gravity at midlength. It is supported by a ball-and-socket joint at A and the two cables under tensions T_1 and T_2. The cable which supports the 2000-lb load leads through a sheave at B and is secured to the vertical x-y plane at F. Calculate the tension T_1 using only one equation of equilibrium.

$$Ans. \ T_1 = 2014 \text{ lb}$$

3/64 The mast weighs 300 lb and is supported by a ball-and-socket joint at A. Calculate the tension T_1 if the horizontal 1000-lb force is applied at F.

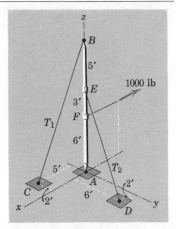

Problem 3/64

3/65 The rigid member ABC is attached to the vertical x-y surface by a ball-and-socket joint at A and is supported by the cables BE and CD. The weight of the member may be neglected compared with the 5000-lb load which it supports. There is one position of D along the horizontal slot through which the cable must be passed and secured for the member to maintain the position shown. Find x.   *Ans.* $x = 7.50$ ft

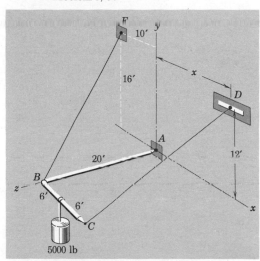

Problem 3/65

3/66 A rectangular sign over a store weighs 200 lb, with the center of gravity in the center of the rectangle. The support against the wall at point C may be treated as a ball-and-socket joint. At corner D support is provided in the y-direction only. Calculate the tensions T_1 and T_2 in the supporting wires, the total force supported at C, and the lateral force R supported at D.

        Ans. $T_1 = 70.7$ lb $T_2 = 87.8$ lb
             $R = 12.9$ lb $C = 156.6$ lb

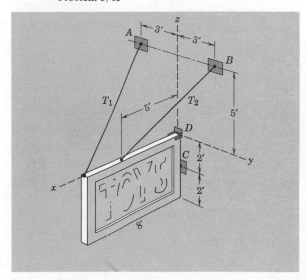

Problem 3/66

◀ **3/67** Develop an alternative set of independent equations that are both necessary and sufficient to establish the equilibrium of a space system of forces concurrent at a point. (See case 3 of Fig. 29.)

◀ **3/68** Develop an alternative set of independent equations that are both necessary and sufficient to establish the equilibrium of a space system of forces concurrent with a line. (See case 4 of Fig. 29.)

▽   ▽   ▽   ▽   ▽

16 Adequacy of Constraint. In the previous article the equations of equilibrium were applied to various bodies that were statically determinate, i.e. that had only the minimum number of supports necessary to establish or maintain an equilibrium position, thus permitting the calculation of all unknown external forces from the equations of equilibrium. It was pointed out that the introduction of more supports than were needed to maintain a fixed equilibrium position gave rise to a condition of redundancy in the supports, in which case the equations of equilibrium were no longer sufficient criteria for the determination of all of the unknown support reactions. The determination of the adequacy or inadequacy of the supports to maintain a body in an equilibrium position is generally clear from inspection, but there are many cases where it is necessary to settle the question of adequacy by analytical criteria. Hence a more detailed examination into the nature of supporting constraints is required.

The position of a body in space is described by its coordinates, measured from some convenient base of reference. The number of independent coordinates required to specify the complete position of a body is known as the number of *degrees of freedom* of the body. Thus a slider confined to move in a fixed slot has a single degree of freedom, since a measurement of its distance along the slot is the only quantity required to specify its position. A body pivoted about a fixed axis has a single degree of freedom which is its angular position about the axis. A rigid body that is confined to move in a given plane can have at the most three degrees of freedom, represented by two linear coordinates in the plane that define the position of any particular point in the body, and an angle that specifies the rotational position of the body about this point. A rigid body in space has six degrees of freedom corresponding to its six possible motions—linear motion in, say, the x-, y-, and z-directions and angular motion about axes in the x-, y-, and z-directions.

To provide for the equilibrium of a body, there must be at least one constraining force or its equivalent for each degree of freedom possessed by the body to secure the body against possible motion due to an imposed imbalance of forces. An equilibrium position is not ensured, however, merely by the existence of the number of constraints equal to the number of degrees of freedom, because the geometrical arrangement of the constraints is also a determining factor, as will now be seen. In the two-dimensional case where the position of the body is confined to a given plane, it is necessary to have at least three constraining reactions to ensure complete fixity of the body. This condition is illustrated in Fig. 33a, where the rectangular body is secured by two links that fix the one corner and by a third link that prevents the body from rotating about the fixed corner. The body is now constrained from all possible movements in the plane resulting from any applied forces (not shown), and its fixity is ensured. Each of the links is treated as a two-force member capable of applying force to the body in the direction of the link only. The constraint provided by the two links at the corner is the model equivalent to a pinned or hinged joint that can offer constraint in the two directions normal to the axis of the pin.

There is no limit to the number of configurations of the restraining links that will maintain fixity of the body. There are, however, two special cases where the configuration of the three restraining links produces only partial fixity, and, hence the equilibrium position is not a perfectly rigid one. The first case is where the links are concurrent, Fig. 33b, and it is seen that this arrangement permits a small initial rotation of the body to take place before the induced angularity of the third link from its initial unloaded position provides the constraining forces necessary to prevent further motion. The second case of partial constraint is where the three links are parallel, an example of which is shown in Fig. 33c. Here a small movement in the vertical direction can occur without initial resistance. Each of the partially constrained bodies illustrated is statically indeterminate.

If a fourth link is added to constrain a two-dimensional body already in a condition of complete fixity with three links, Fig. 33d, there are then more supports than are necessary to maintain the equilibrium position, and link 4 becomes redundant. With the fourth link in place, the body is statically indeterminate.

In the three-dimensional case the schematic model is the rectangular block shown in Fig. 34a, which represents one of an unlimited number of possible configurations of six constraining links that will produce complete fixity of the body when subjected to applied loads (not shown). Here the corner A is securely fixed by the three links 1, 2, and 3, which would correspond to the fixity provided by a ball-and-socket joint. Rotation about axis AB is pre-

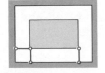

(a) Complete fixity
Adequate constraints

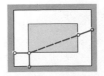

(b) Incomplete fixity
Partial constraints

(c) Incomplete fixity
Partial constraints

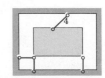

(d) Excessive fixity
Redundant constraint

Figure 33

vented by link 4; rotation about AC is prevented by link 5; and the small remaining rotation possible about AD is prevented by link 6. It is not possible to fix the body completely with less than six restraining links or their equivalent.

As in the two-dimensional case, there are two special situations in which the configuration of the restraining links produces only partial fixity, and hence the equilibrium position is not a perfectly rigid one. An example of the first case is shown in Fig. 34*b*, where the links or their extensions all intersect a common line AE. Since the forces induced in the restraining links all pass through AE, this system of constraints can offer no initial resistance to a moment about AE that could be induced by applied loads. Hence an initial small rotation could occur about AE, and the body would be only partially constrained. Two examples of the second case of partial constraint are illustrated in Figs. 34*c* and *d*, where the constraining links lie in parallel planes. In the *c*-part of the figure the block could experience a small unconstrained movement in the *y*-direction. In the *d*-part of the figure, where all of the constraint forces are parallel, the body could undergo small unconstrained movements

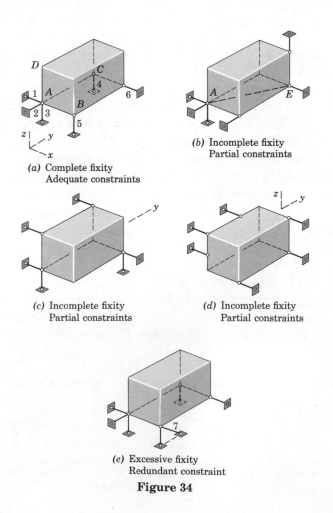

(a) Complete fixity
Adequate constraints

(b) Incomplete fixity
Partial constraints

(c) Incomplete fixity
Partial constraints

(d) Incomplete fixity
Partial constraints

(e) Excessive fixity
Redundant constraint

Figure 34

in both the y- and z-directions, normal to the links, before the induced angularity in the links was sufficient to prevent further motion. With six constraints the bodies represented by Figs. 34b, c, and d are statically indeterminate.

If a seventh constraining link were imposed on a system of six properly placed constraints for complete fixity of a rigid body in space, such as is shown with Fig. 34e, more supports would be provided than would be necessary to maintain the equilibrium position, and link 7 would be redundant. The body would then be statically indeterminate with such a seventh link in place.

With the foregoing illustrations in mind of the constraint configurations that determine fixity, an analytical criterion for determining constraint adequacy will now be developed. Consider the general rigid body in space with six constraining links represented schematically in Fig. 35. As has been pointed out, the position of such a body can be completely determined by specifying the location of any convenient point in the body such as P and the angular orientation of the body about P. This concept is now used to express the possible movement of each of the points where the constraining links attach to the body. The location with respect to the fixed x-y-z reference of the attachment point A for link n, which represents any one of the six links, is merely

$$\rho = \mathbf{R} + \mathbf{r}_n$$

Assume now that the body suffers a small rotation through an angle $\Delta\theta$ about an axis through P. This rotation may be expressed by the vector $\Delta\boldsymbol{\theta} = \mathbf{i}\,\Delta\theta_x + \mathbf{j}\,\Delta\theta_y + \mathbf{k}\,\Delta\theta_z$ where the direction of the vector is taken to be that of the rotation axis in the sense determined by the right-hand rule.* On account of such a rotation, point A will undergo a small arc movement $\Delta\mathbf{r}_n$, shown in the right-hand view of Fig. 35, equal in magnitude to the radius $r_n \sin\alpha$ times the angle $\Delta\theta$ in radians or $\Delta\theta(r_n \sin\alpha)$ where $\Delta\theta$ approaches zero in the limit. In vector form this movement becomes $\Delta\mathbf{r}_n = \Delta\boldsymbol{\theta} \times \mathbf{r}_n$. If, in addition, P suffers a small displacement given by the vector $\Delta\mathbf{R} = \mathbf{i}\,\Delta x + \mathbf{j}\,\Delta y + \mathbf{k}\,\Delta z$, then the corresponding change in the position of point A resulting from

---

*Finite rotations may not be written as a proper vector, but infinitesimal ones may. Since $\Delta\theta$ is a small rotation approaching an infinitesimal rotation in the limit, its representation as a vector is permissible.

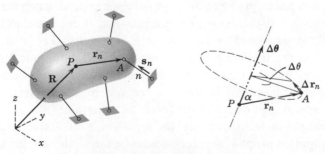

Figure 35

both movements together is

$$\Delta\rho = \Delta\mathbf{R} + \Delta\boldsymbol{\theta} \times \mathbf{r}_n$$

But point A is constrained by link n so that, if A has any movement at all, its motion must be perpendicular to link n, whose length remains unchanged. If $\mathbf{s}_n = \mathbf{i}s_{nx} + \mathbf{j}s_{ny} + \mathbf{k}s_{nz}$ stands for a unit vector in the direction of link n, then the movement $\Delta\rho$ can have no component in the $\mathbf{s}_n$-direction. Thus

$$\Delta\rho \cdot \mathbf{s}_n = 0$$

The restriction on any possible movement of point A now becomes

$$(\Delta\mathbf{R} + \Delta\boldsymbol{\theta} \times \mathbf{r}_n) \cdot \mathbf{s}_n = 0$$

Replacement of the foregoing vectors by their $\mathbf{i}$-, $\mathbf{j}$-, and $\mathbf{k}$-components yields, upon expansion and collection of terms,

$$s_{nx}\,\Delta x + s_{ny}\,\Delta y + s_{nz}\,\Delta z + (r_{ny}s_{nz} - r_{nz}s_{ny})\,\Delta\theta_x + (r_{nz}s_{nx} - r_{nx}s_{nz})\,\Delta\theta_y$$
$$+ (r_{nx}s_{ny} - r_{ny}s_{nx})\,\Delta\theta_z = 0 \qquad (16)$$

where $\qquad\qquad\qquad n = 1, 2, 3, 4, 5, 6$

In the general case, there will be six such linear homogeneous equations, one for each of the attachment points of the six equivalent constraining links. These equations contain the six unknown displacements Δx, Δy, Δz, $\Delta\theta_x$, $\Delta\theta_y$, $\Delta\theta_z$. Clearly the body will be completely constrained in a fixed position if the six displacements all vanish, i.e.

$$\Delta x = \Delta y = \Delta z = \Delta\theta_x = \Delta\theta_y = \Delta\theta_z = 0$$

It is known from the theory of homogeneous linear equations that, if the determinant D of the coefficients of the unknowns in Eqs. 16 is not zero, then the unknowns are all zero. Conversely, if the determinant is zero, then the unknowns need not all be zero, and displacements are possible. Consequently, the test for adequacy of the constraints to provide complete fixity may be stated as

Constraints adequate for complete fixity if $D \neq 0$

Constraints inadequate for complete fixity if $D = 0$

It is seldom necessary to solve a 6-by-6 determinant, since simplification in the equations is usually possible. For instance, if one point in the body is obviously fixed, as would be the case at a ball-and-socket joint, this point can be used as the reference point P, which would automatically eliminate Δx, Δy, and Δz, thereby reducing the determinant to a 3-by-3 array.

For the equilibrium of bodies not secured by fixed constraints, other forces that act are equivalent to the forces of constraint. For example, in an airplane flying at constant velocity, the aerodynamic reactive forces, in contrast to the applied propulsive forces, would be the equivalent of the supporting constraints of a fixed structure.

Sample Problem

3/69   Determine whether the constraints on the cubical block are adequate to ensure a completely fixed position.

Solution. The intersection of links 1 and 2 is chosen both as the reference point P and as the origin of coordinates. Since Δx and Δy are necessarily zero in all six of Eqs. 16 and since $s_{1z} = s_{2z} = 0$ for point P, the only nonzero terms that will appear will be found in the equations written for $n = 3, 4, 5,$ and 6.

The components of $\mathbf{s}_n$ and $\mathbf{r}_n$ are seen by inspection of the figure to be

$(n = 3)$	$s_{3x} = 0$	$s_{3y} = 0$	$s_{3z} = -1$	$r_{3x} = 0$	$r_{3y} = 0$	$r_{3z} = b$
$(n = 4)$	$s_{4x} = 0$	$s_{4y} = -1$	$s_{4z} = 0$	$r_{4x} = 0$	$r_{4y} = b$	$r_{4z} = b$
$(n = 5)$	$s_{5x} = 0$	$s_{5y} = -1$	$s_{5z} = 0$	$r_{5x} = b$	$r_{5y} = b$	$r_{5z} = 0$
$(n = 6)$	$s_{6x} = -1$	$s_{6y} = 0$	$s_{6z} = 0$	$r_{6x} = b$	$r_{6y} = 0$	$r_{6z} = b$

The determinant of the coefficients of the remaining four equations from Eq. 16 becomes

$$
\begin{array}{c}
 \quad (\Delta z) \quad (\Delta\theta_x) \quad (\Delta\theta_y) \quad (\Delta\theta_z) \\
\begin{array}{c}
(n = 3) \\
(n = 4) \\
(n = 5) \\
(n = 6)
\end{array}
\quad D =
\begin{vmatrix}
-1 & 0 & 0 & 0 \\
0 & b & 0 & 0 \\
0 & 0 & 0 & -b \\
0 & 0 & -b & 0
\end{vmatrix}
\end{array}
$$

where the identity of the terms in each row and column is indicated for easy reference. The determinant simplifies to

$$
D = -b \begin{vmatrix} 0 & b \\ b & 0 \end{vmatrix} = -b(-b^2) = b^3
$$

Since $D \neq 0$, the unknowns Δz, $\Delta\theta_x$, $\Delta\theta_y$, and $\Delta\theta_z$, in addition to Δx and Δy, are identically zero, and the cube is now known to be supported in a fixed position. The supports are, therefore, adequately placed to ensure complete constraint.

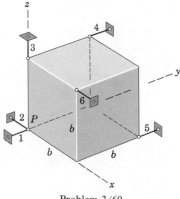

Problem 3/69

Problems

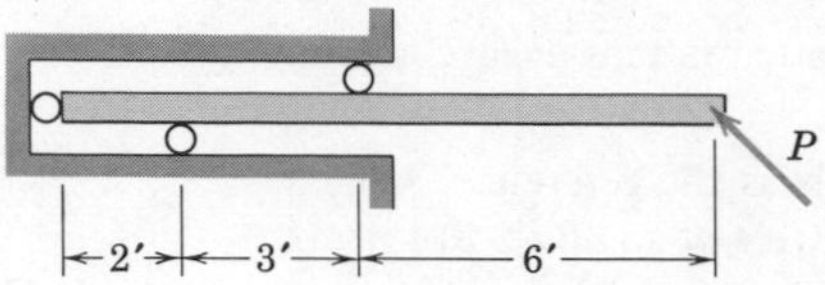

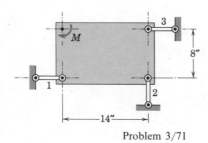

Problem 3/70

3/70   Prove formally that the beam is adequately constrained to support the load shown.

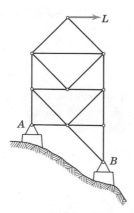

Problem 3/71

3/71   The flat plate is confined to the plane represented and is subjected to the couple $M = 100$ lb-in. Prove formally that the three supporting links hold the plate in a fixed position.

3/72   If the load L and the dimensions of the truss were known, indicate what calculations could be made regarding the forces of constraint at A and B, using only the equations of equilibrium.

Problem 3/72

3/73 The figure shows a series of rectangular plates and their constraints, all confined to the plane of representation. The plates could be subjected to various known loads applied in the plane of the plate. Identify the plates that belong to each of the following categories.

   (A) Complete fixity with minimum number of adequate constraints

   (B) Partial fixity with inadequate constraints

   (C) Complete fixity with redundant constraints

   (D) Partial fixity with redundant constraint

   (E) Partial fixity changed to complete fixity following small movement under load

   (F) Partial fixity with large movement possible

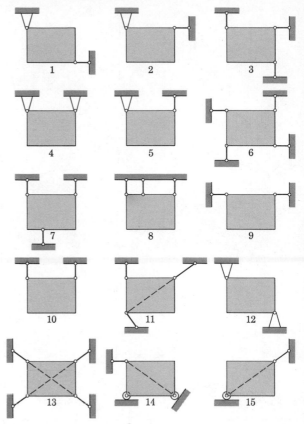

Problem 3/73

3/74 The space truss *ABCDEF,* consisting of a rigid assemblage of 13 interconnected bars, is anchored to the *x-y* plane at its triangular base and subjected to the loads *L* and *P*. Prove whether the constraints are adequate or inadequate to hold the truss in place.

3/75 Work Prob. 3/74 with the *x*-direction support at *B* eliminated and replaced by a supporting link in the *y*-direction at *B*.

              Ans. Support adequate

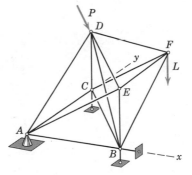

Problem 3/74

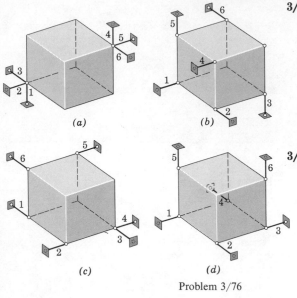

(a)   *(b)*

(c)   *(d)*

Problem 3/76

3/76 By inspection show that each of the cubical blocks is not completely constrained. Identify the movement that could occur under load and indicate one possible relocation of one of the constraints that would ensure complete fixity.

3/77 If the equivalent constraining links for a rigid body all lie in planes that have a common line of intersection, prove that the body is inadequately constrained.

◄ **3/78** Examine analytically the adequacy or inadequacy of the constraints for the cubical block shown. *Ans.* Support inadequate

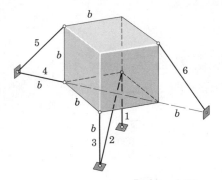

Problem 3/78

4 STRUCTURES

17 Structures. In Chapter 3 attention was centered on the equilibrium of a single rigid body or upon a system of connected members which, when taken as a whole, could be treated as a single rigid body. In such problems a free-body diagram of this single body showing all forces external to the isolated body was drawn prior to the application of the force and moment equations of equilibrium. In this chapter attention is directed toward the determination of the forces internal to a structure, i.e., forces of action and reaction between the connected members. An engineering structure is any connected system of members built to support or transfer forces and to withstand safely the loads applied to it. In the force analysis of structures it is necessary to dismember the structure and to analyze separate free-body diagrams of individual members or combinations of members in order to determine the forces internal to the structure. This analysis calls for very careful observance of Newton's third law which states that each action is accompanied by an equal and opposite reaction. Three types of structures, namely, trusses, frames and machines, and beams under concentrated loading are analyzed. In this treatment only *statically determinate* structures will be considered, i.e., structures for which the statical equations of equilibrium, Eqs. 13 or 14, are sufficient conditions to solve for the unknown forces.

The student who has mastered the basic procedure developed in Chapter 3 of defining unambiguously the body under consideration by constructing a correct free-body diagram will have no difficulty with the analysis of statically determinate structures. The analysis of trusses, frames and machines, and beams under concentrated loads constitutes a straightforward application of the material developed in the previous chapter.

18 Plane Trusses. A framework composed of members joined at their ends to form a rigid structure is known as a truss. Bridges, roof supports, derricks, and other such structures are common examples of trusses. Structural members used are I-beams, channels, angles, bars, and special shapes which are fastened together at their ends by welding, riveted connections, or large bolts or pins. When the members of the truss lie essentially in a single plane, the truss is known as a *plane truss*. Plane trusses, such as those used for bridges, are commonly designed in pairs with one truss panel placed on each side of the bridge and connected together by cross beams that support the roadway and transfer the applied loads to the truss members. Several examples of commonly used trusses that can be analyzed as plane trusses are shown in Fig. 36.

103

The basic element of a plane truss is the triangle. Three bars joined by pins at their ends, Fig. 37a, constitute a rigid frame. On the other hand, four or more bars pin-jointed to form a polygon of as many sides constitute a non-rigid frame. The nonrigid frame in Fig. 37b can be made stable or rigid with an additional diagonal bar joining A and D or B and C and thereby forming two triangles. The structure may be extended by adding additional units of two end-connected bars, such as DE and CE or AF and DF, Fig. 37c, which are pinned to two fixed joints, and in this way the entire structure will remain rigid. The term rigid is used in the sense of noncollapsible and also in the sense of negligible deformation of the members due to induced internal strains.

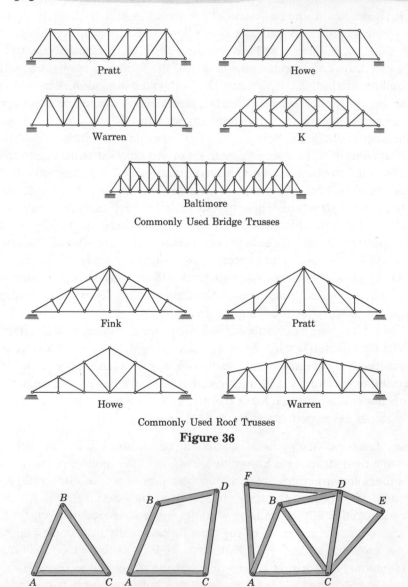

Pratt   Howe

Warren   K

Baltimore

Commonly Used Bridge Trusses

Fink   Pratt

Howe   Warren

Commonly Used Roof Trusses

Figure 36

(a)   *(b)*   *(c)*

Figure 37

Structures that are built from a basic triangle in the manner described are known as *simple trusses*. When more members are present than are needed to prevent collapse, the truss is statically indeterminate. A statically indeterminate truss cannot be analyzed by the equations of equilibrium alone. Additional members or supports which are not necessary for maintaining the equilibrium position are called *redundant*.

The design of a truss involves the determination of the forces in the various members and the selection of appropriate sizes and structural shapes to withstand the forces. Several assumptions are made in the force analysis of simple trusses. First, all members are assumed to be *two-force members*. A two-force member is one in equilibrium under the action of two forces only, as defined in general terms with Fig. 30*a* in Art. 15. For trusses each member is a straight link joining the two points of application of force. The two forces are applied at the ends of the member and are necessarily equal, opposite, and *collinear* for equilibrium. The member may be in tension or compression, as shown in Fig. 38. Note that in representing the equilibrium of a portion of a two-force member the tension T or compression C acting on the cut section is the same for all sections. It is assumed here that the weight of the member is small compared with the force it supports. If it is not, or if the small effect of the weight is to be accounted for, the weight W of the member, if uniform, may be assumed to be replaced by two forces, each $W/2$, acting at each end of the member. These forces, in effect, are treated as loads externally applied to the pin connections. Accounting for the weight of a member in this way gives the correct result for the average tension or compression along the member but will not account for the effect of bending of the member.

When welded or riveted connections are used to join structural members, the assumption of a pin-jointed connection is usually satisfactory if the center lines of the members are concurrent at the joint as in Fig. 39.

It is also assumed in the analysis of simple trusses that all external forces are applied at the pin connections. This condition is satisfied in most trusses. In bridge trusses the deck is usually laid on cross beams that are supported at the joints.

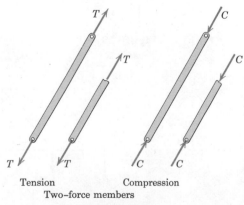

Tension　　　　　　　Compression
Two–force members

Figure 38

Provision for expansion and contraction due to temperature changes and for deformations resulting from applied loads is usually made at one of the supports for large trusses. A roller, rocker, or some kind of slip joint is provided. Trusses and frames wherein such provision is not made are statically indeterminate, as explained in Arts. 15 and 16.

Two methods for the force analysis of simple trusses will be given, and reference will be made to the simple truss shown in Fig. 40*a* for each of the two methods. The free-body diagram of the truss as a whole is shown in Fig. 40*b*. The external reactions are usually determined by computation from the equilibrium equations applied to the truss as a whole before proceeding with the force analysis of the remainder of the truss.

(*a*) *Method of Joints.* This method consists of satisfying the conditions of equilibrium for the forces acting on the connecting pin of each joint. The method therefore deals with the equilibrium of concurrent forces, and only two independent equilibrium equations are involved. The analysis is begun with any joint where at least one known load exists and where not more than two unknown forces are present. Solution may be started with the pin at the left end, and its free-body diagram is shown in Fig. 41. With the joints indicated by letters, the force in each member is designated by the two letters defining the ends of the member. The proper directions of the forces should be evident for this simple case by inspection. The free-body diagrams of portions of members *AF* and *AB* are also shown to indicate clearly the mechanism of the action and reaction. The member *AB* actually makes

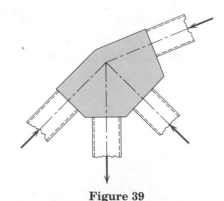

Figure 39

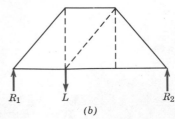

(a)   *(b)*

Figure 40

contact on the left side of the pin, although the force *AB* is drawn from the right side and is shown acting away from the pin. Thus, if the force arrows are consistently drawn on the *same* side of the pin as the member, then tension (such as *AB*) will always be indicated by an arrow *away* from the pin, and compression (such as *AF*) will always be indicated by an arrow *toward* the pin. The magnitude of *AF* is obtained from the equation $\Sigma F_y = 0$, and *AB* is then found from $\Sigma F_x = 0$.

Joint *F* must be analyzed next, since it now contains only two unknowns, *EF* and *BF*. Joints *B*, *C*, *E*, and *D* are subsequently analyzed in that order. The free-body diagram of each joint and its corresponding force polygon which represents graphically the two equilibrium conditions $\Sigma F_x = 0$ and $\Sigma F_y = 0$ are shown in Fig. 42. The numbers indicate the order in which the joints are analyzed. It should be noted that, when joint *D* is finally reached, the computed reaction R_2 must be in equilibrium with the forces in members *CD* and *ED*, determined previously from the two neighboring joints. This requirement will provide a check on the correctness of the work. It should also be noted that isolation of joint *C* quickly discloses the fact that the force in *CE* is zero when the equation $\Sigma F_y = 0$ is applied. The force in this member would not be zero, of course, if an external load were applied at *C*.

It is often convenient to indicate the tension *T* and compression *C* of the various members directly on the original truss diagram by drawing arrows away from the pins for tension and toward the pins for compression. This designation is illustrated at the bottom of Fig. 42.

In some instances it is not possible to assign initially the correct direction of one or both of the unknown forces acting on a given pin. In this event an arbitrary assignment may be made. A negative value from the computation indicates that the assumed direction should be reversed.

If a simple truss has more external supports than are necessary to ensure a stable equilibrium configuration, the truss as a whole is statically indeterminate, and the extra supports constitute *external* redundancy. If the truss has more internal members than are necessary to prevent collapse, then the extra members constitute *internal* redundancy. For a truss that is statically determinate externally, there is a definite relation between the number of its

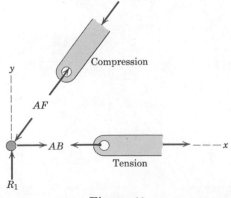

Figure 41

members and the number of its joints necessary for internal stability without redundancy. Since the equilibrium of each joint is specified by two scalar force equations, there are a total of $2j$ such equations for a simple truss with j joints. For the entire truss composed of m two-force members and a maximum of three unknown support reactions, there are a total of $m + 3$ unknowns. Thus, for a simple plane truss composed of triangular elements, the equation $m + 3 = 2j$ will be satisfied if the truss is statically determinate internally.

This relation is a necessary condition for stability but it is not a sufficient condition, since one or more of the m members can be arranged in such a way as not to contribute to a stable configuration of the entire truss. If $m+3>2j$, there are more members than there are independent equations, and the truss is statically indeterminate internally with redundant members present. If

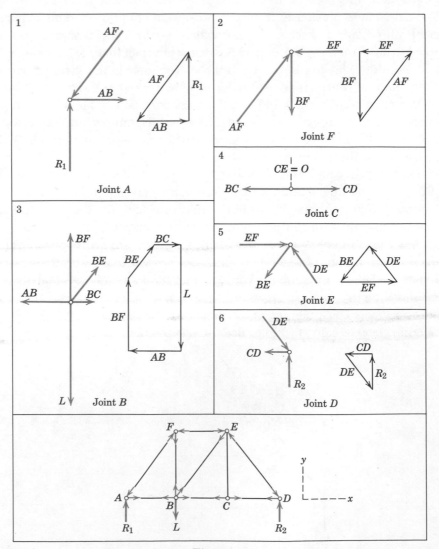

Figure 42

$m + 3 < 2j$, there is a deficiency of internal members, and the truss is unstable and will collapse under load.

The force polygon for each joint, shown in Fig. 42, may be constructed graphically to obtain the unknown forces in the members as an alternative to or as a check on the algebraic calculations using the force equations of equilibrium. If a consistent sequence around each joint, e.g., clockwise, has been used for the addition of the forces, these force polygons may be superimposed on one another to form a composite graphical figure known as the *Maxwell diagram.** The force and its sense may be obtained directly from the diagram. The student who is interested in structures may wish to experiment with this construction and to consult other books dealing more completely with structural analysis for a more detailed description of the Maxwell diagram.

(*b*) *Method of Sections.* In the method of joints, advantage is taken of only two of the three equilibrium equations, since the procedures involve concurrent forces at each joint. The third or moment-equilibrium principle may be used to advantage by selecting an entire section of the truss for the free body in equilibrium under the action of a nonconcurrent system of forces. This *method of sections* has the basic advantage that the force in almost any desired member may be found directly from an analysis of a section which has cut that member. Thus it is not necessary to proceed with the calculation from joint to joint until the member in question has been reached. In choosing a section of the truss it should be noted that in general not more than three members whose forces are unknown may be cut, since there are only three available equilibrium relations which are independent.

SEE NOTES
EXAMPLE 4/24

The method of sections will now be illustrated for the truss in Fig. 40, which was used in the explanation of the previous method. The truss is shown again in Fig. 43*a* for ready reference. The external reactions are first computed as before, considering the truss as a whole. Now let it be desired to determine the force in the member *BE* for example. An imaginary section, indicated by the dotted line, is passed through the truss, cutting it into two parts, Fig. 43*b*. This section has cut three members whose forces are initially unknown. In order for the portion of the truss on each side of the section to remain in equilibrium it is necessary to apply to each cut member the force that was exerted on it by the member cut away. These forces, either tensile or compressive, will always be in the directions of the respective members for simple trusses composed of two-force members. The left-hand section is in equilibrium under the action of the applied load L, the end reaction R_1, and the three forces exerted on the cut members by the right-hand section which has been removed. The forces may usually be drawn with their proper senses by a visual approximation of the equilibrium requirements. Thus in balancing the moments about point *B* for the left-hand section, the force *EF* is clearly to the left, which makes it compressive, since it acts toward the cut section of member *EF*. The load L is greater than the reaction R_1, so that the

---

* The method was published by James Clerk Maxwell in 1864.

force *BE* must be up and to the right to supply the needed upward component for vertical equilibrium. Force *BE* is therefore tensile, since it acts away from the cut section. With the approximate magnitudes of R_1 and L in mind the balance of moments about point E requires that *BC* be to the right. A casual glance at the truss should lead to the same conclusion when it is realized that the lower horizontal member will stretch under the tension caused by bending. The equation of moments about joint B eliminates three forces from the relation, and *EF* may be determined directly. The force *BE* is calculated from the equilibrium equation for the *y*-direction. Finally *BC* may be determined by balancing moments about point E. In this way each of the three unknowns has been determined independently of the other two.

The right-hand section of the truss, Fig. 43*b*, is in equilibrium under the action of R_2 and the same three forces in the cut members applied in the directions opposite to those for the left section. The proper sense for the horizontal forces may easily be seen from the balance of moments about points B and E.

Either section of a truss may be used for the calculations, but the one involving the smaller number of forces will usually yield the simpler solution.

It is essential to understand that in the method of sections an entire portion of the truss is considered a single body in equilibrium. Thus the forces in members internal to the section are not involved in the analysis of the section as a whole. In order to clarify the free body and the forces acting externally on it, the section is preferably passed through the members and not the joints.

The moment equations may be used to great advantage in the method of sections, and a moment center through which as many forces pass as possible should be chosen. It is not always possible to assign an unknown force in the proper sense when the free-body diagram of a section is drawn. With

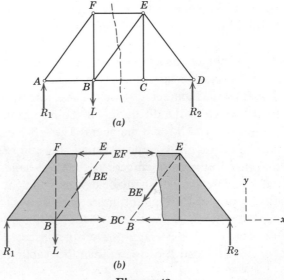

Figure 43

an arbitrary assignment made, a positive answer will verify the assumed sense and a negative result will indicate that the force is in the sense opposite to that assumed. Any system of notation desired may be used, although usually it is found convenient to letter the joints and designate a member and its force by the two letters defining the ends of the member.

Sample Problems

4/1 Compute the force in each member of the loaded cantilever truss by the method of joints.

Solution. If it were not desired to calculate the external reactions at D and E, the analysis for a cantilever truss could begin with the joint at the loaded end. However, this truss will be solved completely, so that the first step will be to compute the external forces at D and E from the free-body diagram of the truss as a whole in the *b*-part of the figure. The equations of equilibrium give

$$T = 8000 \text{ lb} \qquad E_x = 6930 \text{ lb} \qquad E_y = 1000 \text{ lb}$$

In the *c*-part of the figure are drawn the free-body diagrams showing the forces acting on each of the connecting pins. The correctness of the assigned directions of the forces is verified when each joint is considered in sequence. There should be no question about the correct direction of the forces on joint A. Equilibrium requires

$[\Sigma F_y = 0]$	$0.866AB - 3000 = 0$	$AB = 3464 \text{ lb } T$	*Ans.*
$[\Sigma F_x = 0]$	$AC - 0.5(3464) = 0$	$AC = 1732 \text{ lb } C$	*Ans.*

where T stands for tension and C stands for compression.

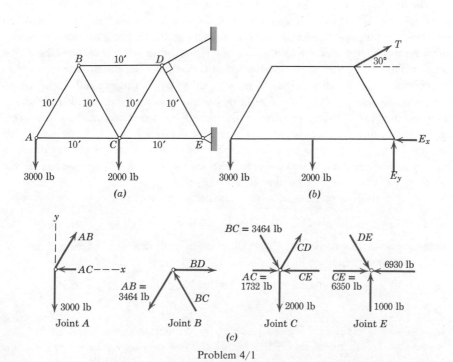

(a)   *(b)*

(c)

Problem 4/1

Joint *B* must be analyzed next, since there are more than two unknown forces on joint *C*. The force *BC* must provide an upward component, in which case *BD* must balance the force to the left. Again the forces are obtained from

$[\Sigma F_y = 0]$ $0.866BC - 0.866(3464) = 0$ $BC = 3464$ lb *C*   *Ans.*

$[\Sigma F_x = 0]$ $BD - 0.5(2)(3464) = 0$ $BD = 3464$ lb *T*   *Ans.*

Joint *C* now contains only two unknowns, and these are found as before:

$[\Sigma F_y = 0]$ $0.866CD - 0.866(3464) - 2000 = 0$

$CD = 5774$ lb *T*   *Ans.*

$[\Sigma F_x = 0]$ $CE - 1732 - 0.5(3464) - 0.5(5774) = 0$

$CE = 6350$ lb *C*   *Ans.*

Finally, from joint *E* there results

$[\Sigma F_y = 0]$ $0.866DE = 1000$ $DE = 1154$ lb *C*   *Ans.*

and the equation $\Sigma F_x = 0$ checks.

4/2 By using the method of sections calculate the force in member *DJ* of the Howe roof truss illustrated. Neglect any horizontal components of force at the supports. The loads are expressed in kips (K) where 1 kip = 1000 lb.

Solution. It is not possible to pass a section through *DJ* without cutting four members whose forces are unknown. Although three of these cut by section 2 are concurrent at *J* and therefore the moment equation about *J* could be used to obtain the fourth, *DE*, the force in *DJ* cannot be obtained from the remaining two equilibrium principles. It is necessary to consider first the adjacent section 1 before considering section 2.

The free-body diagram for section 1 is drawn and includes the reaction of 3.67 kips at *A*, which is previously calculated from the equilibrium of the truss as a whole. In assigning the proper directions for the forces acting on the three cut members a balance of moments about *A* eliminates the effects of *CD* and *JK* and clearly requires that *CJ* be up and to the left. A balance of moments about *C* eliminates the effect of the three forces concurrent at *C* and indicates that *JK* must be to the right to supply sufficient counterclockwise moment. Again it should be fairly obvious that the lower chord is under tension because of the bending tendency of the truss. Although it should also be apparent that the top chord is under compression, the force in *CD* will be arbitrarily assigned as tension. There is no harm in assigning one or more of the forces in the wrong direction as long as the calculations are consistent with the assumption. A negative answer will show the need for reversing the direction of the force.

By the analysis of section 1, *CJ* is obtained from

$[\Sigma M_A = 0]$ $0.707CJ(30) - 2(10) - 2(20) = 0$ $CJ = 2.83$ kips *C*

In this equation the moment of *CJ* is calculated by considering its horizontal and vertical components acting at point *J*. Equilibrium of moments about *J* requires

$[\Sigma M_J = 0]$ $0.894CD(15) + 3.67(30) - 2(10) - 2(20) = 0$

$CD = -3.73$ kips

The moment of *CD* about *J* is calculated here by considering its two components as acting through *D*. The minus sign indicates that *CD* was assigned in the wrong direction.

Hence $$CD = 3.73 \text{ kips } C \qquad \qquad Ans.$$

If desired, the direction of CD may be changed on the free-body diagram and the algebraic sign of CD reversed in the calculations, or else the work may be left as it stands with a note stating the proper direction.

From the free-body diagram of section 2, which now includes the known value of CJ, a balance of moments about G is seen to eliminate DE and JK. Thus

$$[\Sigma M_G = 0] \qquad 30DJ + 2(40) + 2(50) - 3.67(60) - 2.83(0.707)(30) = 0$$

$$DJ = 3.33 \text{ kips } T \qquad \qquad Ans.$$

Again the moment of CJ is determined from its components considered to be acting at J. The answer for DJ is positive, so that the assumed tensile direction was correct. An analysis of the joint D alone also verifies this conclusion.

Although the procedure used here is undoubtedly the shortest for obtaining DJ, the student should consider other possibilities. It may be observed that if for some other problem the 2-kip load at I were not present but the other loads remained the same, the forces in IE and JE (as well as in HF and IF) would be zero. In this event a section through members CD, DJ, JE, and IJ would involve only three unknown forces, and a solution for DJ could be obtained with a single moment equation about A.

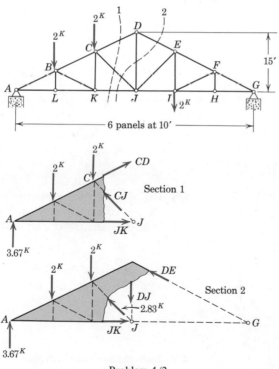

Problem 4/2

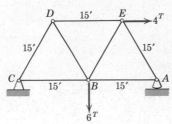

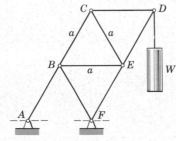

Problem 4/3

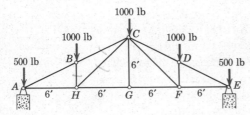

Problem 4/4

Problem 4/5

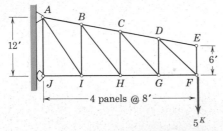

Problem 4/6

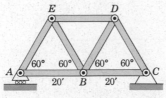

Problem 4/8

Problems

4/3 Calculate the force in each member of the truss shown.

$Ans.$ $AE = 5.46$ tons C, $AB = 2.73$ tons T
$BE = 5.46$ tons T, $DE = 1.46$ tons C
$BD = 1.46$ tons T, $CD = 1.46$ tons C
$BC = 4.73$ tons T

4/4 A snow load transfers the forces shown to the upper joints of a Pratt roof truss. Neglect any horizontal reactions at the supports and compute the forces in members BH, BC, and CH.

4/5 Calculate the forces in members BE and BC for the truss composed of equilateral triangles. Solve for each force from an equilibrium equation which contains that force as the only unknown.

$Ans.$ $BE = W/\sqrt{3}$, T
$BC = W/\sqrt{3}$, T

4/6 Calculate the force supported by member BH of the loaded cantilever truss. Use only one free-body diagram and one equation of equilibrium.

4/7 For the truss of Prob. 4/6 calculate the forces in members EF, DF, and GF.

4/8 Each member of the truss is a uniform 20-ft bar weighing 400 lb. Calculate the average tension or compression in each member due to the weights of the members.

$Ans.$ $AE = CD = 2000/\sqrt{3}$ lb C
$AB = BC = 1000/\sqrt{3}$ lb T
$BE = BD = 800/\sqrt{3}$ lb T
$DE = 1400/\sqrt{3}$ lb C

4/9 Calculate the force in each member of the truss, which is composed of 45-deg right triangles and loaded as shown.

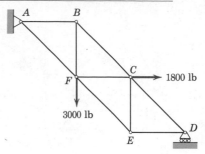

Problem 4/9

4/10 Calculate the forces in members *BH*, *CD*, and *GD* for the truss loaded by the 4- and 6-ton forces.

Ans. BH = 4.71 tons *C*
CD = 0.67 tons *C*
GD = 0

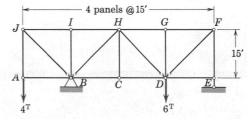

Problem 4/10

4/11 Determine the force in each member of the two trusses that support the 1000-lb load at their common pin *K*.

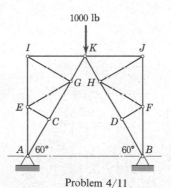

Problem 4/11

4/12 Calculate the forces in members *CF*, *BF*, *BG*, and *FG* for the simple crane truss.

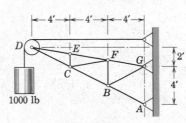

Problem 4/12

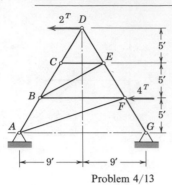

Problem 4/13

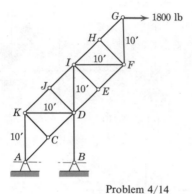

Problem 4/14

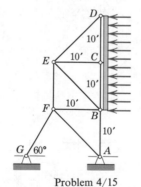

Problem 4/15

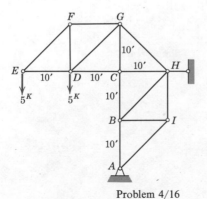

Problem 4/16

4/13 Determine the forces in members *BF* and *AF* for the truss under the action of the horizontal 2- and 4-ton loads.

Ans. BF = 0, *AF* = 3.51 tons *C*

4/14 Calculate the forces in members *DI*, *DE*, and *EI* for the loaded truss shown.

4/15 Calculate the forces in members *BE*, *BC*, *BF*, and *BA* for the signboard truss produced by a horizontal wind load of 800 lb on the sign. A separate analysis of the sign considered as a beam shows that $\frac{5}{8}$ of this load is supported at the midpoint *C*.

Ans. BE = 919 lb *T*, *BC* = 150 lb *T*
BF = 800 lb *C*, *BA* = 800 lb *T*

4/16 Calculate the forces in members *CH*, *CB*, and *GH* for the cantilevered truss. Solve for each force from a moment equation which contains that force as the only unknown.

Ans. CH = 15 kips *C*
CB = 25 kips *C*
GH = 21.2 kips *T*

4/17 Solve for the force in member *CF* of the truss in terms of the load *L*. All triangles are equilateral.

4/18 After solving for the external reaction at *E* for the truss of Prob. 4/17, determine the force in member *BF* from only one additional equation of equilibrium applied to one additional free-body diagram.   *Ans. BF = 3L/5, T*

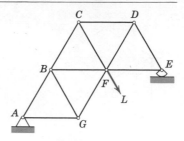

Problem 4/17

4/19 Each of the loaded trusses has supporting constraints which are statically indeterminate. List all members of each truss whose forces are not affected by the indeterminacy of the supports and that may be computed directly by using only the equations of equilibrium. Assume that the loading and dimensions of the trusses are known.

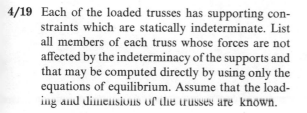

(a)

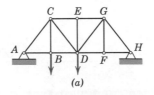

(b)

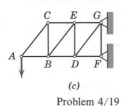

(c)

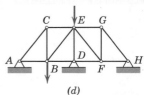

(d)

Problem 4/19

4/20 Verify the fact that each of the trusses contains one or more elements of redundancy, and propose two separate changes, either one of which would remove the redundancy and produce complete statical determinacy. All members can support compression as well as tension.

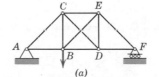

(a)

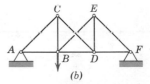

(b)

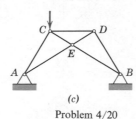

(c)

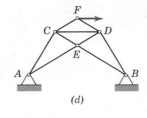

(d)

Problem 4/20

4/21 Verify the fact that each of the loaded trusses shown is unstable internally (nonrigid), and indicate at least two alternative ways to insure internal stability (rigidity) for each truss by the addition of one or more members without introducing redundancy.

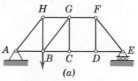

(a)

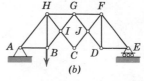

(b)

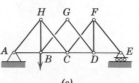

(c)

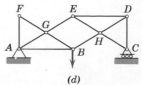

(d)

Problem 4/21

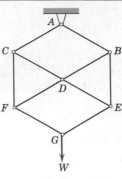

Problem 4/22

4/22 Solve for the forces in members *BE* and *BD* of the truss that supports the load *W*. All interior angles are 60 deg or 120 deg.

Ans. BE = W, T;   BD = W, C

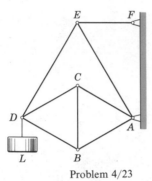

Problem 4/23

4/23 Determine the force in member *AC* in terms of the load *L* supported by the truss. All interior acute angles are either 30 or 60 deg.

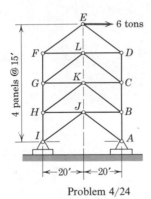

Problem 4/24

4/24 Solve for the force in member *GL* of the loaded tower truss by the method of joints.

Ans. GL = 3.75 tons T

4/25 Both horizontal and vertical components of the external support reactions exist at *A* and *I* for the tower truss of Prob. 4/24. Is the truss statically indeterminate? Explain.

4/26 Determine by inspection the force in member *CJ* for the crane truss. Calculate the force in members *CL* and *CB*.

Ans. CL = 0,   *CB* = 20 tons *C*

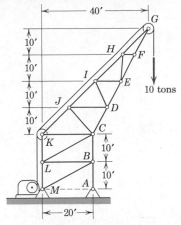

Problem 4/26

4/27 The movable gantry is used to erect and prepare a 500-ton rocket for firing. The primary structure of the gantry is approximated by the symmetrical plane truss shown, which is statically indeterminate. As the gantry is positioning a 60-ton section of the rocket suspended from *A*, strain gage measurements indicate a compressive force of 5 tons in member *AB* and a tensile force of 12 tons in member *CD* due to the 60-ton load. Calculate the corresponding force in members *BF* and *EF*.

Ans. BF = 19.38 tons *C*,   *EF* = 12 tons *T*

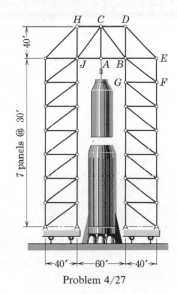

Problem 4/27

4/28 Calculate the forces in members *FC* and *FB* due to the 10-ton load on the crane truss.

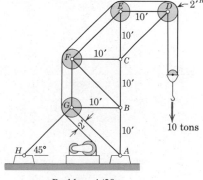

Problem 4/28

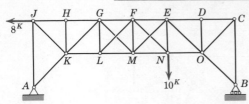

Problem 4/29

4/29   The truss shown is composed of 45-deg right tri-angles. The crossed members in the center two panels are slender tie rods incapable of support-ing compression. Retain the two rods which are under tension and compute the magnitudes of their tensions. Also find the force in member *MN*.

$$Ans. \ FN = 8.48 \text{ kips } T$$
$$GM = 8.48 \text{ kips } T$$
$$MN = 2 \text{ kips } T$$

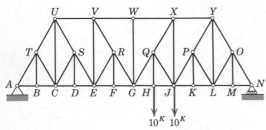

Problem 4/30

4/30   Find the force in member *JQ* for the Baltimore truss where all angles are 30, 60, 90, or 120 deg.

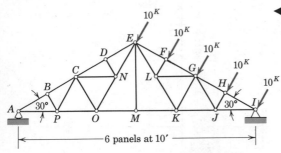

Problem 4/31

◀ **4/31**   Find the forces in members *EF*, *KL*, and *GL* for the Fink truss shown. (*Hint:* Note that the forces in *BP*, *PC*, *DN*, etc., are zero.)

$$Ans. \ EF = 40.4 \text{ kips } C$$
$$KL = 20.0 \text{ kips } T$$
$$GL = 10.0 \text{ kips } T$$

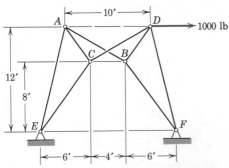

Problem 4/32

◀ **4/32**   The hinged frames *ACE* and *DFB* are connected by two hinged bars, *AB* and *CD*, which cross without being connected. Compute the force in *AB*.   *Ans. AB* = 377 lb *C*

◄ 4/33 In the traveling bridge crane shown all crossed members are slender tie rods incapable of supporting compression. Determine the forces in members *DF* and *EF* and find the horizontal reaction on the truss at *A*. Show that if *CF* = 0, *DE* = 0 also.

$$Ans.\ DF = 76.8\ tons\ C$$
$$EF = 36.4\ tons\ C$$
$$A_x = 10.1\ tons\ to\ the\ right$$

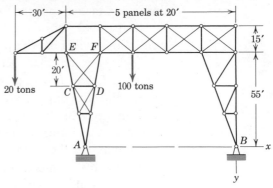

Problem 4/33

19 Space Trusses. A space truss is the three-dimensional counterpart of the plane truss described in the previous article. The idealized space truss consists of rigid links connected at their ends by ball-and-socket joints. It was seen that a triangle of pin-connected bars forms the basic noncollapsible unit for the plane truss. A space truss, on the other hand, requires six bars joined at their ends to form the edges of a tetrahedron for the basic noncollapsible unit. In Fig. 44*a* the two bars *AD* and *BD* joined at *D* require a third support *CD* to keep the triangle *ABD* from rotating about *AB*. In Fig. 44*b* the supporting base is replaced by three more bars *AB*, *BC*, and *AC* to form a tetrahedron independent of the foundation for its own rigidity. Added units to the structure may be formed with three additional concurrent bars whose

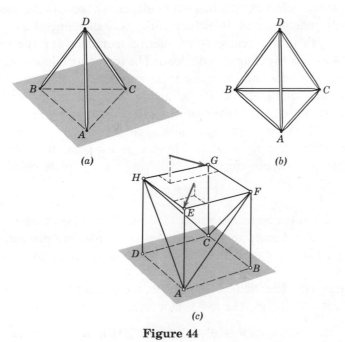

Figure 44

ends are attached to three fixed joints on the existing structure. Thus in Fig. 44c the bars *AF*, *BF*, and *CF* are attached to the foundation and therefore fix point *F* in space. Likewise point *H* is fixed in space by the bars *AH*, *DH*, and *CH*. The three additional bars *CG*, *FG*, and *HG* are attached to the three fixed points *C*, *F*, and *H* and therefore fix *G* in space. Point *E* is similarly established. The structure is entirely rigid, and the two applied loads shown will result in forces in all of the members.

Ideally there must be point support, such as represented by a ball-and-socket joint, for the connections of a space truss in order that there be no bending in the members. Again, as in riveted and welded connections for plane trusses, if the center lines of joined members intersect at a point, the assumption of two-force members under simple tension and compression may be justified.

For a space truss which is supported externally in such a way that it is statically determinate as an entire unit, a relationship exists between the number of its joints and the number of its members necessary for internal stability without redundancy. Since the equilibrium of each joint is specified by three scalar force equations, there are a total of $3j$ such equations for a simple space truss with j joints. For the entire truss composed of m members there are m unknowns plus six unknown support reactions in the general case of a statically determinate space structure. Thus, for a simple space truss composed of tetrahedral elements, the equation $m + 6 = 3j$ will be satisfied if the truss is statically determinate internally. Again, as in the case of the plane truss, this relation is a necessary condition for stability, but it is not a sufficient condition, since one or more of the m members can be arranged in such a way as not to contribute to a stable configuration of the entire truss. If $m + 6 > 3j$, there are more members than there are independent equations, and the truss is statically indeterminate internally with redundant members present. If $m + 6 < 3j$, there is a deficiency of internal members, and the truss is unstable and subject to collapse under load. The foregoing relationship between the number of joints and the number of members for a space truss is very helpful in the preliminary design for such a truss, since the configuration is not nearly so obvious as in the case of a plane truss, where the geometry for statical determinacy is generally quite apparent.

The method of joints developed in Art. 18 for plane trusses may be extended directly to space trusses by satisfying the complete vector equation

$$\Sigma \mathbf{F} = 0$$

for each joint. It is necessary to start at some joint where at least one known force acts and not more than three unknown forces are present. Adjacent joints upon which not more than three unknown forces act may be analyzed in turn.

The method of sections developed in the previous article may also be applied to space trusses. The two vector equations

$$\Sigma \mathbf{F} = 0 \qquad \text{and} \qquad \Sigma \mathbf{M} = 0$$

must be satisfied for any section of the truss, where the zero moment sum will hold for all moment axes. Since the two vector equations are equivalent to six scalar equations, it follows that a section should in general not be passed through more than six members whose forces are unknown. The method of sections for space trusses is not widely used, however, because a moment axis can seldom be found which eliminates all but one unknown as in the case of plane trusses.

Vector notation for expressing the terms in the force and moment equations for space trusses may often be used to advantage and is used in the sample problem that follows.

Sample Problem

4/34 The space truss shown in the *a*-part of the figure is secured to a vertical wall by a ball-and-socket joint at A, by links in the x- and y-directions at F, and by a link in the y direction at E, thus providing six statically determinate constraints. Joints B, D, and C in that order are seen to be fixed. The truss carries a vertical load L at D and a vertical load $2L$ at C. Determine the forces in members ED and FB.

Solution. It is observed first that the truss as a whole has $j = 6$ joints and $m = 12$ members, so that the relation $m + 6 = 3j$ holds. Thus there are a sufficient number of members to provide a stable configuration, as is accomplished by the arrangement shown.

The support reactions A_y, A_z, E_y, and F_y are easily assigned in their correct directions by inspection. The reactions F_x and A_x may also be assigned correctly by observing the moment requirement about axes through F and A in the y-direction. These reactions are shown on the free-body diagram of a section of the truss in the *b*-part of the figure. The support reactions are obtained from the truss as a whole as follows.

$$[\Sigma M_{FE} = 0] \qquad A_y \frac{a\sqrt{3}}{2} - La - 2La = 0, \qquad A_y = 2\sqrt{3}L$$

$$[\Sigma M_{F_z} = 0] \qquad 2\sqrt{3}L\frac{a}{2} - E_ya = 0, \qquad E_y = \sqrt{3}L$$

$$[\Sigma F_y = 0] \qquad \sqrt{3}L + F_y - 2\sqrt{3}L = 0, \qquad F_y = \sqrt{3}L$$

$$[\Sigma M_{AB} = 0] \qquad L\frac{a}{4} - 2L\frac{a}{4} + F_x\frac{a\sqrt{3}}{2} = 0, \qquad F_x = \frac{L}{2\sqrt{3}}$$

$$[\Sigma F_x = 0] \qquad A_x = \frac{L}{2\sqrt{3}}$$

$$[\Sigma F_z = 0] \qquad A_z = 3L$$

A joint-by-joint analysis of the truss may begin with either joint A or joint C, upon either of which only three unknowns act. In the case of joint C, an incidental observation may be made that $2L$, DC, and BC all lie in the same vertical plane, so that F_{EC} must be zero. With F_{DC} determined from joint C, the force in ED can be obtained by the equilibrium of forces on joint D. The geometry of this particular truss, however, permits a determination of the force F_{ED} from a single moment equation taken about the axis AB for the free-body diagram of the truss section shown in the *b*-part of the figure. All unknown forces in the cut members for the section are shown in the positive sense as tensile forces. Thus a negative sign upon computation will signify a compressive force.

The lines of action of F_{FB}, F_{EB}, F_{FD}, and F_{EC} all intersect the moment axis AB, so that the only moments about AB are due to F_{ED} and F_x. The distance $\overline{ED}$ is $a\sqrt{(3/4)^2 + 1^2 + 3/4^2} = a\sqrt{7}/2$, and the direction cosines of $\mathbf{F}_{ED}$ are $l = 3/(2\sqrt{7})$, $m = 2/\sqrt{7}$, $n = \sqrt{3}/(2\sqrt{7})$. Thus the force in ED is

$$\mathbf{F}_{ED} = \frac{F_{ED}}{2\sqrt{7}}(3\mathbf{i} + 4\mathbf{j} + \sqrt{3}\mathbf{k})$$

and the vector from A to E is

$$\mathbf{r}_{AE} = -\frac{a}{2}(\mathbf{i} + \sqrt{3}\mathbf{k})$$

The moment equation becomes

$$[\Sigma M_{AB} = 0] \qquad \mathbf{r}_{AE} \times \mathbf{F}_{ED} \cdot \mathbf{j} + F_x\frac{a\sqrt{3}}{2} = 0$$

Thus

$$-\frac{a}{2}(\mathbf{i} + \sqrt{3}\mathbf{k}) \times \frac{F_{ED}}{2\sqrt{7}}(3\mathbf{i} + 4\mathbf{j} + \sqrt{3}\mathbf{k}) \cdot \mathbf{j} + \frac{L}{2\sqrt{3}}\frac{a\sqrt{3}}{2} = 0$$

which upon expansion gives

$$F_{ED} = \frac{\sqrt{7}L}{2\sqrt{3}} \qquad\qquad\qquad Ans.$$

To reduce the number of unknowns at joint F to three, the force F_{FA} is obtained by an analysis of joint A in the x-z plane. Thus the free-body diagram of joint A in the c-part of the figure gives

$$[\Sigma F_{x'} = 0] \qquad F_{FA}\frac{\sqrt{3}}{2} + \frac{L}{2\sqrt{3}}\frac{\sqrt{3}}{2} - \frac{3L}{2} = 0, \qquad F_{FA} = \frac{5L}{2\sqrt{3}}$$

The unknown forces acting on joint F as shown in the d-part of the figure are expressed in vector form as

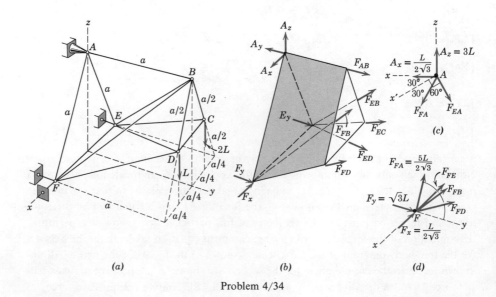

(a)   (b)   (d)

Problem 4/34

$$\mathbf{F}_{FD} = \frac{F_{FD}}{2\sqrt{5}}(-\mathbf{i} + 4\mathbf{j} + \sqrt{3}\mathbf{k}), \qquad \mathbf{F}_{FB} = \frac{F_{FB}}{2\sqrt{2}}(-\mathbf{i} + 2\mathbf{j} + \sqrt{3}\mathbf{k})$$

$$\mathbf{F}_{FA} = \frac{5L}{4\sqrt{3}}(-\mathbf{i} + \sqrt{3}\mathbf{k}), \qquad \mathbf{F}_{FE} = F_{FE}(-\mathbf{i})$$

Equilibrium of joint F requires

$$[\Sigma\mathbf{F} = 0] \qquad \left(-\frac{F_{FD}}{2\sqrt{5}} - \frac{F_{FB}}{2\sqrt{2}} - \frac{5L}{4\sqrt{3}} - F_{FE} - \frac{L}{2\sqrt{3}}\right)\mathbf{i}$$

$$+ \left(\frac{2F_{FD}}{\sqrt{5}} + \frac{F_{FB}}{\sqrt{2}} + L\sqrt{3}\right)\mathbf{j} + \left(\frac{\sqrt{3}F_{FD}}{2\sqrt{5}} + \frac{\sqrt{3}F_{FB}}{2\sqrt{2}} + \frac{5L}{4}\right)\mathbf{k} = 0$$

The coefficients of $\mathbf{i}$, $\mathbf{j}$, and $\mathbf{k}$ must be zero, and the simultaneous solution of the three equations gives

$$F_{FD} = -\frac{\sqrt{5}L}{2\sqrt{3}} \qquad F_{FE} = -\frac{L}{2\sqrt{3}} \qquad F_{FB} = -\frac{2\sqrt{2}L}{\sqrt{3}} \qquad \textit{Ans.}$$

The negative signs indicate compression since the positive directions were chosen as tensile (away from the joint). The remaining unknown forces may be determined by continuing the analysis with joint B.

Problems

4/35 The space truss is built upon the triangular base *ABC*. The locations of joints *D* and *E* are established by the links shown. Show that this configuration is internally stable. Also replace link *AE* by a different link that will preserve the rigidity of the truss.

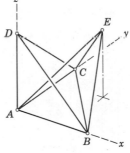

Problem 4/35

4/36 The prismatic space truss has a horizontal base *ADE* and a parallel top face *BCF* in the shape of equal equilateral triangles which are connected by three equal vertical legs and braced by three diagonal members as shown. Show that this truss represents a stable configuration.

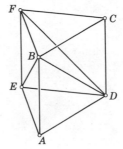

Problem 4/36

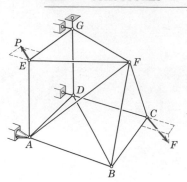

Problem 4/37

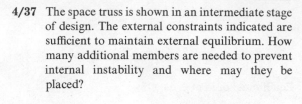

4/37 The space truss is shown in an intermediate stage of design. The external constraints indicated are sufficient to maintain external equilibrium. How many additional members are needed to prevent internal instability and where may they be placed?

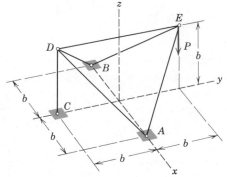

Problem 4/38

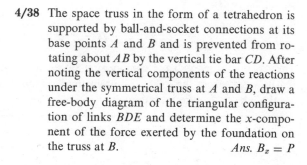

4/38 The space truss in the form of a tetrahedron is supported by ball-and-socket connections at its base points A and B and is prevented from rotating about AB by the vertical tie bar CD. After noting the vertical components of the reactions under the symmetrical truss at A and B, draw a free-body diagram of the triangular configuration of links BDE and determine the x-component of the force exerted by the foundation on the truss at B. *Ans.* $B_x = P$

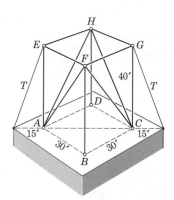

Problem 4/39

4/39 The rectangular space truss 40 ft in height is erected on a horizontal square base 30 ft on a side. Guy wires are attached to the structure at E and G as shown and are tightened until the tension T in each wire is 1800 lb. Calculate the compression C in each of the similar diagonal members. *Ans.* $C = 745$ lb

4/40 The tetrahedral space truss has a horizontal base
ABC in the form of an isosceles triangle and legs
AD, BD, and CD which support the weight W
from point D. Each vertex of the base is suspended by a vertical wire from overhead supports. Calculate the force induced in member
AC.

$$Ans. \ F_{AC} = \frac{5W}{54}, \quad C$$

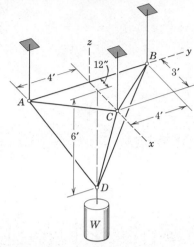

Problem 4/40

4/41 A space truss is constructed in the form of a cube
with six diagonal members shown. Verify that the
truss is internally stable. If the truss is subjected
to the compressive forces P applied at F and D
along the diagonal FD, determine the forces in
members FE and EG.

$$Ans. \ F_{FE} = P/\sqrt{3}, \ C; \qquad F_{EG} = P/\sqrt{6}, \ T$$

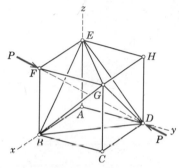

Problem 4/41

4/42 Each of the landing struts for a lunar spacecraft
is a space truss symmetrical about the vertical
x-z plane as shown. For a landing force of $F =$
480 lb, calculate the corresponding force in member BE. The assumption of static equilibrium for
the truss is permissible if the mass of the truss is
very small. Assume equal loads in the symmetrically placed members.

$$Ans. \ F_{BE} = 306 \ lb \ T$$

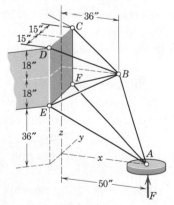

Problem 4/42

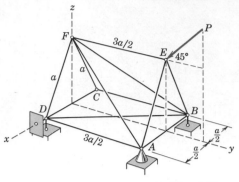

Problem 4/43

4/43 For the space truss shown check the sufficiency of the supports and also the number and arrangement of the members to ensure statical determinacy both external and internal. By inspection determine the forces in DC, CB, and CF. Calculate the force in member AF and the x-component of the reaction on the truss at D.

$$Ans. \ F_{AF} = \frac{\sqrt{13}}{3\sqrt{2}} P, \ T$$

$$D_x = \frac{P}{3\sqrt{2}}, \quad (-x\text{-direction})$$

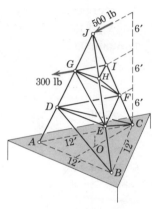

Problem 4/44

◀ **4/44** A space tower is erected on an equilateral triangular base and forms a symmetrical pyramid with its three 6-ft-high truss sections. The tower supports the horizontal 500-lb force, which is parallel to the base center line OC, and the horizontal 300-lb force applied along the line of GI. Check the internal rigidity of the truss and calculate the force in member GF. Determine by inspection from the appropriate free bodies the forces in members GI, HI, GE, and HF.

$$Ans. \ F_{GF} = 321 \ \text{lb} \ T$$

20 Frames and Machines. Structures and mechanisms composed of joined members any one of which has more than two forces acting on it cannot be analyzed by the methods developed for simple trusses. Such members are multiforce members (three or more forces), and in general the forces will *not* be in the directions of the members. In the previous chapter the equilibrium of multiforce bodies was discussed and illustrated, but attention was focused on the equilibrium of a *single* rigid body. In this present article attention is focused on the equilibrium of *interconnected* rigid bodies which contain multiforce members. Although most such bodies may be analyzed as two-dimensional systems, there are numerous examples of frames and machines that are three-dimensional.

The forces acting on each member of a connected system are found by isolating the member with a free-body diagram and applying the established equations of equilibrium. The *principle of action and reaction* must be carefully observed when representing the forces of interaction on the separate free-body diagrams. If the structure contains more members or supports than are necessary to prevent collapse, then, as in the case of trusses, the problem is statically indeterminate, and the principles of equilibrium, although necessary, are not sufficient for solution.

If the frame or machine constitutes a rigid unit by itself when removed from its supports, the analysis is best begun by establishing all the forces external to the structure considered as a single rigid body. The structure is then dismembered and the equilibrium of each part is considered. The equilibrium equations for the several parts will be related through the terms involving the forces of interaction. If the structure is not a rigid unit by itself but depends on its external supports for rigidity, then it is usually necessary to consider first the equilibrium of some portion of the system that itself is inherently rigid.

It will be found that in most cases the analysis of frames and machines is facilitated by representing the forces in terms of their rectangular components. This is particularly so when the dimensions of the parts are given in mutually perpendicular directions. The advantage of this representation is that the calculation of moment arms is accordingly simplified. In some three-dimensional problems, particularly when moments are evaluated about axes that are not parallel to the coordinate axes, the use of vector notation is of advantage.

It is not always possible to assign every force or its components in the proper sense when drawing the free-body diagrams, and it becomes necessary to make an arbitrary assignment. In any event it is *absolutely necessary* that a force be *consistently* represented on the diagrams for interacting bodies which involve the force in question. Thus for two bodies connected by the pin A, Fig. 45a, the components when separated must be consistently represented in the *opposite* directions. For a ball-and-socket connection between members of a space frame, the action-and-reaction principle must be applied to all three components as shown in Fig. 45b. The assigned directions may prove to be wrong when the algebraic signs of the components are determined upon calculation. If A_x, for instance, should turn out to be negative, it is actually acting in the direction opposite to that originally represented. Accordingly it would be necessary to reverse the direction of the force on *both* members and to reverse the sign of this force term in the equations. Or the representation may be left as originally made, and the proper sense of the force will be understood from the negative sign.

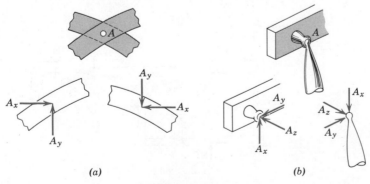

(a)   (b)

Figure 45

Finally, situations occasionally arise where it is necessary to solve two or more equations simultaneously in order to separate the unknowns. In most instances, however, simultaneous solutions may be avoided by careful choice of the member or group of members for the free-body diagram and by a careful choice of moment axes which will eliminate undesired terms from the equations. The method of solution described in the foregoing paragraphs is illustrated in the following sample problems.

Sample Problems

4/45 The frame supports the 400-lb load in the manner shown in part *a* of the figure. Neglect the weights of the members compared with the forces induced by the load and compute the horizontal and vertical components of all forces acting on each of the members.

Solution. It is observed first that the three supporting members that constitute the frame form a rigid assembly that can be analyzed as a single unit. It is also observed that the arrangement of the external supports makes the frame statically determinate.

The free-body diagram of the entire frame is drawn in part *b* of the figure, and the external reactions determined. Thus

$$[\Sigma M_A = 0] \qquad 11(400) - 10D = 0, \qquad D = 440 \text{ lb}$$

$$[\Sigma F_x = 0] \qquad A_x - 440 = 0, \qquad A_x = 440 \text{ lb}$$

$$[\Sigma F_y = 0] \qquad A_y - 400 = 0, \qquad A_y = 400 \text{ lb}$$

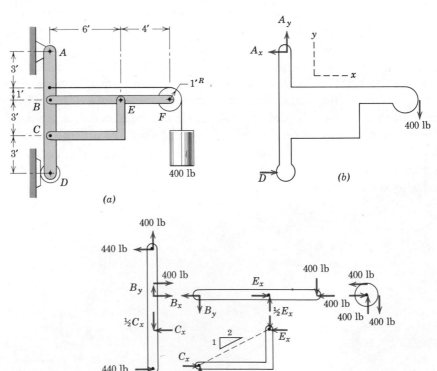

Problem 4/45

Next the frame is dismembered and a free-body diagram of each member is drawn in part *c* of the figure. The diagrams are arranged in their approximate relative position to aid in keeping track of the common forces of interaction. The external reactions just obtained are entered onto the diagram for *AD*. Other known forces are the 400-lb forces exerted by the shaft of the pulley on the member *BF*, as obtained from the free-body diagram of the pulley. The cable tension of 400 lb is also shown acting on *AD* at its attachment point.

Next, the components of all unknown forces are shown on the diagrams. Here it is observed that *CE* is a two-force member, so that the direction of the line joining the two points of application of force, and not the shape of the member, determines the direction of the force and hence the ratio of the force components acting at *C* and *E*. These components have equal and opposite reactions, which are shown on *BF* at *E* and on *AD* at *C*. The positive sense of the components at *B* may not be recognized at first glance and so may be arbitrarily but consistently assigned.

The solution may proceed by using a moment equation about *B* or *E* for member *BF* followed by the two force equations. Thus

$[\Sigma M_B = 0]$ $\qquad$ $400(10) - \frac{1}{2}E_x(6) = 0,$ $\qquad$ $E_x = 1333$ lb $\qquad$ *Ans.*

$[\Sigma F_y = 0]$ $\qquad$ $B_y + 400 - 1333/2 = 0,$ $\qquad$ $B_y = 267$ lb $\qquad$ *Ans.*

$[\Sigma F_x = 0]$ $\qquad$ $B_x + 400 - 1333 = 0,$ $\qquad$ $B_x = 933$ lb $\qquad$ *Ans.*

Positive numerical values of the unknowns mean that their directions were correctly assumed on the free-body diagrams. The value of $C_x = E_x = 1333$ lb obtained by inspection of the free-body diagram of *CE* is now entered onto the diagram for *AD*, along with the values of B_x and B_y just determined. The equations of equilibrium may now be applied to member *AD* as a check, since all the forces acting on it have already been computed. The equations give

$[\Sigma M_C = 0]$ $\qquad$ $440(7) + 440(3) - 400(4) - 933(3) = 0$

$[\Sigma F_x = 0]$ $\qquad$ $440 - 1333 + 933 + 400 - 440 = 0$

$[\Sigma F_y = 0]$ $\qquad$ $-1333/2 + 267 + 400 = 0$

which verify the previous calculations.

4/46 The loader has a capacity of 4 yd³ and is handling dirt that weighs 80 lb/ft³. For the particular position shown, where the arm *EB* is horizontal, find the compressive force in the hydraulic piston rod *JL* and the total shear force supported by the pin at *A*. The machine is symmetrical about a central vertical plane in the fore-and-aft direction and has two sets of the linkages shown. The weights of the members may be neglected compared with the forces they support.

Solution. The weight of the load of dirt supported by the set of links shown is one-half of the total or $\frac{1}{2}(80)(27)(4) = 4320$ lb. The free-body diagrams of the several members are shown in the *b*-part of the figure, where the directions of the force components either are assigned correctly by observation or are assumed. In each case the principle of action and reaction has been consistently observed. The hydraulic cylinders *BK* and *JL* act as two-force members so that

$$B_y = B_x \tan \alpha = \tfrac{70}{40}B_x = 1.75B_x$$
$$L_y = L_x \tan \beta = \tfrac{12}{64}L_x = 0.1875L_x$$

The force L in JL and the reaction at A require a determination of the other forces shown on the free-body diagrams. For the scoop

$[\Sigma M_E = 0]$ $\qquad\qquad$ $4320(18) - 36F_x = 0,$ $\qquad$ $F_x = 2160$ lb

$[\Sigma F_x = 0]$ $\qquad\qquad$ $E_x - 2160 = 0,$ $\qquad\qquad$ $E_x = 2160$ lb

$[\Sigma F_y = 0]$ $\qquad\qquad$ $E_y - 4320 = 0,$ $\qquad\qquad$ $E_y = 4320$ lb

From the link DF it is easily seen that

$$D_x = F_x = 2160 \text{ lb}, \qquad G_x = 2(2160) = 4320 \text{ lb}$$

Similarly from link LH,

$$L_x = 4320 \text{ lb}, \qquad C_x = 2(4320) = 8640 \text{ lb}$$

and

$[\Sigma F_y = 0]$ $\quad$ $C_y - L_y = 0,$ $\quad$ $C_y = 0.1875L_x,$ $\quad$ $C_y = 0.1875(4320) = 810$ lb

From the free-body diagram of EBA it is seen that B_x and B_y may be eliminated by summing moments about K and B. Hence

$[\Sigma M_K = 0]$ $\quad$ $52A_x + (2160 + 2160 - 8640)70 + 810(64) - 4320(120) = 0$

$\qquad\qquad\qquad$ $A_x = 14{,}790$ lb

$[\Sigma M_B = 0]$ $\qquad$ $40A_y + 810(24) - 14790(18) - 4320(80) = 0$

$\qquad\qquad\qquad$ $A_y = 14{,}810$ lb

and $\qquad\qquad$ $A = \sqrt{(14{,}790)^2 + (14{,}810)^2} = 20{,}920$ lb $\qquad\qquad\qquad$ *Ans.*

The force in the cylinder JL is

$$L = \sqrt{(4320)^2 + (810)^2} = 4400 \text{ lb} \qquad\qquad\qquad \textit{Ans.}$$

If desired the force at B may now be computed by

$[\Sigma F_x = 0]$ $\quad$ $2160 + 2160 - 8640 - B_x + 14{,}790 = 0,$ $\quad$ $B_x = 10{,}470$ lb

and

$$B_y = 1.75B_x, \qquad B_y = 1.75(10{,}470) = 18{,}320 \text{ lb}$$

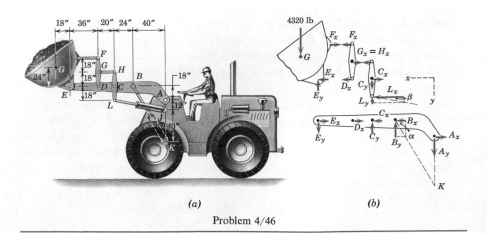

(a) $\qquad\qquad\qquad\qquad\qquad\qquad\qquad$ *(b)*

Problem 4/46

4/47 The A-frame shown in the *a*-part of the figure is hinged about the *y*-axis through points *E* and *F*, which cannot offer restraint in the *y*-direction. The connections at *A*, *B*, *C*, *D*, and *G* may be treated as ball-and-socket joints. Link *CG* is the only two-force member. Compute the *x*-, *y*-, and *z*-components of all forces acting on each member of the frame. The weights of the members may be neglected compared with the loads transmitted.

Solution. The free-body diagram of the frame as a whole is shown in the *b*-part of the figure and is used to compute the external reactions.

$$[\Sigma M_y = 0] \qquad G\frac{4}{\sqrt{2}} - 500(4) = 0, \qquad G = 500\sqrt{2} \text{ lb}$$

$$[\Sigma M_{F_x} = 0] \qquad 500(4) - 8E_z = 0, \qquad E_z = 250 \text{ lb}$$

$$[\Sigma F_z = 0] \qquad F_z + 250 - 500 = 0, \qquad F_z = 250 \text{ lb}$$

$$[\Sigma M_{F_z} = 0] \qquad 500(4) - 500(2) - 8E_x = 0, \qquad E_x = 125 \text{ lb}$$

$$[\Sigma F_x = 0] \qquad 500 + F_x - 125 - 500 = 0, \qquad F_x = 125 \text{ lb}$$

Each of the three multiforce members of the frame is isolated in the *c*-part of the figure with its free-body diagram with the computed support reactions included. The correct sense of each component is not necessarily apparent at the outset, and no harm is done if a component is assigned in the wrong sense as long as the action and reaction principle is not violated. By inspection, the equilibrium of the member *BD* requires

$$B_x = D_x = 250 \text{ lb}, \qquad B_z = D_z = 250 \text{ lb} \qquad \qquad Ans.$$

With two of the three components of *D* known, members *ADF* and *ABE* may be analyzed. Hence for *ADF*

$$[\Sigma F_z = 0] \qquad A_z + 250 - 250 = 0, \qquad A_z = 0 \qquad Ans.$$

$$[\Sigma F_x = 0] \qquad A_x + 250 + 125 - 500 = 0, \qquad A_x = 125 \text{ lb} \qquad Ans.$$

$$[\Sigma M_{D_x} = 0] \qquad 250(2) - 4A_y = 0, \qquad A_y = 125 \text{ lb} \qquad Ans.$$

$$[\Sigma F_y = 0] \qquad 125 - D_y = 0, \qquad D_y = 125 \text{ lb} \qquad Ans.$$

From *BD* it is now clear that $\qquad\qquad B_y = 125 \text{ lb} \qquad Ans.$

As a check, the equilibrium of *ABE* may be verified with the known force components.

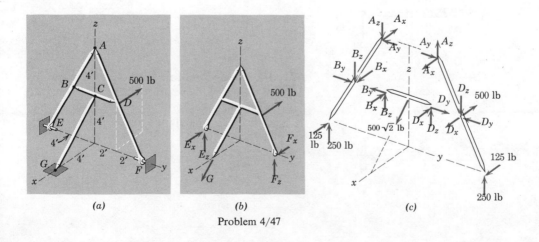

(a)   (b)   (c)

Problem 4/47

It should be observed that the sequence of the solution was chosen to eliminate coupled equations that require simultaneous solutions. A little thought will disclose the most favorable sequence or alternative sequences of steps in any given problem. In more involved problems it is not always feasible to avoid simultaneous solutions to the equilibrium equations.

Problems

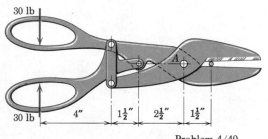

Problem 4/48

4/48 Compute the force supported by pin *A* of the frame loaded by the 800-lb-in. couple.

Ans. A $= 20\sqrt{2}$ lb

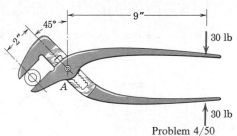

Problem 4/49

4/49 Compound-lever snips, shown in the figure, are often used in place of regular tinners' snips when large cutting forces are required. For the gripping force of 30 lb, what is the cutting force *P* at a distance of $1\frac{1}{2}$ in. along the blade from the pin at *A*?

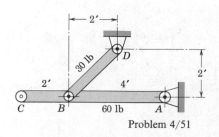

Problem 4/50

4/50 Compute the force supported by the pin at *A* for the channel-lock pliers under a grip of 30 lb.

Ans. A $= 157.6$ lb

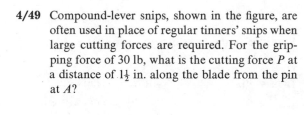

Problem 4/51

4/51 The two uniform bars are supported in the vertical plane as shown. If *AC* weighs 60 lb and *BD* weighs 30 lb, each with the center of gravity at its midlength, compute the force supported by the pin at *A*.

4/52 The force P holds the spring-loaded frame in the equilibrium position shown. For this position determine the expression for the tension T in the spring and the force supported by the pin at B.

$$Ans. \quad T = P \operatorname{ctn} \frac{\theta}{2}, \quad B = P \operatorname{csc} \frac{\theta}{2}$$

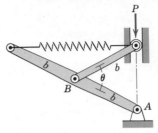

Problem 4/52

4/53 Calculate the total force supported by the pin at A under the action of the 100-lb force and the 200-lb-ft couple.

$$Ans. \quad A = 80.2 \text{ lb}$$

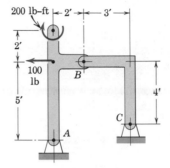

Problem 4/53

4/54 Calculate the force supported by the pin at C for the frame loaded by the 60-lb force and the 300-lb-in. couple.

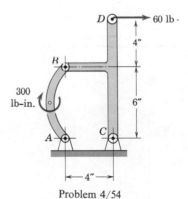

Problem 4/54

4/55 The speed reducer consists of the input shaft A and attached pinion B, which drives gear C and its output shaft D with a 2:1 reduction. The center of gravity of the 60-lb unit is at G. If a clockwise input torque of 400 lb-in. is applied to shaft A and if the output shaft D drives a machine at a constant speed, determine the net forces exerted *on* the base flange of the reducer at E and F by the combined action of the bolts and the supporting foundation.

$$Ans. \quad E = 27.3 \text{ lb down}, \quad F = 87.3 \text{ lb up}$$

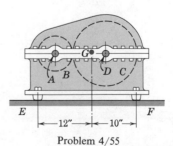

Problem 4/55

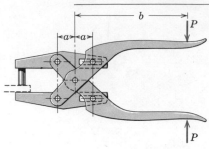

Problem 4/56

4/56 For the paper punch shown find the punching force Q corresponding to a hand grip P.

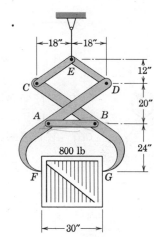

Problem 4/57

4/57 Compute the force supported by the pin at C for the frame subjected to the 300-lb-ft couple.

Ans. $C = 83.9$ lb

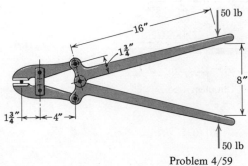

Problem 4/58

4/58 Compute the force acting in the link AB of the lifting tongs which cross without touching.

Problem 4/59

4/59 Determine the force Q exerted on each side of the bolt by the cutters if the handles are subjected to loads of 50 lb. *Ans. $Q = 4050$ lb*

4/60 The toggle pliers are used for a variety of clamping purposes. For the handle position given by $\alpha = 10$ deg and for a handle grip of $P = 30$ lb calculate the clamping force C produced.

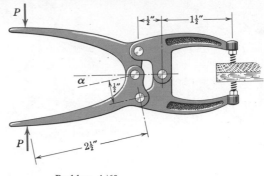

Problem 4/60

4/61 The lower jaw D of the toggle press slides with negligible frictional resistance along the fixed vertical column. Calculate the compressive force R exerted on the cylinder E and the force supported by the pin at A if a force $F = 40$ lb is applied to the handle at an angle of $\theta = 75$ deg.

Ans. $R = 193.2$ lb, $A = 166.7$ lb

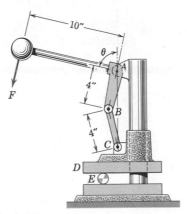

Problem 4/61

4/62 A double-axle suspension for use on small trucks is shown in the figure. The weight of the central frame F is 80 lb, and the weight of each wheel and attached link is 70 lb with center of gravity 27 in. from the vertical center line. For a load of $L = 2400$ lb transmitted to the frame F, compute the total shear force supported by the pin at A. If each spring has an uncompressed length of 10 in. and a stiffness of 640 lb/in., what is its compressed length l under the specified loads?

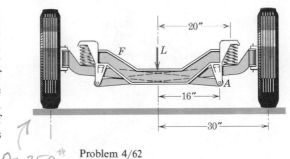

$A = 350^{\#}$
$S = 1590^{\#}$
$l = 7.52''$

Problem 4/62

4/63 Neglect the weights of the members and compute the x- and y-components of all forces on each of the two members of the machine resulting from the application of the 30-lb-ft couple.

$A_x = 0$, $A_y = 27.7$
$C_y = 27.7$
$B_x = 36.9$, $B_y = 27.7$
$D_x = 36.9$

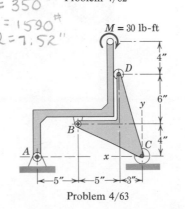

Problem 4/63

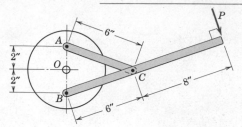

Problem 4/64

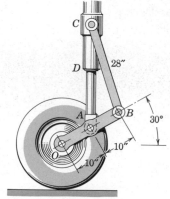

Problem 4/65

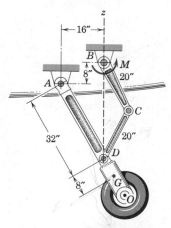

Problem 4/66

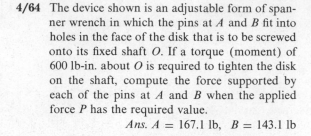

4/64 The device shown is an adjustable form of spanner wrench in which the pins at *A* and *B* fit into holes in the face of the disk that is to be screwed onto its fixed shaft *O*. If a torque (moment) of 600 lb-in. about *O* is required to tighten the disk on the shaft, compute the force supported by each of the pins at *A* and *B* when the applied force *P* has the required value.

Ans. A = 167.1 lb, *B* = 143.1 lb

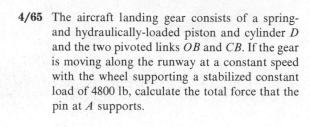

4/65 The aircraft landing gear consists of a spring- and hydraulically-loaded piston and cylinder *D* and the two pivoted links *OB* and *CB*. If the gear is moving along the runway at a constant speed with the wheel supporting a stabilized constant load of 4800 lb, calculate the total force that the pin at *A* supports.

4/66 The nose-wheel assembly is raised by the application of a torque *M* to link *BC* through the shaft at *B*. If the arm and wheel *AO* have a combined weight of 94 lb with center of gravity at *G*, find the value of *M* necessary to lift the wheel when *D* is directly under *B*. Neglect the weights of *BC* and *DC*. *Ans. M* = 1130 lb-in.

4/67 An anti-torque wrench is designed for use by a crewman of a spacecraft where he has no stable platform against which to push as he tightens a bolt. The pin A fits into an adjacent hole in the structure which contains the bolt to be turned. By successive oscillations of the gear and handle unit the socket turns in one direction through the action of a ratchet mechanism. The reaction against the pin A provides the "anti-torque" characteristic of the tool. For a gripping force of $P = 30$ lb determine the torque M transmitted to the bolt and the external reaction R against the pin A normal to the line AB. (One side of the tool is used for tightening and the opposite side for loosening a bolt.)

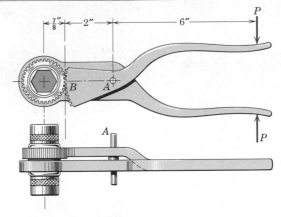

Problem 4/67

4/68 In the special position shown for the log hoist, booms AF and EG are at right angles to one another and AF is perpendicular to AB. If the hoist is handling a log weighing 4800 lb, compute the force supported by the pins at A and D in this one position due to the weight of the log.

　　　　Ans. $A = 34{,}000$ lb,　$D = 17{,}100$ lb

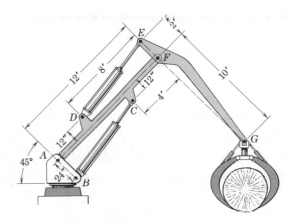

Problem 4/68

4/69 Determine the shearing force Q applied to the bar if a 40-lb force is applied to the handle for $\theta = 30$ deg. For a given applied force what value of θ gives the greatest shear?

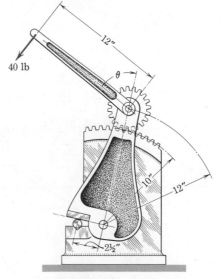

Problem 4/69

4/70 Determine the punching force P in terms of the gripping force F for the rivet squeezer shown.

$$\textit{Ans. } P = \frac{2Fe}{c\left(1 - \dfrac{a}{b}\right)}$$

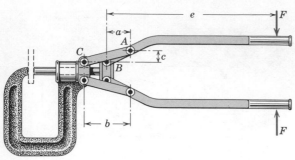

Problem 4/70

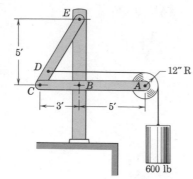

Problem 4/71

4/71 The fixed post supports the two members CE and AC together with the pulley and its load. Determine the force acting on the pin at B for application of the 600-lb weight. $\textit{Ans. } B = 1755$ lb

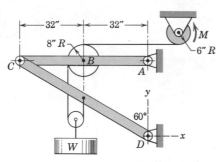

Problem 4/72

4/72 A torque M of 200 lb-ft must be applied to the shaft of the hoisting drum to support the load W suspended from the frame as shown. Calculate the x- and y-components of all forces acting on members AC and CD. The weights of the members and pulleys are small compared with W and may be neglected.

$A_x = 294 \quad A_y = 200$
$D_x = 694 \quad D_y = 600$

4/73 Determine the force supported by the pin at C for the loaded frame. $\textit{Ans. } C = 216$ lb

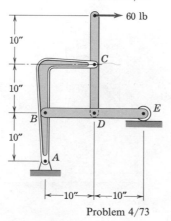

Problem 4/73

4/74 A pneumatic cylinder pivoted at F operates the lever AB of the quick-acting toggle clamp for holding work in position while it is being machined. For an air pressure of 60 lb/in.^{2} above atmospheric pressure against the 2-in.-diameter piston, determine the clamping force at G for the position $\alpha = 10$ deg. For this position the piston rod is perpendicular to AB.

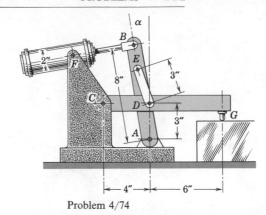

Problem 4/74

4/75 Compute the x- and y-components of all forces acting on members AC and DE for application of the 300-lb load. The two pulleys are fastened together as an integral unit.

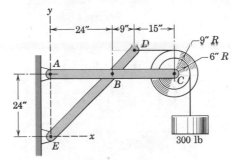

Problem 4/75

4/76 Calculate the force supported by the pin at A for the hoisting rig with the crank in the vertical position shown. The geometry is such that the cable between the hoisting drum at D and the pulley at C is very nearly parallel to BC and may be so assumed. The force P on the handle is just sufficient to support the load.   *Ans.* $A = 814$ lb

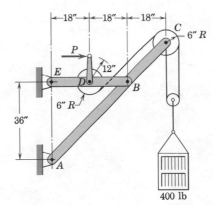

Problem 4/76

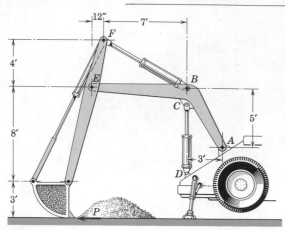

Problem 4/77

4/77 The back hoe is controlled by the three hydraulic cylinders, and, in the particular position shown, the hoe can apply a horizontal force $P = 2000$ lb. Neglect the weights of the members and compute the forces supported by the pins at A and E.

Ans. $A = 4470$ lb, $E = 7350$ lb

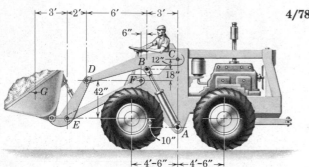

Problem 4/78

4/78 The rubber-tired loader weighs 20,000 lb empty and handles 2 yd³ of dirt with center of gravity at G. Duplicate linkages exist on the two sides of the loader. If the loader handles dirt weighing 80 lb/ft³, determine for the particular position shown the force P on the piston of each hydraulic cylinder AB and the forces on the pins C and E on one side of the machine.

Ans. $P = 8520$ lb
$C = 8800$ lb
$E = 4650$ lb

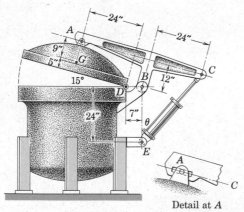

Detail at A

Problem 4/79

4/79 The lid of the large vulcanizer is hinged at D and is raised and lowered by the hydraulic cylinder CE. A short, smooth slot at A in the direction of AC (see detail) allows for the slight difference in radius of the arcs made by A about D and B as the lid and control arm pivot. In the 15-deg position shown, the lower surface of the lid and the arm AC are parallel. The lid weighs 880 lb, with its center of gravity at G. The weights of the arm and cylinder are small by comparison and may be neglected. Calculate the force supported by each of the hinge pins at D and B for the 15-deg position. (From the given geometry, the angle θ equals 41°49′.)

Ans. $D = 233$ lb, $B = 2290$ lb

4/80 Determine the force acting on member ABC at the connection A for the loaded space frame shown. Each connection may be treated as a ball-and-socket joint.

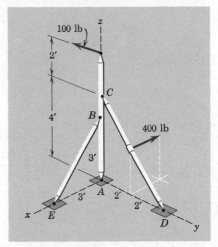

Problem 4/80

4/81 Compute the total force acting on the member BD at D for the space frame loaded by the two forces shown. Each of the connections may be treated as a ball-and-socket joint.

Ans. $D = 490$ lb

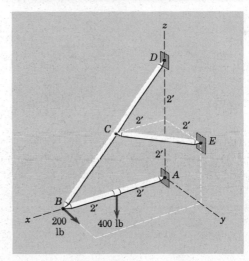

Problem 4/81

4/82 The frame shown rests on a smooth horizontal surface, and the connections at D, E, F, G, and H act as ball-and-socket joints. Determine the total force acting on the connection at D produced by the 500-lb load applied at F.

Ans. $D = 257$ lb

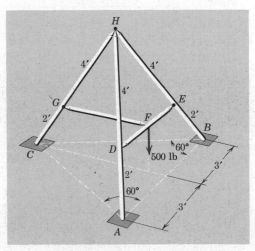

Problem 4/82

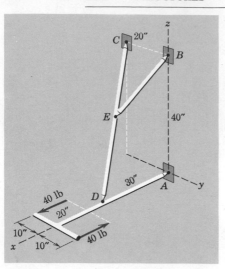

Problem 4/83

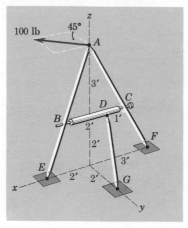

Problem 4/84

◀ **4/83** Determine the force in member *EB* produced by the action of the applied couple on the space frame. Also compute the components of force acting on *CD* at *D*. Point *E* is midway between *C* and *D*, and connections *A*, *B*, *C*, *D*, and *E* may be treated as ball-and-socket joints.

Ans. $F_{EB} = 71.8$ lb compression
$D = D_y = 26.7$ lb

◀ **4/84** The cross member *BC* of the *A*-frame is restrained from motion in the *x*-direction by a small collar at *C*, but no *x*-constraint exists at *B*. The remaining joints *A*, *D*, *E*, *F*, and *G* act as ball-and-socket connections. Calculate the force in *DG* from only one equation of equilibrium. Also determine the *x*-, *y*-, and *z*-components of the force that acts on *AE* at *E*. Assume that the positive directions of these components agree with the positive coordinate directions. Neglect the weights of the members.

Ans. $F_{DG} = 260$ lb, $E_x = -56.6$ lb
$E_y = -35.4$ lb, $E_z = 176.8$ lb

21 Beams with Concentrated Loads. Structural members that offer resistance to bending caused by applied loads are known as beams. Most beams are long prismatical bars, and the loads are usually applied normal to the axes of the bars. Beams are undoubtedly the most important of all structural members, and the basic theory underlying their design must be thoroughly understood. The analysis of the load-carrying capacities of beams consists, first, in establishing the equilibrium requirements of the beam as a whole and any portion of it considered separately. Second, the relations between the resulting forces and the accompanying internal resistance of the beam to support these forces are established. The first part of this analysis requires the application of the principles of statics, while the second part of the problem involves the strength characteristics of the material and is usually treated under the heading of the mechanics of solids or the mechanics of materials. This article concerns the first aspect of the problem only.

Beams supported in such a way that the external support reactions can be calculated by the methods of statics alone are called *statically determinate* beams. A beam that has more supports than are necessary to provide equilibrium is said to be *statically indeterminate,* and it is necessary to consider the load-deformation properties of the beam in addition to the equations of statical equilibrium to determine the support reactions. In Fig. 46 are shown examples of both types of beams. In this article only statically determinate beams are analyzed.

In the analysis of simple trusses the resultant of the forces acting on the section of a cut member was a single force, either tension or compression, in the direction of the member. Bending, shearing, and twisting of the member were either negligible or absent. In the analysis of beams the resultant of the forces acting on a transverse section of the beam cannot, in general, be represented in terms of a single force but must be expressed by a force **F** and a couple **M** as shown in Fig. 47*a*. It should be recalled from Chapter 2 that the resultant of any force system is expressible as a resultant force acting at any point and a corresponding couple. In some beams the actual distribution of force over a cross section is exceedingly complex and requires for a complete description extended analysis involving the load-deformation properties of the beam materials. In the present discussion only the *resultant*

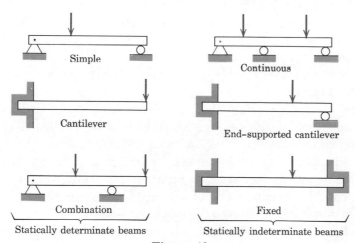

Figure 46

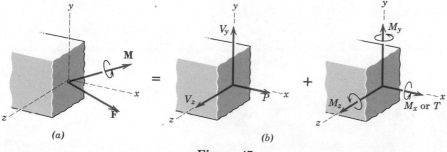

Figure 47

of the force distribution across any section of a statically determinate beam will be considered.

In Fig. 47*b* the force **F** and couple **M** acting on the cut section of one portion of the beam are shown separately in terms of their three components. The components of **F** are a tensile force P and two shearing forces V_y and V_z. The components of **M** are a torsional moment T and two bending moments M_y and M_z. The bending moment M_y is due to forces applied in the x-z plane, and M_z is due to forces applied in the x-y plane. The torsional moment M_x or T results from moments applied to the beam about the x-axis which tend to twist the beam about this axis.

It should be noted that the components of **F** and **M** in Fig. 47 are taken in the positive directions of the coordinate axes for a right-handed orthogonal x-y-z system. This notation for internal forces is being used with increasing frequency for the designation and analysis of stresses in solids and is the one employed in this book.

If the forces applied to the beam are coplanar, such as in the x-y plane of Fig. 47, then $V_z = 0$, $M_y = 0$, and $T = 0$. The beams shown in Fig. 46 are of this type where bending in the plane of the figure alone is involved.

In Fig. 48 are shown two adjoining portions of the same beam subjected to coplanar forces in the x-y plane of the paper. Only the shear $V_y = V$, the tensile force P, and the bending moment $M_z = M$ remain. The forces are represented as acting through the center of the cross section. The *average shear stress* over the cross section is V/A, where A is the area of the cross section. The force P is a tensile force as shown, and the *average tensile stress* over the cross section is P/A. Distributed forces or stresses will be discussed further in Chapter 5.

It should be noted carefully that the positive directions of V, P, and M are reversed on the two sections in Fig. 48 by the principle of action and reaction. This designation for the positive directions of these actions agrees with the notation adopted in Fig. 47 and should be used consistently in the problem work.* It is frequently impossible to ascertain at a glance whether the shear and moment on a certain section of a loaded beam are positive or negative. For this reason it will be found advisable to represent V and M in their positive directions on the free-body diagrams and let the algebraic signs of the calculated values indicate the proper direction.

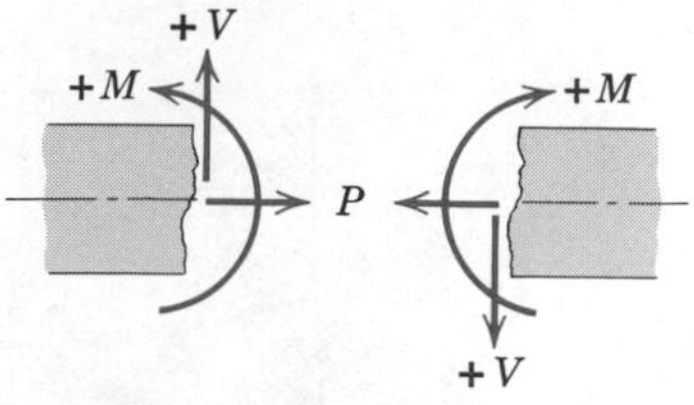

Figure 48

* A notation that employs the opposite sense for shear to that adopted here will be observed particularly in the older literature.

As an aid to the physical interpretation of the bending couple M consider the beam shown in Fig. 49 bent by the two equal and opposite moments applied at the ends. The cross section of the beam is taken to be that of an H-section with a very narrow center web and heavy top and bottom flanges. For this beam the load carried by the small web may be neglected compared with that carried by the two flanges. It should be perfectly clear that the upper flange of the beam is shortened and is under compression while the lower flange is lengthened and is under tension. The resultant of the two forces, one tensile and the other compressive, acting on any section is a couple and has the value of the bending moment on the section. If a beam of some other cross section were loaded in the same way, the distribution of force over the cross section would be different, but the resultant would be the same couple.

The variation of shear force V and bending moment M over the length of a beam provides information necessary for the design analysis of the beam. In particular, the maximum magnitude of the bending moment is usually the primary consideration in the design or selection of a beam, and its value and position should be determined. The variations in shear and moment are best shown graphically, and the expressions for V and M when plotted against distance along the beam give the *shear-force* and *bending-moment diagrams* for the beam.

The first step in the determination of the shear and moment relations is to establish the values of all external reactions on the beam by applying the equations of equilibrium to a free-body diagram of the beam as a whole. Next, a portion of the beam, either to the right or to the left of an arbitrary transverse section, is isolated with a free-body diagram, and the equations of equilibrium are applied to this isolated portion of the beam. These equations will yield expressions for the shear force V and bending moment M acting at the cut section on the part of the beam isolated. The part of the beam which involves the smaller number of forces, either to the right or to the left of the arbitrary section, usually yields the simpler solution. A transverse section that coincides with the location of a concentrated load should not be chosen, as such a position represents a point of discontinuity in the variation of shear and bending moment. Finally, it is important to note that the calculations for V and M on each section chosen should be consistent with the positive convention illustrated in Fig. 48. For horizontal beams where the coordinate axes are not specified, it will be assumed that the positive direction of the hori-

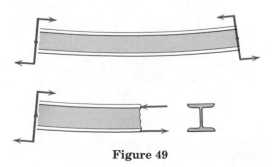

Figure 49

zontal axis is to the right and that of the vertical axis is up. This assumption is consistent with the notation of Fig. 48.

There are many possible combinations of loading and supports. In each case the method and procedure described here and illustrated in the sample problems that follow will always work.

Sample Problems

4/85 Draw the shear-force and bending-moment diagrams for the beam shown in the *a*-part of the illustration and determine the maximum magnitude of the bending moment and its location.

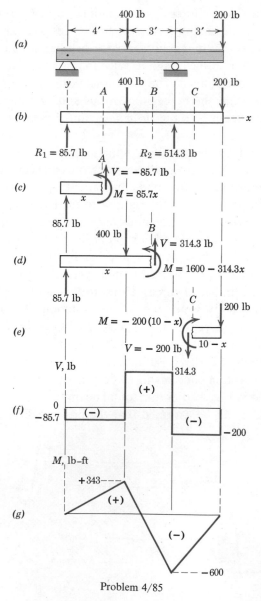

Problem 4/85

Solution. From the free-body diagram of the beam in part *b* the reactions are

$[\Sigma M_{R_1} = 0]$ $7R_2 - 4(400) - 10(200) = 0,$ $R_2 = 514.3$ lb

$[\Sigma F_y = 0]$ $R_1 + 514.3 - 400 - 200 = 0,$ $R_1 = 85.7$ lb

The free-body diagram of the part of the beam to the left of section *A* is shown in part *c* of the figure. The shear *V* and moment *M* are shown in their positive directions. Equilibrium requires

$[\Sigma F_y = 0]$ $V = -85.7$ lb

$[\Sigma M_A = 0]$ $M = 85.7x$

These values hold between $x = 0$ and $x = 4$ ft.

The next interval is analyzed by the free-body diagram of the entire portion of the beam to the left of section *B*. Again *V* and *M* are shown in their positive directions, part *d*. Equilibrium gives

$[\Sigma F_y = 0]$ $V - 400 + 85.7 = 0,$ $V = 314.3$ lb

$[\Sigma M_B = 0]$ $M + 400(x - 4) - 85.7x = 0,$ $M = 1600 - 314.3x$

Finally, from the simpler of the two diagrams for section *C*, shown in part *e* of the figure, literal application of the equilibrium equations gives

$[\Sigma F_y = 0]$ $V = -200$ lb

$[\Sigma M_C = 0]$ $M + 200(10 - x) = 0,$ $M = -200(10 - x)$

It should be noted that *M* and *V* are shown in their positive directions on the free-body diagram.

In parts *f* and *g* are plotted the shear-force and bending-moment diagrams for the entire beam. The maximum magnitude of the moment and its position are clearly

$$|M| = 600 \text{ lb-ft at } x = 7 \text{ ft} \qquad Ans.$$

4/86 The beam is built into a foundation at *A* and supports the three forces and the couple shown. Construct the bending-moment diagrams in the *x-z* and *x-y* planes and determine the resultant bending moment at $x = 2$ in. What is the torsional moment *T* about the *x*-axis?

Solution. The positive sense for shear and moment is first established as shown in the separate diagrams. Next a portion of the beam is isolated in the two views represented. In each view the bending moment and shear are shown acting in the positive sense. In the *x-z* view a moment equation about the cut section gives for $x < 7$ in.

$[\Sigma M = 0]$ $M_y - 100(3) + 60(11 - x) = 0,$ $M_y = -360 + 60x$ lb-in.

For $x > 7$ in. an isolated portion of the beam gives $M_y = -60(11 - x)$ lb-in. The variation of M_y with *x* is shown in the figure, which discloses the discontinuity at $x=7$ in. Also, it is clear from the figure that the shear is constant at $V_z = -60$ lb and that the axial force is $P = -100$ lb (compression) on the foundation side of the shaft.

For the *x-y* view of the free-body diagram, a moment sum about the cut section gives for $x < 7$ in.

$[\Sigma M = 0]$ $M_z + 50(11 - x) - 300 = 0,$ $M_z = -250 + 50x$ lb-in.

For $x > 7$ in. an isolated portion of the beam gives $M_z = -50(11 - x)$ lb-in. The variation of M_z with x is shown in the lower figure, and a discontinuity in the bending moment is observed at the point where the 300 lb-in. couple is applied. Clearly the shear is $V_y = -50$ lb for the entire length of the beam. The torsional moment (not shown on the free-body diagrams) is $T = 50(3) = 150$ lb-in. over the entire length of the beam.

At $x = 2$ in. the bending moments are

$$M_y = -240 \text{ lb-in.} \qquad M_z = -150 \text{ lb-in.}$$

so that the resultant bending moment is

$$M = \sqrt{(-240)^2 + (-150)^2} = 283 \text{ lb-in.} \qquad Ans.$$

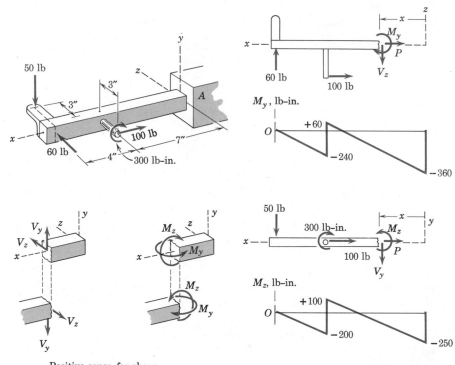

Positive sense for shear
force and bending moment

Problem 4/86

Problems

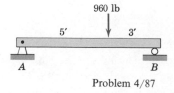

Problem 4/87

4/87 Draw the shear and moment diagrams for the simply supported beam and determine the maximum moment M. *Ans.* $M_{max} = 1800$ lb-ft

4/88 Draw the shear and moment diagrams for the beam shown.

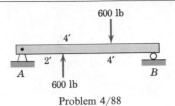

Problem 4/88

4/89 Draw the shear and moment diagrams for the cantilever beam. Observe the sign convention consistent with the axes indicated.

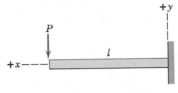

Problem 4/89

4/90 Draw the shear and moment diagrams for the beam subjected to the end couple. What is the moment M at a section 5 in. to the right of B?
Ans. $M = -1440$ lb-in.

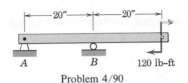

Problem 4/90

4/91 Draw the shear and moment diagrams for the beam loaded at its center by the couple C.

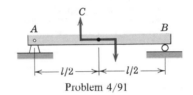

Problem 4/91

4/92 The angle strut is welded to the beam AB and supports the 100-lb load. Draw the shear and moment diagrams for the beam.

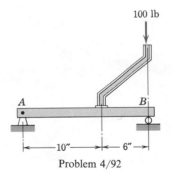

Problem 4/92

4/93 Draw the shear and moment diagrams for the beam shown and find the bending moment M at section C.   *Ans. $M = -1667$ lb-ft*

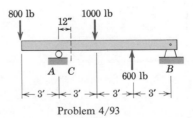

Problem 4/93

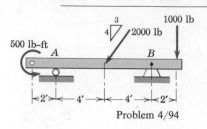

Problem 4/94

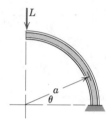

Problem 4/95

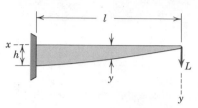

Problem 4/96

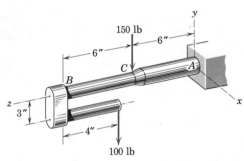

Problem 4/97

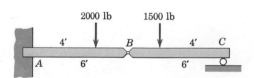

Problem 4/98

4/94 Draw the shear and moment diagrams for the beam loaded as shown. Determine the bending moment M with the maximum magnitude.

Ans. $M = -2000$ lb-ft

4/95 The resistance of a beam of uniform width to bending is found to be proportional to the square of the beam depth y. For the cantilever beam shown the depth is h at the support. Find the required depth y as a function of the length x in order for all sections to offer the same resistance to bending as the section at the support.

Ans. $y = h\sqrt{x/l}$

4/96 A curved cantilever beam has the form of a quarter-circular arc. Determine the expressions for the shear V (perpendicular to the arc) and the bending moment M as functions of θ.

4/97 Construct the bending-moment diagram for the cantilevered shaft AB of the rigid unit shown.

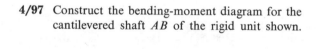

4/98 Construct the moment diagram for the two beams loaded as shown and connected by a hinged joint at B.

4/99 The 18-in. shaft is rigidly held in the fixed collar at A and carries the two loads shown. Determine the resultant bending moment in the shaft at A and express it in vector notation.

Ans. $\mathbf{M}_A = 360\mathbf{i} - 624\mathbf{j}$ lb-in.

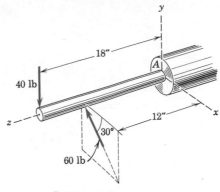

40 lb

18″

60 lb

30°

12″

Problem 4/99

4/100 Write expressions for the torsional moment T and bending moment M in the curved quarter-circular beam under the end load L. Use a notation consistent with the right-handed r-θ-z coordinate system.

Ans. $M_r = -La\cos\theta, \quad T = -La(1 - \sin\theta)$

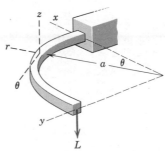

L

Problem 4/100

4/101 Sketch the family of bending-moment diagrams for the upper support beam for various values of overhang x where $l < b$.

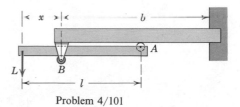

Problem 4/101

4/102 Determine the maximum bending moment M and the corresponding value of x in the crane beam and indicate the section where this moment acts.

Ans. $M_A = \dfrac{L}{4l}(l - a)^2, \quad x = \dfrac{a + l}{2}$

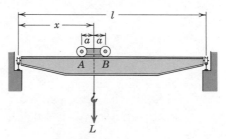

Problem 4/102

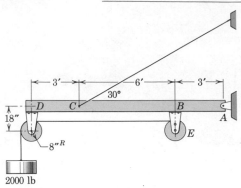

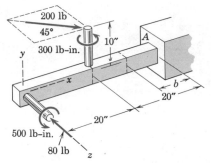

Problem 4/103

4/103 The beam AD is supported and loaded as shown. The cable that holds the 2000-lb load is wrapped around the drum at E, and the drum is locked to its bracket and cannot rotate. Neglect the dimensions of the flanges that attach the brackets to the beam at B and D and construct the shear-force and bending-moment diagrams for the beam. Determine the bending moment M with the greatest magnitude and the distance x from the pivot A at which M occurs.

Ans. $M = -108{,}000$ lb-in. at $x = 9$ ft

4/104 Draw the moment diagram for the beam in each of the coordinate planes x-y and x-z and write the vector expression for the total moment (torsional plus bending) $\mathbf{M}$ acting on the left-hand portion of the beam at the section $b = 10$ in.

Ans. $\mathbf{M} = -1414\mathbf{i} + 686\mathbf{j} + 914\mathbf{k}$ lb-in.

Problem 4/104

5 DISTRIBUTED FORCES

22 Introduction. In the preceding chapters all forces have been assumed to be concentrated and have been represented by single vector arrows at the points of application or along the unique lines of action of the forces. Actually, no "concentrated" forces exist in the exact sense, since every real force applied to a body is distributed over some finite area or volume. In the case of force applied to a surface it is clear that when the dimensions of the area over which the force is distributed are negligible compared with the other dimensions of the body, the concept of a concentrated force raises no question. On the other hand force may be applied over a surface area of considerable size, and the variation in the distribution of the force acting over the area must be accounted for. There is also force which is distributed over the volume of a body, such as the earth's gravitational force or electric and magnetic forces. When a force is distributed over a region, the distribution is known as a *force field.* The field may be represented by the distribution of force along a line, over an area, or throughout a volume.

A distributed force is measured at any point by its intensity. The intensity of force that is distributed over an area is known as *pressure* or *stress* and is measured as force per unit area on which it acts (lb/in.^{2}). The term *pressure* is usually used to denote the intensity of distributed force due to the action of fluids, whereas the term *stress* is more generally used to denote internal distributed force in solids. Force distributed over the volume of a body is known as *body force* and is measured as force per unit of volume (lb/in.^{3}).

In this chapter the resultants and equilibrium of various distributed force systems are described. Problems of this type involve a continuum of change over the region involved, and here the calculus is the appropriate analytical tool for the analysis of force fields.

23 Center of Gravity. The distributed force most commonly encountered is the force of attraction of the earth. This body force is distributed over all parts of every object in the earth's field of influence. The resultant of this distribution of body force is known as the *weight* of the body, and it is necessary to determine its magnitude and position for bodies whose weights are appreciable.

Consider a three-dimensional body of any size, shape, and weight. If it is suspended, as shown in Fig. 50, by a cord from any point such as A, the body will be in equilibrium under the action of the tension in the cord and the resultant of the gravity or body forces acting on all the particles. This resultant is clearly collinear with the cord, and it will be assumed that its

155

position will be marked, say, by drilling a hole of negligible size along its
line of action. The experiment is repeated by suspension at other points such
as *B* and *C*, and in each case the line of action of the resultant is marked.
For all practical purposes these lines of action will be concurrent at a point
which is known as the *center of gravity* or *center of mass* of the body. An exact
analysis, however, would take into account the fact that the directions of
the gravity forces for the various particles of the body differ slightly because
they converge toward the center of attraction of the earth. Also, since the
particles are at different distances from the earth, the intensity of the earth's
force field is not exactly constant over the body. These considerations lead
to the conclusion that the lines of action of the gravity force resultants in
the experiments just described will not quite be concurrent, and therefore
no unique center of gravity exists in the exact sense. This condition is of no
practical importance as long as we deal with bodies whose dimensions are
small compared with those of the earth. We therefore assume a uniform
and parallel field of force due to the earth's gravitational attraction, and
this condition results in the concept of a unique center of gravity.

To express mathematically the location of the center of gravity *G* of any
body, Fig. 51*a*, an equation may be written that states by Varignon's theorem
that the moment about any axis of the resultant *W* of the gravitational forces
equals the sum of the moments about the same axis of the gravitational
forces *dW* acting on all particles considered as infinitesimal elements of the
body. The resultant of the gravitational forces acting on all elements is the

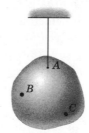

Figure 50

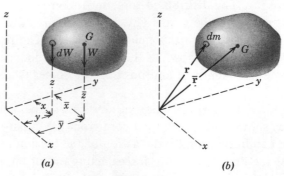

(a)   (b)

Figure 51

weight of the body and is given by the sum $W = \int dW$. If the moment principle is applied about the y-axis, for example, the moment about this axis of the elemental weight is $x\,dW$, and the sum of these moments for all elements of the body is $\int x\,dW$. This sum of moments must equal the moment of the sum $W\bar{x}$. Thus for all three axes the moment expressions become

$$\bar{x} = \frac{\int x\,dW}{W} \qquad \bar{y} = \frac{\int y\,dW}{W} \qquad \bar{z} = \frac{\int z\,dW}{W} \qquad (17)$$

The numerator of each expression represents the *sum of the moments,* and the product of W and the corresponding coordinate of G represents the *moment of the sum.* The third equation is obtained by revolving the body and reference frame 90 deg about a horizontal axis so that the z-axis is horizontal.

Equations 17 may be expressed in vector form with the aid of Fig. 51*b*, where the elemental mass and the center of gravity G are located by their respective position vectors $\mathbf{r} = \mathbf{i}x + \mathbf{j}y + \mathbf{k}z$ and $\bar{\mathbf{r}} = \mathbf{i}\bar{x} + \mathbf{j}\bar{y} + \mathbf{k}\bar{z}$. Thus Eqs. 17 are the components of the single vector equation

$$\bar{\mathbf{r}} = \frac{\int \mathbf{r}\,dW}{W} \qquad \text{or} \qquad \bar{\mathbf{r}} = \frac{\int \mathbf{r}\,dm}{m} \qquad (18)$$

where the mass replaces the weight by means of the relation $W = mg$ in the second equation.

This second equation defines the position of the *center of mass,* which is clearly the same point as the center of gravity as long as the gravity field is treated as uniform and parallel. It is meaningless to speak of the center of gravity of a body that is removed from the earth's gravitational field, since no gravitational forces would act on the body. It would, however, still possess its unique center of mass. It is quite proper to use the term center of gravity whenever reference is made to the effect of gravitational forces on a body. The term center of mass, on the other hand, is more properly used when reference is made to the influence of the distribution of mass on the dynamic response of a body to unbalanced forces. This class of problems is discussed at length in the companion volume *Dynamics.*

In most problems the calculation of the position of the center of gravity may be simplified by an intelligent choice of reference axes. In general the axes should be placed so as to simplify the equations of the boundaries as much as possible. Thus polar coordinates will be useful for bodies having circular boundaries. Another important clue may be taken from considerations of symmetry. Whenever there exists a line or plane of symmetry, a coordinate axis or plane should be chosen to coincide with this line or plane. The center of gravity will always lie on such a line or plane, since the moments due to symmetrically located elements will always cancel, and the body can be considered as composed of pairs of these elements. Thus the axis of a right circular cone is such a line of symmetry, and the central vertical plane in the fore-and-aft direction of a ship's hull is such a plane of symmetry.

The proper choice of the differential element for the integration process is important. Whenever possible a first-order differential element of volume should be selected, so that only one integration will be required to cover the entire figure. Regardless of the element used, the coordinate x, y, or z that is multiplied by dW in the numerator of Eqs. 17 is the coordinate of the *center of gravity of the element*. When a first-order element is not convenient, a second-order quantity should be selected in preference to a third-order element. As an example of these statements consider the calculation of the center of gravity of a right circular cone. An element $dx\,dy\,dz$ could be chosen and a triple integration carried out. The process would be relatively laborious. On the other hand if a first-order element defined, for example, by a coaxial circular slice of differential thickness were selected, only one simple integration would be required in moving the element from vertex to base. Other considerations in the selection of the element and integration limits are best shown by actual examples, several of which follow the next article.

The specific weight γ of a body is its weight per unit volume. Thus the weight of a differential element of volume dV becomes $dW = \gamma\,dV$. In the event that γ is not constant throughout the body but can be expressed as a function of the coordinates of the body, it will be necessary to account for this variation in the calculation of both the numerators and denominators of Eqs. 17. These expressions would then be written

$$\bar{x} = \frac{\int x\gamma\,dV}{\int \gamma\,dV} \qquad \bar{y} = \frac{\int y\gamma\,dV}{\int \gamma\,dV} \qquad \bar{z} = \frac{\int z\gamma\,dV}{\int \gamma\,dV} \tag{19}$$

24   Centroids of Lines, Areas, and Volumes. Whenever the specific weight γ of a body is uniform throughout, it will be a constant factor in both the numerators and denominators of Eqs. 19 and will therefore cancel. The expressions that remain define a purely geometrical property of the body, since any reference to its physical properties is absent. The term *centroid* is used when the calculation concerns a geometrical shape only. When speaking of an actual physical body, the term *center of gravity* or *center of mass* is used. If the specific weight is uniform throughout the body, the positions of the centroid and center of gravity are identical, whereas if the specific weight varies, these two points will, in general, not coincide.

In the case of a slender rod or wire of length L, cross-sectional area A, and specific weight γ, Fig. 52a, the body approximates a line segment, and $dW = \gamma A\,dL$. If γ and A are constant over the length of the rod, the coordinates of the center of gravity also become the coordinates of the centroid C of the line segment which, from Eqs. 17, may be written

$$\bar{x} = \frac{\int x\,dL}{L} \qquad \bar{y} = \frac{\int y\,dL}{L} \qquad \bar{z} = \frac{\int z\,dL}{L} \tag{20}$$

It should be noted that, in general, the centroid C will not lie on the line. If the rod lies in a single plane, such as the x-y plane, only two coordinates will require calculation.

When a body of specific weight γ has a small thickness t and approximates a surface of area A, Fig. 52b, then $dW = \gamma t \, dA$. Again, if γ and t are constant over the entire area, the coordinates of the center of gravity of the body also become the coordinates of the centroid C of the surface area, and from Eqs. 17 may be written

$$\bar{x} = \frac{\int x \, dA}{A} \qquad \bar{y} = \frac{\int y \, dA}{A} \qquad \bar{z} = \frac{\int z \, dA}{A} \qquad (21)$$

If the surface area is curved, as illustrated in Fig. 52b with the shell segment, all three coordinates will be involved. Here again the centroid C for the curved surface will in general not lie on the surface. If the area is a flat surface in, say, the x-y plane, only the coordinates in that plane will be unknown.

For a general body of volume V and weight γ per unit of volume, the element has a weight $dW = \gamma \, dV$. The specific weight γ cancels if it is constant over the entire volume, and the coordinates of the center of gravity also become the coordinates of the centroid C of the body. From Eqs. 17 or 19 they become

$$\bar{x} = \frac{\int x \, dV}{V} \qquad \bar{y} = \frac{\int y \, dV}{V} \qquad \bar{z} = \frac{\int z \, dV}{V} \qquad (22)$$

A summary of the centroidal coordinates for some of the commonly used shapes is given in Tables C4 and C5, Appendix C.

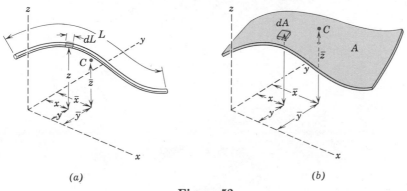

(a)   (b)

Figure 52

Sample Problems

5/1   *Centroid of a Circular Arc.* Locate the centroid of a circular arc as shown in the figure.

Solution. Choosing the x-axis as the axis of symmetry makes $\bar{y} = 0$. A differential element of arc has a length $dL = r\, d\theta$, and the x-coordinate of the element is $r\cos\theta$. Applying the first of Eqs. 20 gives

$$[L\bar{x} = \int x\, dL]\qquad\qquad (2\alpha r)\bar{x} = \int_{-\alpha}^{\alpha} (r\cos\theta) r\, d\theta$$

$$2\alpha r\bar{x} = 2r^2 \sin\alpha$$

$$\bar{x} = \frac{r\sin\alpha}{\alpha}\qquad\qquad\qquad Ans.$$

For $2\alpha = \pi/2$, $\bar{x} = 0.900r$; for a semicircular arc $2\alpha = \pi$, which gives $\bar{x} = 2r/\pi$.

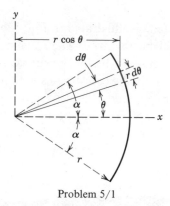

Problem 5/1

5/2   *Centroid of a Triangular Area.* Locate the centroid of the area of a triangle of base b and altitude h.

Solution. The x-axis is taken to coincide with the base. A differential strip of area $x\, dy$ is chosen. By similar triangles $x/(h - y) = b/h$. Applying the second of Eqs. 21 gives

$$[A\bar{y} = \int y\, dA]\qquad\qquad \frac{bh}{2}\bar{y} = \int_0^h y\,\frac{b(h - y)}{h}\, dy = \frac{bh^2}{6}$$

and

$$\bar{y} = \frac{h}{3}\qquad\qquad\qquad Ans.$$

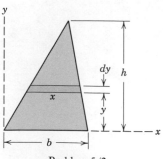

Problem 5/2

This same result holds with respect to either of the other two sides of the triangle considered as a new base with corresponding new altitude. Thus it can be said that the centroid lies at the intersection of the medians, since the distance of this point from any side is one third the altitude of the triangle with that side considered as base.

It should be noted that the differential strip is a trapezoid, and its area becomes $x\,dy$ only in the limit for which dy approaches zero. In setting up the expression for the elemental area, then, the small triangular end sections are disregarded, since they involve the higher-order product $dx\,dy$.

5/3 *Centroid of the Area of a Circular Sector.* Locate the centroid of the area of a circular sector with respect to its vertex.

Solution I. The x-axis is chosen as the axis of symmetry, and $\bar{y}$ is therefore automatically zero. The area may be covered by moving an element in the form of a segment of the circular ring, shown in the figure, from the center to the outer periphery. The radius of the ring is r_0 and its thickness is dr_0. In the first of Eqs. 21 the coordinate x is the coordinate of the *centroid* of the element dA. From Prob. 5/1 this distance is $r_0 \sin \alpha / \alpha$, where r_0 replaces r. Thus the first of Eqs. 21 gives

$$[A\bar{x} = \int x\,dA] \qquad \frac{2\alpha}{2\pi}(\pi r^2)\bar{x} = \int_0^r \left(\frac{r_0 \sin \alpha}{\alpha}\right)(2r_0\alpha\,dr_0)$$

$$r^2\alpha\bar{x} = \tfrac{2}{3}r^3 \sin \alpha$$

$$\bar{x} = \frac{2}{3}\frac{r \sin \alpha}{\alpha} \qquad\qquad\qquad Ans.$$

For a semicircular area $2\alpha = \pi$ and $\bar{x} = 4r/3\pi$.

Solution II. The area may also be covered by swinging a triangle of differential area about the vertex and through the total angle of the sector. This triangle, shown in the illustration, has an area $dA = (r/2)(r\,d\theta)$, where higher-order terms are neglected. Again the x-coordinate of dA is measured to the centroid of the element, and from Prob. 5/2 this coordinate is seen to be $2r/3$ multiplied by $\cos \theta$. Applying the first of Eqs. 21 gives

$$[A\bar{x} = \int x\,dA] \qquad (r^2\alpha)\bar{x} = \int_{-\alpha}^{\alpha} (\tfrac{2}{3}r \cos \theta)(\tfrac{1}{2}r^2\,d\theta)$$

$$r^2\alpha\bar{x} = \tfrac{2}{3}r^3 \sin \alpha$$

and as before

$$\bar{x} = \frac{2}{3}\frac{r \sin \alpha}{\alpha} \qquad\qquad\qquad Ans.$$

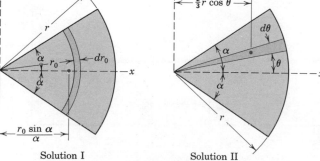

Solution I   Solution II

Problem 5/3

It should be noted that, if a second-order element $r \, dr \, d\theta$ is chosen, one integration with respect to θ would yield the ring with which *Solution I* began. On the other hand integration with respect to r_0 initially would give the triangular element with which *Solution II* began.

5/4 Locate the centroid of the area under the curve $x = ky^3$ from $x = 0$ to $x = a$.

Solution I. A vertical element of area $dA = y \, dx$ is chosen as shown in the left part of the figure. The x-coordinate of the centroid is found from the first of Eqs. 21. Thus

$$[A\bar{x} = \int x \, dA] \qquad\qquad \bar{x} \int_0^a y \, dx = \int_0^a xy \, dx$$

Substituting $y = (x/k)^{1/3}$ and $k = a/b^3$ and integrating give

$$\frac{3ab}{4} \bar{x} = \frac{3a^2b}{7}, \qquad \bar{x} = \tfrac{4}{7}a \qquad\qquad\qquad Ans.$$

In solving for $\bar{y}$ by the second of Eqs. 21 the coordinate y is *not* the coordinate to the curve $x = ky^3$ but is the y-distance to the centroid of the element dA. Hence the value $y/2$, which locates the centroid of the rectangular element, must be used. The moment principle becomes

$$[A\bar{y} = \int y \, dA] \qquad\qquad \frac{3ab}{4}\bar{y} = \int_0^a \left(\frac{y}{2}\right) y \, dx$$

Substituting $y = b(x/a)^{1/3}$ and integrating give

$$\frac{3ab}{4}\bar{y} = \frac{3ab^2}{10}, \qquad \bar{y} = \tfrac{2}{5}b \qquad\qquad\qquad Ans.$$

Solution II. The horizontal element of area shown in the right-hand part of the figure may be employed in place of the vertical element. In calculating $\int x \, dA$ the x-coordinate of the centroid of the element must be used for "x". This distance is $x + (a - x)/2 = (a + x)/2$. Hence

$$[A\bar{x} = \int x \, dA] \qquad \bar{x} \int_0^b (a - x) \, dy = \int_0^b \left(\frac{a + x}{2}\right)(a - x) \, dy$$

The value of $\bar{y}$ is found from

$$[A\bar{y} = \int y \, dA] \qquad \bar{y} \int_0^b (a - x) \, dy = \int_0^b y(a - x) \, dy$$

The evaluation of these integrals will check the previous results for $\bar{x}$ and $\bar{y}$.

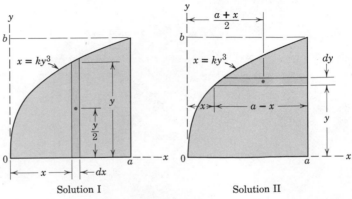

Solution I   Solution II

Problem 5/4

5/5 *Hemispherical Volume.* Locate the centroid of the volume of a hemisphere of radius r with respect to its base.

Solution. With the axes chosen as shown in the figure $\bar{x} = \bar{z} = 0$ by symmetry. The most convenient element is a circular slice of thickness dy parallel to the x-z plane. Since the hemisphere intersects the y-z plane in the circle $y^2 + z^2 = r^2$, the radius of the circular slice is $z = +\sqrt{r^2 - y^2}$. The volume of the elemental slice becomes

$$dV = \pi(r^2 - y^2)\,dy$$

The second of Eqs. 22 requires

$$[V\bar{y} = \int y\,dV] \qquad \bar{y}\int_0^r \pi(r^2 - y^2)\,dy = \int_0^r y\pi(r^2 - y^2)\,dy$$

Integrating gives

$$\tfrac{2}{3}\pi r^3 \bar{y} = \tfrac{1}{4}\pi r^4, \qquad \bar{y} = \tfrac{3}{8}r \qquad\qquad\qquad Ans.$$

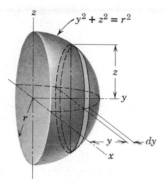

Problem 5/5

Problems

5/6 Locate the centroid of the shaded area shown.
Ans. $\bar{x} = 2.09, \quad \bar{y} = 1.43$

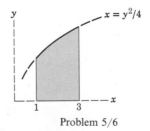

Problem 5/6

5/7 Locate the center of gravity of the wire formed into the circular arc shown. Solve by direct integration and check by using the results of Sample Prob. 5/1.

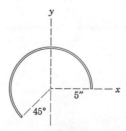

Problem 5/7

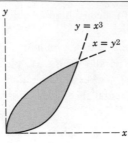

Problem 5/8

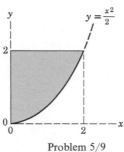

Problem 5/9

Problem 5/11

Problem 5/12

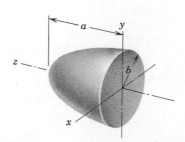

Problem 5/13

5/8 Locate the centroid of the shaded area between the two curves. *Ans.* $\bar{x} = \frac{12}{25}$, $\bar{y} = \frac{3}{7}$

5/9 Locate the centroid of the shaded area above the parabola by using a horizontal strip of area dA in the integration.

5/10 Work Prob. 5/9 by using a vertical strip of area dA in the integration.

5/11 Specify the coordinates of the center of gravity of the circular shell section by direct reference to the results of Sample Prob. 5/1.

5/12 Find the distance $\bar{z}$ from the vertex of the right circular cone to the centroid of its volume. *Ans.* $\bar{z} = \frac{3}{4}h$

5/13 A parabola is revolved about the z-axis to obtain the paraboloid shown. Locate the centroid of its volume with respect to the base $z = 0$.

Ans. $\bar{z} = \frac{a}{3}$

5/14 Locate the centroid of the area of the elliptical quadrant.

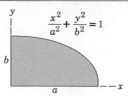

$$\frac{x^2}{a^2} + \frac{y^2}{b^2} = 1$$

Problem 5/14

5/15 Determine the distance $\bar{z}$ from the base of any cone or pyramid of altitude h to the centroid of its volume.

Ans. $\bar{z} = \dfrac{h}{4}$

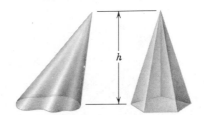

Problem 5/15

5/16 Determine the distance h from the centroid of the lateral area of any cone or pyramid of altitude h to the base of the figure.

5/17 Locate the center of gravity of the homogeneous hemispherical shell of radius r and negligible wall thickness.

Ans. $\bar{z} = \dfrac{r}{2}$

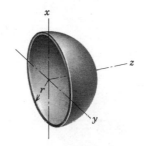

Problem 5/17

5/18 Locate the centroid of the area shown in the figure by direct integration.

Ans. $\bar{x} = \dfrac{10 - 3\pi}{4 - \pi} \dfrac{a}{3}$

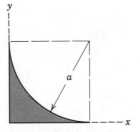

Problem 5/18

5/19 Locate the centroid of the shaded area between the ellipse and the circle.

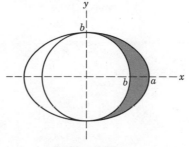

Problem 5/19

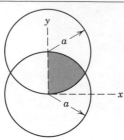

Problem 5/20

5/20 Determine the x-coordinate of the centroid of the shaded area shown.   *Ans.* $\bar{x} = 0.339a$

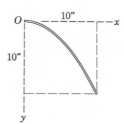

Problem 5/21

5/21 The slender rod has a uniform cross section and is bent into the shape of a parabolic arc with the vertex at the origin. Determine the coordinates of the center of gravity of the rod.

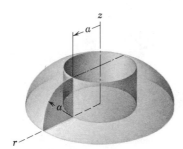

Problem 5/22

5/22 Determine the z-coordinate of the center of gravity of the solid obtained by revolving the quarter-circular area about the z-axis.

$$Ans. \ \bar{z} = \frac{11a}{2(4 + 3\pi)}$$

Problem 5/23

5/23 Determine the z-coordinate of the centroid of the volume generated by revolving the quarter-circular area about the z-axis through 90 deg.

5/24 A portion of a thin-walled right circular tube of radius r is formed by cutting off a length h with a plane which passes through the base diameter AB. Determine the coordinates of the center of gravity of the portion of the shell.

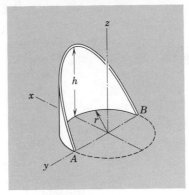

Problem 5/24

5/25 Locate the centroid of the conical wedge generated by revolving the right triangle of altitude a about its base b through an angle θ.

$$Ans. \; \bar{r} = \frac{a \sin (\theta/2)}{\theta}, \quad \bar{z} = \frac{b}{4}$$

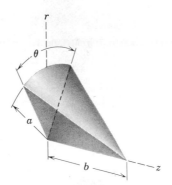

Problem 5/25

5/26 Locate the center of gravity of the thin spherical shell that is formed by cutting out one-eighth of a complete spherical shell.

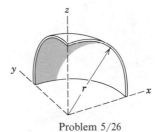

Problem 5/26

5/27 The specific weight of a sphere decreases uniformly with radial distance from μ_0 at the center to one-half this value at the surface $r = a$. If the sphere is cut on a diametral plane as shown, determine the distance $\bar{z}$ from this base plane to the center of gravity of either hemisphere. (*Hint:* Use the results of Prob. 5/17 in setting up the element of volume.)   *Ans.* $\bar{z} = \frac{9}{25}a$

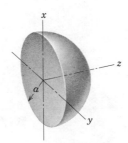

Problem 5/27

◀ **5/28** Determine the position of the center of gravity of the thin conical shell shown.

Problem 5/28

◀ **5/29** Determine the distance $\bar{r}$ to the center of gravity of the solid spherical wedge shown.

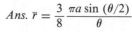

$$Ans. \;\; \bar{r} = \frac{3}{8} \frac{\pi a \sin (\theta/2)}{\theta}$$

Problem 5/29

◀ **5/30** A cylindrical body of radius r and height h is divided in half by a diagonal slice as shown. Locate the center of gravity of the part indicated.

$$Ans. \;\; \bar{x} = \tfrac{1}{4}r, \quad \bar{z} = \tfrac{5}{16}h$$

Problem 5/30

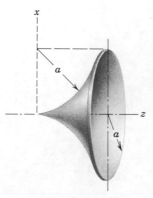

◀ **5/31** Locate the center of gravity of the bell-shaped shell of uniform but negligible thickness.

$$Ans. \;\; \bar{z} = \frac{a}{\pi - 2}$$

◀ **5/32** Determine the position of the centroid of the volume within the bell-shaped shell of Prob. 5/31.

$$Ans. \;\; \bar{z} = \frac{a}{2(10 - 3\pi)}$$

Problem 5/31

5/33 Locate the center of gravity G of the steel half ring. (*Hint:* Choose an element of volume in the form of a cylindrical shell whose intersection with the plane of the ends is shown in section.)

$$Ans. \ \bar{r} = \frac{a^2 + 4R^2}{2\pi R}$$

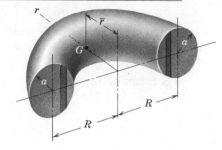

Problem 5/33

---

25 Composite Bodies and Figures; Approximations. When a body or figure can be conveniently divided into several parts of simple shape, the principle of Varignon may be used if each part is treated as a finite element of the whole. Thus for a body whose parts have weights W_1, W_2, W_3, ..., and whose separate coordinates of the centers of gravity of these parts in, say, the x-direction are $\bar{x}_1, \bar{x}_2, \bar{x}_3, \ldots$, the moment principle gives

$$(W_1 + W_2 + W_3 + \cdots)\overline{X} = W_1\bar{x}_1 + W_2\bar{x}_2 + W_3\bar{x}_3 + \cdots$$

where $\overline{X}$ is the x-coordinate of the center of gravity of the whole. Similar relations hold for the other two coordinate directions. These sums may be expressed in condensed form and written as

$$\overline{X} = \frac{\Sigma W\bar{x}}{\Sigma W} \qquad \overline{Y} = \frac{\Sigma W\bar{y}}{\Sigma W} \qquad \overline{Z} = \frac{\Sigma W\bar{z}}{\Sigma W} \qquad \text{(23)}$$

Analogous relations hold for composite lines, areas, and volumes, where the W's are replaced by L's, A's, and V's, respectively. It should be pointed out that if a hole or cavity is considered to be one of the component parts of a composite body or figure, the corresponding weight or area represented by the cavity or hole is considered a negative quantity.

Frequently in practice the boundaries of an area or volume are not expressible in terms of the simple geometrical shapes or in shapes that can be represented mathematically. For such cases it is necessary to resort to a method of approximation.

Consider the problem of locating the centroid C of the irregular area shown in Fig. 53. The area may be divided into strips of width Δx and

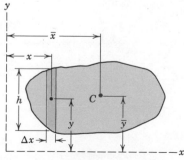

Figure 53

variable height h. The area A of each strip, such as the one shown in color, is multiplied by the coordinates x and y to its *centroid* to obtain the moments of the element of area. The sum of the moments for all strips divided by the total area of the strips will give the corresponding centroidal component. A systematic tabulation of the results will permit an orderly evaluation of the total area ΣA, the sums ΣAx and ΣAy, and the results

$$\bar{x} = \frac{\Sigma Ax}{\Sigma A} \qquad \bar{y} = \frac{\Sigma Ay}{\Sigma A}$$

The accuracy of the approximation will be increased by decreasing the widths of the strips used. In all cases the average height of the strip should be estimated in approximating the areas. Although it is usually of advantage to use elements of constant width, it is not necessary to do so. In fact elements of any size and shape which approximate the given area to satisfactory accuracy may be used.

In locating the centroid of an irregular volume the problem may be reduced to one of determining the centroid of an area. It is necessary only to plot a curve representing the magnitude of the areas of cross sections of the volume normal to a desired axis. The position along the axis of the centroid of the figure defined by this curve of areas will be the corresponding position of the volume centroid.

Sample Problem

5/34 Locate the center of gravity of the bracket-and-shaft combination. The vertical face is made from sheet metal which weighs 5 lb/ft², the material of the horizontal base weighs 8 lb/ft², and the steel shaft has a specific weight of 0.283 lb/in.³

Solution. The composite body may be considered as composed of the five elements shown in the right-hand part of the illustration. The triangular part will be taken as a negative area. For the reference axes indicated it is clear by symmetry that the x-coordinate of the center of gravity is zero.

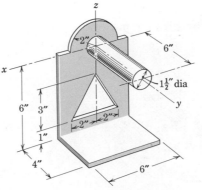
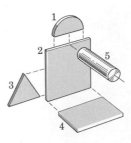

Problem 5/34

The weight W of each part and the coordinates $\bar{y}$ and $\bar{z}$ of the respective centers of gravity are determined. The terms involved in applying Eqs. 23 are best handled in the form of a table as follows:

Part	W, lb	$\bar{y}$, in.	$\bar{z}$, in.	$W\bar{y}$, lb-in.	$W\bar{z}$, lb-in.
1	0.218	0	0.849	0	0.185
2	1.25	0	−3.00	0	−3.75
3	−0.208	0	−4.00	0	0.832
4	1.33	2.00	−6.00	2.66	−7.98
5	3.00	3.00	0	9.00	0
Totals	5.59			11.66	−10.71

Equations 23 are now applied and the results are

$$\left[\bar{Y} = \frac{\Sigma W\bar{y}}{\Sigma W}\right] \qquad \bar{Y} = \frac{11.66}{5.59} = 2.08 \text{ in.} \qquad\qquad Ans.$$

$$\left[\bar{Z} = \frac{\Sigma W\bar{z}}{\Sigma W}\right] \qquad \bar{Z} = \frac{-10.71}{5.59} = -1.92 \text{ in.} \qquad\qquad Ans.$$

Problems

5/35 Compute the dimension b that will place the centroid of the shaded area 7 in. above the base.
   Ans. $b = 8$ in.

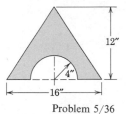

Problem 5/35

5/36 The triangular plate has a semicircular notch cut from its base as shown. Compute the distance $\bar{Y}$ from the base to the center of gravity of the plate.

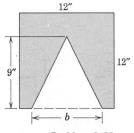

Problem 5/36

5/37 Determine the coordinates of the centroid for the shaded area bounded by the circular arc and the two straight lines.
   Ans. $\bar{X} = 5.00$ in., $\bar{Y} = 4.27$ in.

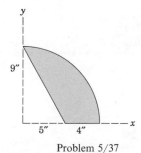

Problem 5/37

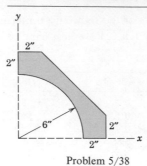

Problem 5/38

5/38 Calculate the coordinates of the centroid of the shaded area.

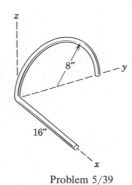

Problem 5/39

5/39 Calculate the coordinates of the center of gravity of the slender rod bent into the shape shown.
Ans. $\overline{X} = 3.11$ in., $\overline{Y} = 4.89$ in., $\overline{Z} = 3.11$ in.

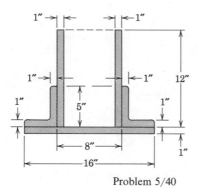

Problem 5/40

5/40 Determine the distance $\overline{H}$ from the bottom of the base plate to the centroid of the build-up structural section shown.

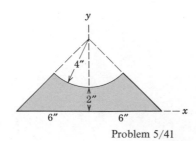

Problem 5/41

5/41 Compute the distance $\overline{Y}$ from the *x*-axis to the centroid of the shaded area.
Ans. $\overline{Y} = 1.14$ in.

5/42 A homogeneous solid of revolution, shown in section, consists of a frustum of a right circular cone containing a cylindrical hole of 8-in. diameter. Calculate the height $\overline{Z}$ of the center of gravity of the solid above its base.

Ans. $\overline{Z} = 4.97$ in.

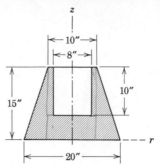

Problem 5/42

5/43 Calculate the coordinates of the center of gravity of the solid semicircular cylinder with the prismatic wedge removed.

Problem 5/43

5/44 The hemispherical shell and its semicircular base are formed from the same piece of sheet metal of small thickness. Use the results of Prob. 5/17 and calculate the coordinates of the mass center of the shell and base combined.

Ans. $\overline{X} = 0.475r, \quad \overline{Y} = r/3$

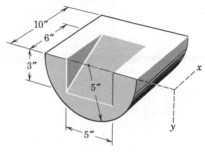

Problem 5/44

5/45 The semicircular disk and square plate are made from steel plate weighing 40 lb/ft² and are welded to the steel shaft, which weighs 4.6 lb per foot of length. Calculate the coordinates of the center of gravity of the assembly.

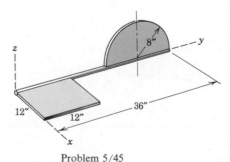

Problem 5/45

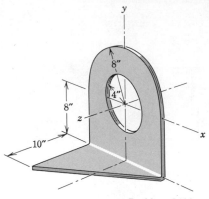

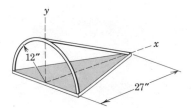

Problem 5/46

5/46 Determine the coordinates of the center of gravity of the bracket, which is made from a plate of uniform thickness.

Ans. $\overline{X} = -0.83$ in.
$\overline{Y} = -3.14$ in.
$\overline{Z} = 1.03$ in.

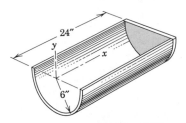

Problem 5/47

5/47 A tubular framework consisting of a semicircular member and a brace is welded to a flat triangular plate as shown. If the tubular material weighs 5 lb per foot of length and the base material weighs 20 lb per square foot of area, calculate the coordinates of the center of gravity of the combined base and frame.

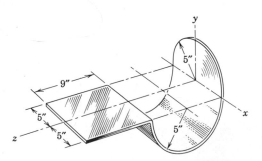

Problem 5/48

5/48 Determine the position of the center of gravity of the cylindrical shell with a closed semicircular end. The shell is made from sheet metal weighing 4.8 lb/ft², and the end is made from metal plate weighing 7.2 lb/ft².

Ans. $\overline{X} = 13.90$ in., $\overline{Y} = -3.62$ in.

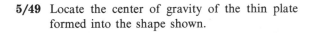

Problem 5/49

5/49 Locate the center of gravity of the thin plate formed into the shape shown.

5/50 As an example of the accuracy involved in graphical approximations calculate the percentage error e in determining the x-coordinate of the centroid of the triangular area by using the five approximating rectangles of width $a/5$ in place of the triangle.

Ans. $e = 1.00$ per cent low

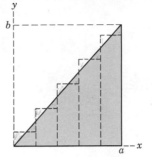

Problem 5/50

5/51 A sheet metal pattern has the shape shown. With the aid of the superimposed grid determine the coordinates of the centroid.

Ans. $\overline{X} = 6.14$ in., $\overline{Y} = 3.92$ in.

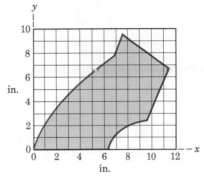

Problem 5/51

5/52 The center of buoyancy B of a ship's hull is the centroid of its submerged volume. The underwater cross-sectional areas A of the transverse sections of the tugboat hull shown are tabulated for every ten feet of waterline length. Plot these values and determine to the nearest foot the distance $\bar{x}$ of B aft of point A.

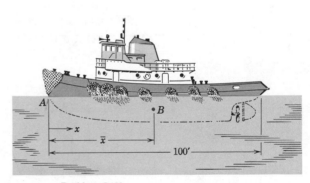

Problem 5/52

x, ft	A, ft^2	x, ft	A, ft^2
0	0	50	270
10	76	60	256
20	170	70	210
30	238	80	135
40	266	90	55
		100	0

Ans. $\bar{x} = 49$ ft

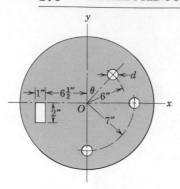

Problem 5/53

5/53 The circular disk contains two 1-in.-diameter holes and the rectangular slot shown. Determine the diameter d and angular position θ of a hole to be drilled in the disk at a radius of 6 in. to ensure balanced rotation of the disk about its center O.   *Ans.* $d = 1.550$ in., $\theta = 48°35'$

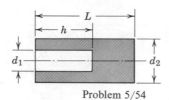

Problem 5/54

5/54 A homogeneous charge of explosive is to be formed as a circular cylinder of length L and diameter d_2 with an axial hole of diameter d_1 and depth h as shown in section. Determine the value of h which will place the center of gravity of the final charge a maximum distance from the open end.

$$Ans.\ h = \left(\frac{d_2}{d_1}\right)^2 L\left[1 - \sqrt{1 - \left(\frac{d_1}{d_2}\right)^2}\right]$$

26 Theorems of Pappus.* A very simple method exists for calculating the surface area generated by revolving a plane curve about a nonintersecting axis in the plane of the curve. In Fig. 54 the line segment of length L in the x-y plane generates a surface when revolved about the x-axis. An element of this surface is the ring generated by dL. The area of this ring is

$$dA = 2\pi y\, dL$$

and the total area is then

$$A = 2\pi \int y\, dL$$

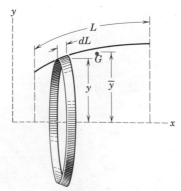

Figure 54

* Attributed to Pappus of Alexandria, a Greek geometer who lived in the third century A.D. The theorems often bear the name of Guldinus (Paul Guldin, 1577–1643), who claimed original authorship, although the works of Pappus were apparently known to him.

But since $\bar{y}L = \int y\, dL$, the area becomes

▶ $$A = 2\pi\bar{y}L \tag{24}$$

where $\bar{y}$ is the y-coordinate of the centroid for the line of length L. Thus the generated area is the same as the lateral area of a right circular cylinder of length L and radius $\bar{y}$.

In the case of a volume generated by revolving an area about a non-intersecting line in its plane an equally simple relation exists for finding the volume. An element of the volume generated by revolving the area A about the x-axis, Fig. 55, is the elemental ring of cross section dA and radius y. The volume of the element is $dV = 2\pi y\, dA$, and the total volume is

$$V = 2\pi \int y\, dA$$

But since $\bar{y}A = \int y\, dA$, the volume becomes

$$V = 2\pi\bar{y}A \tag{25}$$

where $\bar{y}$ is the y-coordinate of the centroid of the revolved area A. Thus the generated volume is obtained by multiplying the generating area by the circumference of the circular path described by its centroid.

The two theorems of Pappus, expressed by Eqs. 24 and 25, not only are useful in determining areas and volumes of generation, but they are also employed to find the centroids of plane curves and plane areas when the corresponding areas and volumes due to revolution of these figures about a nonintersecting axis are known. Dividing the area or volume by 2π times the corresponding line segment length or plane area will give the distance from the centroid to the axis of revolution.

In the event that a line or an area is revolved through an angle θ less than 2π, the generated surface or volume may be found by replacing 2π by θ in Eqs. 24 and 25. Thus

$$A = \theta\bar{y}L \quad \text{and} \quad V = \theta\bar{y}A$$

where θ is expressed in radians.

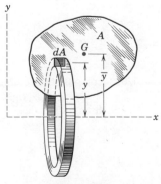

Figure 55

Problems

5/55 Determine the volume V and lateral area A of a right circular cone of base radius r and altitude h by the method of this article.

5/56 From the known surface area $A = 4\pi r^2$ of a sphere of radius r, determine the radial distance $\bar{r}$ to the centroid of the semicircular arc used to generate the surface.

5/57 From the known volume $V = \frac{4}{3}\pi r^3$ of a sphere of radius r, determine the radial distance $\bar{r}$ to the centroid of the semicircular area used to generate the sphere.

5/58 Use the notation of the half torus of Prob. 5/33 and determine the volume V and the surface area A of a complete torus.

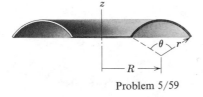

Problem 5/59

5/59 A shell structure has the form of a surface obtained by revolving the circular arc through 360 deg about the z-axis. Determine the surface area of one side of the complete shell.

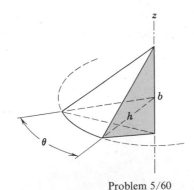

Problem 5/60

5/60 The shaded triangle of base b and altitude h is revolved about its base through an angle θ to generate a portion of a complete solid of revolution. Write the expression for the volume V of the solid generated.

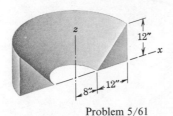

Problem 5/61

5/61 Compute the volume V of the solid generated by revolving the right triangle about the z-axis through 180 deg. *Ans.* $V = 3619$ in.[3]

5/62 Determine the volume V generated by revolving the quarter-circular area about the z-axis through an angle of 90 deg.

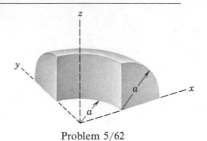

Problem 5/62

5/63 Determine the volume V generated by revolving the quarter-circular area about the z-axis through 90 deg.   *Ans.* $V = \dfrac{\pi a^3}{12}(3\pi - 2)$

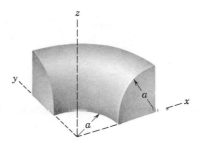

Problem 5/63

5/64 Determine the surface area of one side of the bell-shaped shell of Prob. 5/31 by the method of this article.

5/65 Compute the volume V and total surface area A of the complete ring whose square cross section is shown.

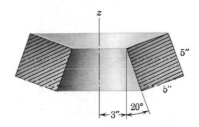

Problem 5/65

5/66 Calculate the weight W and total surface area A of the wheel generated by revolving the shaded section about the z-axis. The material of the wheel weighs 450 lb/ft³.

Ans. $W = 673$ lb, $A = 1838$ in.²

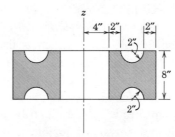

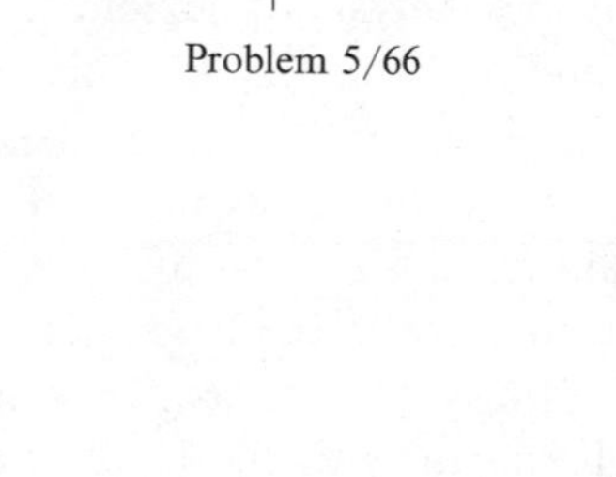

Problem 5/66

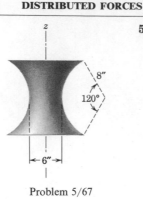

Problem 5/67

5/67 A surface is generated by revolving the circular arc of 8-in. radius and subtended angle of 120 deg completely about the z-axis. The diameter of the neck is 6 in. Determine the area A generated.   *Ans. $A = 462$ in.^{2}*

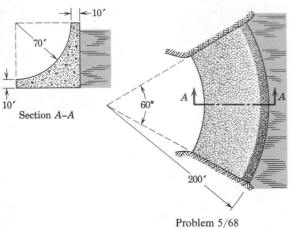

Section A–A

Problem 5/68

5/68 Calculate the weight W in tons of concrete required to construct the arched dam shown. Concrete weighs 150 lb/ft^3.

Ans. $W = 35.18(10^3)$ tons

27 Flexible Cables

(*a*) *General Relationships.* One important type of structural member is the flexible cable which is used in suspension bridges, transmission lines, messenger cables for supporting heavy trolley or telephone lines, and many other applications. In the design of these structures it is necessary to know the relations involving the tension, span, sag, and length of the cables. These quantities are determined by examining the cable as a body in equilibrium. In the analysis of flexible cables it is assumed that any resistance offered to bending is negligible. This assumption means that the force in the cable is always in the direction of the cable.

Flexible cables may support a series of distinct concentrated loads, as shown in Fig. 56a, or they may support loads that are continuously distributed over the length of the cable, as indicated by the variable-intensity loading *w* in Fig. 56b. In some instances the weight of the cable is negligible compared with the loads it supports, and in other cases the weight of the cable may be an appreciable load or the sole load, in which case it cannot be neglected. Regardless of which of these conditions is present, the equilibrium requirements of the cable may be formulated in the same manner.

If the variable and continuous load applied to the cable of Fig. 56b is expressed as w pounds per foot of horizontal length x, then the resultant R of the vertical loading is

$$R = \int dR \qquad R = \int w \, dx$$

where the integration is taken over the desired interval. The position of R is found from the moment principle, so that

$$R\bar{x} = \int x \, dR \qquad \bar{x} = \frac{\int x \, dR}{R}$$

The elemental load $dR = w \, dx$ is represented by an elemental strip of vertical length w and width dx of the shaded area of the loading diagram, and R is represented by the total area. It follows from the foregoing expressions that R passes through the *centroid* of the shaded area.

The equilibrium condition of the cable will be satisfied if each infinitesimal element of the cable is in equilibrium. The free-body diagram of a differential element is shown in Fig. 56c. The tension in the cable at the general position x is T, and the cable makes an angle θ with the horizontal x-direction. At the section $x + dx$ the tension is $T + dT$, and the angle is $\theta + d\theta$. Note that the changes in both T and θ are taken positive with positive change in x. The vertical load $w \, dx$ completes the free-body diagram. The equilibrium of vertical and horizontal forces requires, respectively, that

$$(T + dT) \sin (\theta + d\theta) - T \sin \theta + w \, dx$$
$$(T + dT) \cos (\theta + d\theta) = T \cos \theta$$

The trigonometric expansion for the sine and cosine of the sum of two angles and the substitutions $\sin d\theta = d\theta$ and $\cos d\theta = 1$, which hold in the limit as $d\theta$ approaches zero, yield upon simplification

$$d(T \sin \theta) = w \, dx \qquad \text{and} \qquad d(T \cos \theta) = 0$$

The second relation expresses the fact that the horizontal component of T remains unchanged, which fact is clear from the free-body diagram. If this constant horizontal force is $T_0 = T \cos \theta$, substitution into the first equation

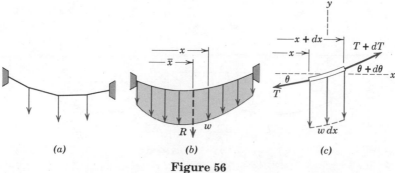

(a) (b) (c)

Figure 56

gives $d(T_0 \tan \theta) = w \, dx$. But $\tan \theta = dy/dx$, so that the equilibrium equation may be written in the form

$$\frac{d^2y}{dx^2} = \frac{w}{T_0} \qquad (26)$$

Equation 26 is the *differential equation* for the flexible cable. The solution to the equation is that functional relation $y = f(x)$ which satisfies the equation and also satisfies the conditions at the fixed ends of the cable, called *boundary conditions*. This relationship defines the shape of the cable.

The general procedure for obtaining the defining differential equation illustrated here applies to a large number of other problems which will be illustrated in subsequent articles. In each case the governing principles are applied to a general differential element of material, and the resulting expressions yield the differential equations that define the behavior of the body at each point. Note that in the present problem there is only the single independent variable x, and consequently the variables y and T are functions of x only, thus yielding an *ordinary differential equation*. When the dependent variables depend on two or more independent coordinates, a *partial differential equation* results. Several examples of partial differential equations are given in the optional section of Art. 31 on the equilibrium of internal stresses.

The differential relationship, Eq. 26, will now be solved for two important and limiting cases of cable loading.

(*b*) *Parabolic Cable.* When the intensity of vertical loading w is constant, the description closely approximates a suspension bridge where the uniform weight of the roadway may be expressed by the constant w. The weight of the cable is not distributed uniformly with the horizontal but is relatively small and is neglected. For this limiting case it will be proved that the cable hangs in a parabolic arc. Figure 57 shows such a suspension bridge of span L and sag h with origin of coordinates taken at the midpoint of the span. With both w and T_0 constant, Eq. 26 may be integrated once with respect to x to obtain

$$\frac{dy}{dx} = \frac{wx}{T_0} + C$$

where C is a constant of integration. For the coordinate axes chosen, $dy/dx = 0$ when $x = 0$, so that $C = 0$. Hence

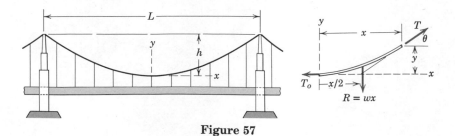

Figure 57

$$\frac{dy}{dx} = \frac{wx}{T_0}$$

which defines the slope of the curve as a function of x. One further integration yields

$$\int_0^y dy = \int_0^x \frac{wx}{T_0} dx \quad \text{or} \quad y = \frac{wx^2}{2T_0} \tag{27}$$

The student should make certain that he can obtain the identical results with the indefinite integral together with the evaluation of the constant of integration. Equation 27 gives the shape of the cable, which is clearly a vertical parabola. The constant horizontal component of cable tension becomes the cable tension at the origin.

Inserting the corresponding values $x = L/2$ and $y = h$ in Eq. 27 gives

$$T_0 = \frac{wL^2}{8h} \quad \text{and} \quad y = \frac{4hx^2}{L^2}$$

The tension T is found from a free-body diagram of a finite portion of the cable, shown in Fig. 57, which requires that $T = \sqrt{T_0^2 + w^2 x^2}$. Elimination of T_0 gives

$$T = w \sqrt{x^2 + \left(\frac{L^2}{8h}\right)^2} \tag{28}$$

The maximum tension occurs when $x = L/2$ and is

$$T_{\max} = \frac{wL}{2} \sqrt{1 + \frac{L^2}{16h^2}}$$

The length S of the complete cable is obtained from the differential relation $dS = \sqrt{(dx)^2 + (dy)^2}$. Thus

$$\frac{S}{2} = \int_0^{L/2} \sqrt{1 + \left(\frac{dy}{dx}\right)^2} \, dx = \int_0^{L/2} \sqrt{1 + \left(\frac{wx}{T_0}\right)^2} \, dx$$

For convenience in computation this expression is changed to a convergent series and is then integrated term by term. From the expansion

$$(1 + x)^n = 1 + nx + \frac{n(n-1)}{2!} x^2 + \frac{n(n-1)(n-2)}{3!} x^3 + \cdots$$

the integral may be written as

$$S = 2 \int_0^{L/2} \left(1 + \frac{w^2 x^2}{2T_0^2} - \frac{w^4 x^4}{8T_0^4} + \cdots\right) dx$$

$$= L \left(1 + \frac{w^2 L^2}{24 T_0^2} - \frac{w^4 L^4}{640 T_0^4} + \cdots\right)$$

Substitution of $w/T_0 = 8h/L^2$ yields

$$S = L \left[1 + \frac{8}{3}\left(\frac{h}{L}\right)^2 - \frac{32}{5}\left(\frac{h}{L}\right)^4 + \cdots\right] \tag{29}$$

When the properties of this series are examined, it is found that it converges for all values of $h/L \leq 1/4$. In most cases h is much smaller than $L/4$, so that the three terms of Eq. 29 give a sufficiently accurate approximation.

(*c*) *Catenary Cable.* Consider now a uniform cable, Fig. 58, suspended at two points in the same horizontal plane and hanging under the action of its own weight only. The free-body diagram of a finite portion of the cable of length s is shown in the right-hand part of the figure. This free-body diagram differs from that in Fig. 57 in that the total vertical force supported is equal to the weight of the section of cable of length s in place of the load distributed uniformly with respect to the horizontal. If the cable weighs μ pounds per foot of its own length, the resultant R of the load is $R = \mu s$, and the incremental vertical load $w\, dx$ of Fig. 56c is replaced by $\mu\, ds$. Thus the differential relation, Eq. 26, for the cable becomes

$$\frac{d^2y}{dx^2} = \frac{\mu}{T_0}\frac{ds}{dx} \tag{30}$$

Since $s = f(x, y)$, it is necessary to change this equation to one containing only the two variables.

Substituting the identity $(ds)^2 = (dx)^2 + (dy)^2$ yields

$$\frac{d^2y}{dx^2} = \frac{\mu}{T_0}\sqrt{1 + \left(\frac{dy}{dx}\right)^2} \tag{31}$$

Equation 31 is the differential equation of the curve (catenary) assumed by the cable. Solution of this equation is facilitated by the substitution $p = dy/dx$, which gives

$$\frac{dp}{\sqrt{1 + p^2}} = \frac{\mu}{T_0}\, dx$$

Integrating this equation produces

$$\log\left(p + \sqrt{1 + p^2}\right) = \frac{\mu}{T_0}x + C$$

The constant C is zero since $dy/dx = p = 0$ when $x = 0$. Substituting $p = dy/dx$, changing to exponential form, and clearing the equation of the radical give

$$\frac{dy}{dx} = \frac{e^{\mu x/T_0} - e^{-\mu x/T_0}}{2} = \sinh\frac{\mu x}{T_0}$$

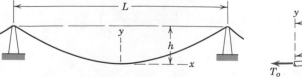

Figure 58

where the hyperbolic function* is introduced for convenience. The slope may be integrated to obtain

$$y = \frac{T_0}{\mu} \cosh \frac{\mu x}{T_0} + K$$

The integration constant K is evaluated from the boundary condition $x = 0$ when $y = 0$. This substitution requires that $K = -T_0/\mu$, and hence

$$y = \frac{T_0}{\mu} \left(\cosh \frac{\mu x}{T_0} - 1 \right) \tag{32}$$

Equation 32 is the equation of the curve (catenary) assumed by the cable hanging under the action of its weight only.

From the free-body diagram in Fig. 58 it is seen that $dy/dx = \tan \theta = \mu s/T_0$. Thus, from the previous expression for the slope,

$$s = \frac{T_0}{\mu} \sinh \frac{\mu x}{T_0} \tag{33}$$

The tension T in the cable is obtained from the equilibrium triangle of forces in Fig. 58. Thus

$$T^2 = \mu^2 s^2 + T_0^2$$

which, upon combination with Eq. 33, becomes

$$T^2 = T_0^2 \left(1 + \sinh^2 \frac{\mu x}{T_0} \right) = T_0^2 \cosh^2 \frac{\mu x}{T_0}$$

or

$$T = T_0 \cosh \frac{\mu x}{T_0} \tag{34}$$

The tension may also be expressed in terms of y with the aid of Eq. 32, which, when substituted into Eq. 34, gives

$$T = T_0 + \mu y \tag{35}$$

Equation 35 shows that the increment in cable tension from that at the lowest position depends only on μy.

Most problems dealing with the catenary involve solutions of Eqs. 32 through 35, which may be handled graphically or solved by computer. The graphical procedure is illustrated in the sample problem following this article.

The solution of catenary problems where the sag-to-span ratio is small may be approximated by the relations developed for the parabolic cable. A small sag-to-span ratio means a tight cable, and the uniform distribution of weight along the cable is not much different from the same load intensity distributed uniformly along the horizontal.

Many problems dealing with both the catenary and parabolic cable

* See Table C3, Appendix C.

involve suspension points that are not on the same level. In such cases the relations may be applied to the part of the cable on each side of the lowest point.

Sample Problem

5/69 A cable weighing 8 lb/ft is suspended between two points on the same level and 1000 ft apart. If the sag is 200 ft, find the length of the cable and the maximum tension.

Solution. Equations 33 and 34 for the cable length and tension both involve the minimum tension T_0 which must be found from Eq. 32. Thus, for $x = 500$ ft, $y = 200$ ft, and $\mu = 8$ lb/ft,

$$200 = \frac{T_0}{8}\left[\cosh\frac{(8)(500)}{T_0} - 1\right] \quad \text{or} \quad \frac{1600}{T_0} = \cosh\frac{4000}{T_0} - 1$$

This equation is most easily solved graphically. The expression on each side of the equals sign is computed and plotted as a function of various values of T_0. The intersection of the two curves establishes the equality and determines the correct value of T_0. This plot is shown in the figure accompanying this problem and yields the solution

$$T_0 = 5250 \text{ lb}$$

The maximum tension occurs for maximum y and from Eq. 35 is

$$T = 5250 + (8)(200) = 6850 \text{ lb} \qquad\qquad Ans.$$

From Eq. 33 the total length of the cable becomes

$$2s = 2\frac{5250}{8}\sinh\frac{(8)(500)}{5250} = 1100 \text{ ft} \qquad\qquad Ans.$$

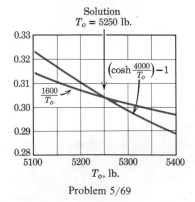

Problem 5/69

Problems

5/70 A cable carrying a load distributed uniformly along the horizontal attaches to the support at A with a horizontal tension of 3000 lb. If the tangent to the cable at the attachment B makes an angle of 50 deg with the vertical, determine the height h of B above A and the horizontally distributed unit load w.

Ans. $h = 41.95$ ft, $w = 25.17$ lb/ft

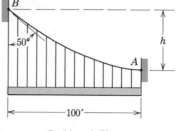

Problem 5/70

5/71 The cable from A to B carries a load of 240,000 lb distributed uniformly along the horizontal. The slope of the cable is zero at A, and its weight is small compared with the load it supports. Calculate the maximum tension T in the cable.

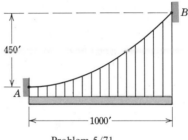

Problem 5/71

5/72 Expand Eq. 32 in a power series for cosh $(\mu x/T_0)$ and show that the equation for the parabola, Eq. 27, is obtained by taking only the first two terms in the series. (See Table C3, Appendix C, for the series expansion of the hyperbolic function.)

5/73 The two cables of a suspension bridge with a span of 4000 ft and a sag of 600 ft support a vertical load of $160(10^6)$ lb distributed uniformly with respect to the horizontal distance between its towers. Compute the tension T_0 in each cable at mid-span and the angle θ made by the cables with the horizontal as they approach the support at the top of either tower.

Ans. $T_0 = 66.7(10^6)$ lb, $\theta = 30°58'$

5/74 The cable supports a total load of 60,000 lb distributed uniformly with respect to the horizontal. If the slope of the cable is 1/3 at mid-span and is zero at its right-hand end, determine the tension T in the cable at position A. The weight of the cable may be neglected compared with the load it supports.

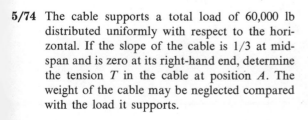

Problem 5/74

5/75 The cable of a suspension bridge with a span of 2000 ft is 300 ft below the top of the supporting towers at the quarter-span position. Calculate the total length S of each cable between the two identical towers.   *Ans. S = 2193 ft*

5/76 A cable of negligible weight supports a load of 40 lb/ft uniformly distributed with respect to the horizontal. The cable is supported at two points on the same horizontal line 120 ft apart and has a sag of 30 ft. With the aid of the principle of the concurrency of three forces in equilibrium, determine graphically the tension T at the support and the angle θ made by T with the horizontal. Also determine graphically the height b above the cable at mid-span of a point on the cable 40 ft from the mid-span position measured horizontally.

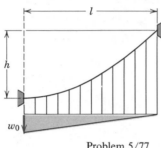

5/77 A cable of negligible weight is suspended from the fixed points shown and has a zero slope at its lower end. If the cable supports a unit load w which decreases uniformly with x from w_0 to zero as indicated, determine the equation of the curve assumed by the cable.

Problem 5/77

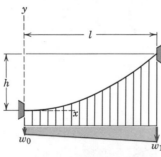

5/78 The light cable has zero slope at its lower point of attachment and supports the unit load which varies linearly with x from w_0 to w_1 as shown. Derive the expression for the tension T_0 in the cable at the origin.

Problem 5/78

5/79 The light cable is suspended from two points a distance L apart and on the same horizontal line. If the load per unit length in the horizontal direction supported by the cable varies from w_0 at the center to w_1 at the ends in accordance with the relation $w = a + bx^2$, derive the equation for the sag h of the cable in terms of the mid-span tension T_0.   *Ans.* $h = \dfrac{L^2}{48T_0}(5w_0 + w_1)$

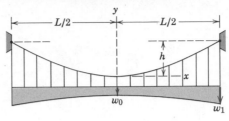

Problem 5/79

5/80 A cable 300 ft long is suspended between two points on the same level 280 ft apart. If the cable supports a large load uniformly distributed with respect to the horizontal direction, find the sag h in the cable from Eq. 29 by successive approximations.

5/81 A cable supports a load of 50 lb/ft uniformly distributed with respect to the horizontal and is suspended from the two fixed points located as shown. Determine the maximum and minimum tensions T and T_0 in the cable.
    Ans. $T = 3630$ lb, $T_0 = 2146$ lb

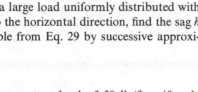

50 lb /ft

Problem 5/81

5/82 A flexible cable is secured to point A and passes over a small pulley at B, which is 600 ft higher than A. If it requires a tension $T = 12{,}000$ lb at B to make $\alpha = 0$ at A, determine the weight μ of the cable per foot of its length.

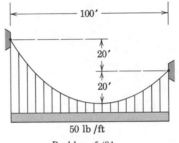

Problem 5/82

5/83 A rope 40 ft in length is suspended between two points which are separated by a horizontal distance of 10 ft. Compute the distance h to the lowest part of the loop.   *Ans.* $h = 18.5$ ft

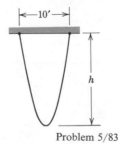

Problem 5/83

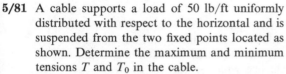

5/84 A cable hanging under the action of its own weight is suspended between two points on the same level and 400 ft apart. If the sag is 100 ft, determine the total length S of the cable. What error is involved if the length is computed from the expression for the parabolic cable?

5/85 A power line is suspended from two towers 500 ft apart on the same horizontal line. The cable weighs 12.20 lb/ft of length and has a sag of 80 ft at mid span. If the cable can support a maximum tension of 12,000 lb, determine the weight w' of ice per foot which can form on the cable without exceeding the maximum cable tension.   *Ans.* $w' = 12.70$ lb of ice per ft

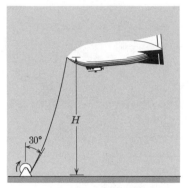

◄ **5/86** The blimp is moored to the ground winch in a gentle wind with 200 ft of $\frac{1}{2}$-in. cable which weighs 0.38 lb/ft. A torque of 300 lb-ft is required on the drum to start winding in the cable. At this condition the cable makes an angle of 30 deg with the vertical as it approaches the winch. Calculate the height H of the blimp. The diameter of the drum is 18 in.

Ans. $H = 177$ ft

Problem 5/86

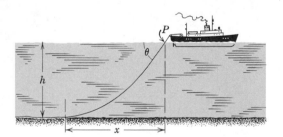

Problem 5/87

◄ **5/87** While repairs on its cable are being made, the propellers of the cable-laying ship exert a forward thrust of 60,000 lb to hold a fixed position in calm water. The ocean depth at this location is h ft, and the cable is observed to make an angle of $\theta = 60$ deg with the horizontal where it enters the water at point P. The cable weighs 15 lb/ft and has a cross-sectional area of 7.10 in.^{2} Salt water weighs 64 lb/ft^3. Calculate the length s of cable from point P to the point where the cable leaves the ocean floor and find the horizontal distance x from this point to a point directly below P. Determine the depth h of the ocean. Also compute the torque M on the 6-ft-diameter cable-laying drum to prevent it from turning. (*Note:* The vertical loading is the difference between the cable weight and the weight of the water displaced.)

Ans. $s = 8775$ ft, $x = 6672$ ft
$h = 5066$ ft, $M = 360,000$ lb-ft

28 Beams with Distributed Loads. Analysis of the shear forces and bending moments in beams under concentrated loads was developed in Art. 21. This development will now be extended to the case of beams subjected to loads that vary continuously over the beam or over portions of the beam. The conventions used for shear V and moment M and the same basic equilibrium method of analysis developed in Art. 21 will also be used for distributed forces. Additionally, there are several relations involving V and M which may be established for any beam with distributed loads and which will aid greatly in the construction of the shear and moment distributions. Figure 59 represents a portion of a loaded beam, and an element dx of the beam is isolated. The loading p represents the force per unit length of beam. At the location x the shear V and moment M acting on the element are drawn in their positive directions. On the opposite side of the element where the coordinate is $x + dx$ these quantities are also shown in their positive directions but must be labeled $V + dV$ and $M + dM$, since the change in V and M with x is required. The applied loading p may be considered constant over the length of the element, since this length is a differential quantity and the effect of any change in p disappears in the limit compared with the effect of p. Representation of the forces and moments and their incremental changes on a differential element of the beam, Fig. 59, follows a procedure analogous to that used for defining the equilibrium of a differential element of the flexible cable.

Equilibrium of the element requires that the sum of the vertical forces be zero. Thus

$$V + p\,dx - (V + dV) = 0$$

or

▶
$$p = \frac{dV}{dx} \tag{36}$$

It is clear from Eq. 36 that the slope of the shear diagram must everywhere be equal to the value of the applied loading. Equation 36 holds on either side of a concentrated load but not at the concentrated load by reason of the discontinuity produced by the abrupt change in shear.

Equilibrium of the element in Fig. 59 also requires that the moment sum

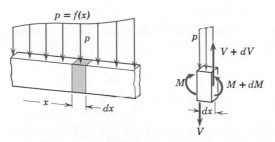

Figure 59

be zero. Taking moments about the left side of the element gives

$$M + p\,dx\,\frac{dx}{2} - (V + dV)\,dx - (M + dM) = 0$$

The two M's cancel, and the terms $p(dx)^2/2$ and $dV\,dx$ may be dropped, since they are differentials of higher order than those that remain. This leaves merely

▶
$$V = -\frac{dM}{dx} \qquad (37)$$

which expresses the fact that the shear everywhere is equal to the negative of the slope of the moment curve. Or the moment M may be considered the negative of the integral of the shear. In integral form Eq. 37 is written

$$\int_{M_0}^{M} dM = -\int_{x_0}^{x} V\,dx$$

or

$$M = M_0 - (\text{area under shear diagram from } x_0 \text{ to } x)$$

In this expression M_0 is the bending moment at x_0 and M is the bending moment at x. For beams where there is no externally applied moment M_0 at $x_0 = 0$, the total moment at any section equals the negative of the area under the shear diagram up to that section. Summing up the area under the shear diagram is usually the simplest way to construct the moment diagram.

When V passes through zero and is a continuous function of x with $dV/dx \neq 0$, the bending moment M will be a maximum or a minimum, since $dM/dx = 0$ at such a point. Critical values of M also occur when V crosses the zero axis discontinuously, as seen in Art. 21 for beams under concentrated loads.

It should be noted from Eqs. 36 and 37 that the degree of V in x is one higher than that of p. Also M is of one higher degree in x than is V. Furthermore M is two degrees in x higher than p. Thus for a beam loaded by $p = kx$ which is of the first degree in x, the shear V is of the second degree in x and the bending moment M is of the third degree in x.

Equations 36 and 37 may be combined to yield

▶
$$\frac{d^2M}{dx^2} = -p \qquad (38)$$

Thus, if p is a known function of x, the moment M may be obtained by two integrations, provided that the limits of integration are properly evaluated each time. This method is usable only if p is a continuous function of x. When p is a discontinuous function of x it is possible to introduce a special set of expressions called *singularity functions* which permit writing analytical expressions for shear V and moment M over a range of discontinuities. These functions are not discussed in this book. It is always possible to construct the moment diagram directly from the areas under the shear diagram.

When bending in a beam occurs in more than a single plane, a separate analysis in each plane may be carried out. The results are then combined vectorially, as described in Art. 21 for beams under concentrated loads and as shown in Fig. 47 and in Sample Prob. 4/86.

Sample Problem

5/88   Draw the shear-force and bending-moment diagrams for the loaded beam shown at the top of the figure and determine the maximum moment M and its location x from the left end.

    Solution. The support reactions are most easily obtained by considering the resultants of the distributed loads as shown on the free-body diagram of the beam as a whole. The first interval of the beam is analyzed from the free-body diagram of the section for $x < 4$ ft. A vertical summation of forces for this section gives

$$[\Sigma F_y = 0] \qquad\qquad V = 12.5x^2 - 247$$

and a moment summation about the cut section yields

$$[\Sigma M = 0] \qquad\qquad M + (12.5x^2)\frac{x}{3} - 247x = 0$$

$$M = 247x - 4.167x^3$$

These values of V and M hold for $0 < x < 4$ ft and are plotted for that interval in the shear and moment diagrams shown.

    From the free-body diagram covering the section for which $4 < x < 8$ ft, equilibrium in the vertical direction requires

$$[\Sigma F_y = 0] \qquad\qquad V - 100(x - 4) - 200 + 247 = 0$$

$$V = 100x - 447$$

A moment sum about the cut section gives

$$[\Sigma M = 0] \qquad M + 100(x - 4)\frac{x - 4}{2} + 200[x - \tfrac{2}{3}(4)] - 247x = 0$$

$$M = -266.7 + 447x - 50x^2$$

These values of V and M are plotted on the shear and moment diagrams for the interval $4 < x < 8$ ft.

    The analysis of the remainder of the beam is continued from the free-body diagram of the portion of the beam to the right of a section in the next interval. It should be noted that V and M are represented in their positive directions. A vertical force summation requires

$$[\Sigma F_y = 0] \qquad\qquad V - 653 + 300 = 0, \qquad V = 353 \text{ lb}$$

and a moment summation about the section yields

$$[\Sigma M = 0] \qquad\qquad M + 300(12 - x) - 653(12 - x - 2) = 0$$

$$M = 2930 - 353x$$

These values of V and M are plotted on the shear and moment diagrams for the interval $8 < x < 10$ ft.

    The last interval may be analyzed by inspection. The shear is constant at $+300$ lb, and the moment follows a straight-line relation beginning with zero at the right end of the beam.

The maximum moment occurs at $x = 4.47$ ft, where the shear curve crosses the zero axis, and the magnitude of M is obtained for this value of x by substitution into the expression for M for the second interval. The maximum moment is

$$M = 731 \text{ lb-ft} \qquad\qquad \textit{Ans.}$$

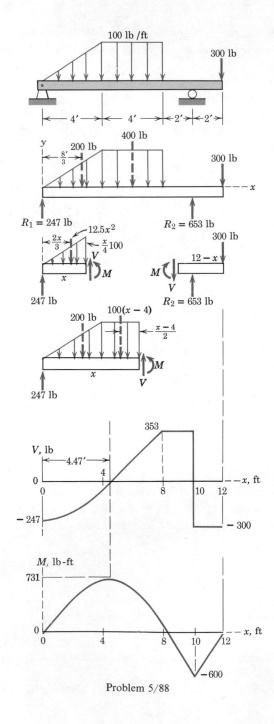

Problem 5/88

The derivative $dM/dx = 0$ for this second interval will check the value $x = 4.47$ ft. Although the shear crosses the zero axis again at $x = 10$ ft, the magnitude of the bending moment is less than that at $x = 4.47$ ft.

Although in this sample solution the moment M was determined by a direct application of the moment equilibrium equation, M may be obtained from the shear diagram as explained in the statements following Eq. 37. Thus, the moment M at any section x equals the negative of the area under the shear diagram to the left of x. For $x < 4$ ft, for instance, the moment is

$$[\Delta M = -\int V\, dx] \qquad M - 0 = -\int_0^x (12.5x^2 - 247)\, dx, \qquad M = 247x - 4.167x^3$$

which agrees with the results already computed.

Problems

5/89 Draw the shear and moment diagrams for the uniformly loaded beam and find the maximum bending moment M.   *Ans.* $M = \dfrac{wl^2}{8}$

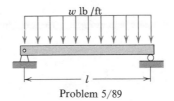

Problem 5/89

5/90 Draw the shear and moment diagrams for the cantilever beam with the uniform load w per unit of length.

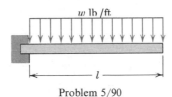

Problem 5/90

5/91 Draw the shear and moment diagrams for the beam, which supports the unit load of 50 lb per foot of beam length distributed over its mid-section.

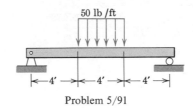

Problem 5/91

5/92 Draw the shear and moment diagrams for the loaded beam and find the maximum magnitude M of the bending moment.   *Ans.* $M = \frac{5}{6}Pl$

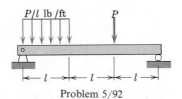

Problem 5/92

5/93 Draw the shear and moment diagrams for the cantilever beam loaded as shown.

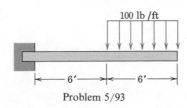

Problem 5/93

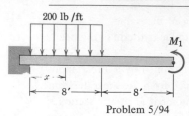

200 lb /ft

M_1

8'   8'

Problem 5/94

5/94 Draw the shear and moment diagrams for the loaded cantilever beam where the end couple M_1 is adjusted so as to produce zero moment at the fixed end of the beam. Find the bending moment M at $x = 4$ ft.   *Ans.* $M = 4800$ lb-ft

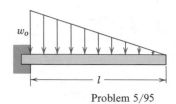

w_o

l

Problem 5/95

5/95 Draw the shear and moment diagrams for the cantilever beam with the linear loading and find the maximum magnitude M of the bending moment.

$$Ans.\ M = \frac{w_0 l^2}{6}$$

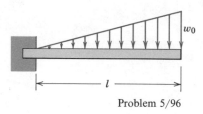

w_0

l

Problem 5/96

5/96 Draw the shear and moment diagrams for the linearly loaded cantilever beam.

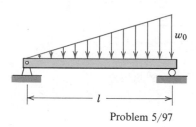

w_0

l

Problem 5/97

5/97 Draw the shear and moment diagrams for the linearly loaded simple beam. The load intensity at the right-hand end is w_0 pounds per foot of beam length. Write the expression for the magnitude M of the maximum bending moment and the distance x, measured to the right from the left end, to the point where the maximum bending moment occurs.

$$Ans.\ M = \frac{w_0 l^2}{9\sqrt{3}} \quad at \quad x = l/\sqrt{3}$$

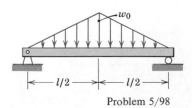

w_0

$l/2$ $l/2$

Problem 5/98

5/98 Draw the shear and moment diagrams for the linearly loaded simple beam shown. Determine the magnitude M of the maximum bending moment. $Ans.\ M = \dfrac{w_0 l^2}{12}$

5/99 Draw the shear and moment diagrams for the beam with the linearly varying loads and find the maximum bending moment M.

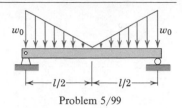

Problem 5/99

5/100 Draw the shear and moment diagrams for the beam shown. Determine the distance b, measured from the left end, to the point where the bending moment is zero between the supports.

Ans. $b = 4.5$ ft

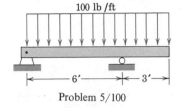

Problem 5/100

5/101 Draw the shear and moment diagrams for the loaded beam and determine the bending moment M that induces the maximum compression in the upper fibers of the beam.

Ans. $M = 46.8$ lb-ft

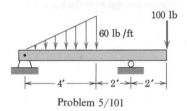

Problem 5/101

5/102 Determine the bending moment M in terms of x in the pyramidal cantilever beam which supports its own weight. The specific weight of the beam material is γ.

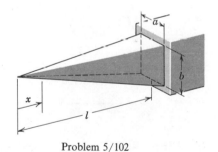

Problem 5/102

5/103 The curved cantilever beam in the form of a quarter-circular arc supports a load of w lb/ft applied along the curve of the beam on its upper surface. Determine the torsional moment T and bending moment M in the beam as functions of θ.

Ans. $T = wr^2\left(\dfrac{\pi}{2} - \theta - \cos\theta\right)$

$M = -wr^2(1 - \sin\theta)$

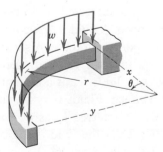

Problem 5/103

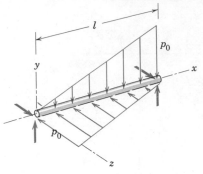

◀ **5/104** The end-supported shaft is subjected to the linearly varying loads in mutually perpendicular planes. Determine the expression for the resultant bending moment M in the shaft.

$$Ans. \ M = \frac{p_0}{6l} x(l - x) \sqrt{5l^2 - 2lx + 2x^2}$$

Problem 5/104

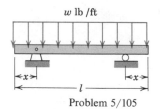

◀ **5/105** The beam supports a uniformly distributed load as shown. Determine the location x of the two supports so as to minimize the maximum bending moment M in the beam. What is M?

$$Ans. \ M = 0.0215wl^2, \quad x = 0.207l$$

Problem 5/105

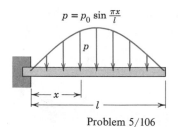

◀ **5/106** The cantilever beam supports the half-sine-wave loading $p = p_0 \sin(\pi x/l)$ per foot of length. Determine the bending moment M in terms of x by direct use of Eq. 38.

$$Ans. \ M = \frac{p_0 l}{\pi} \left(\frac{l}{\pi} \sin \frac{\pi x}{l} + x - l \right)$$

Problem 5/106

29 Fluid Statics. In the work so far, attention has been directed primarily to the action of forces between rigid bodies. In this article the equilibrium of bodies subjected to forces due to the action of fluid pressures will be developed. A fluid is any continuous substance which, when at rest, is unable to support shear force. A shear force is one tangent to the surface upon which it acts and is developed when differential velocities exist between adjacent layers of fluids. Thus a fluid at rest can exert only normal forces on a bounding surface. Fluids may be either gaseous or liquid. The statics of fluids is generally referred to as *hydrostatics* when the fluid is a liquid and as *aerostatics* when the fluid is a gas.

(*a*) *Fluid Pressure.* The pressure at any given point in a fluid is the same in all directions (Pascal's law). This fact may be shown by considering the equilibrium of an infinitesimal triangular prism of fluid as shown in Fig. 60. The fluid pressures normal to the faces of the element are taken to be p_1, p_2, p_3, and p_4 as shown. The equilibrium of forces in the x- and y-directions requires that

$$p_2 \, dx \, dz = p_3 \, ds \, dz \cos \theta \qquad p_1 \, dy \, dz = p_3 \, ds \, dz \sin \theta$$

Since $ds \sin \theta = dy$ and $ds \cos \theta = dx$, these equations require that

$$p_1 = p_2 = p_3 = p$$

By rotating the element through 90 deg it is found that p_4 is also equal to the other pressures. Hence the pressure at any point in a fluid is the same in all directions. In this analysis it is unnecessary to account for the weight of the fluid element since, when the specific weight γ (weight per unit volume) of the fluid is multiplied by the volume of the element, a differential quantity of third order results which may be neglected when passing to the limit compared with the second-order pressure force terms.

In all fluids at rest the pressure is a function of the vertical dimension. To determine this function a change in the vertical dimension must be considered and account of the weight of the fluid must be taken. Figure 61 shows a differential element of a vertical fluid column of cross-sectional area dA. The positive direction of vertical measurement h is taken down. The pressure on the upper face is p, and that on the lower face is p plus the change in p, or $p + dp$. The weight of the element equals its specific weight γ multiplied by the volume. The normal forces on the lateral surface have nothing to do with the balance of forces in the vertical direction and are not shown. Equilibrium of the fluid element in the h-direction requires

$$p \, dA + \gamma \, dA \, dh - (p + dp) \, dA = 0$$
$$dp = \gamma \, dh \tag{39}$$

This differential relation shows that the pressure in a fluid increases with depth or decreases with increased elevation. Equation 39 holds for both

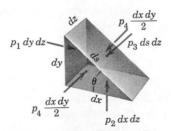

Figure 60

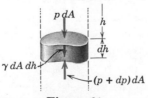

Figure 61

liquids and gases and is in accord with common observation of air and water pressures.

Fluids that are essentially incompressible are called liquids, and it follows that for most practical purposes their specific weight may be considered constant for every part of the liquid.† With γ a constant Eq. 39 may be integrated as it stands, and the result is

$$▶ \qquad p = p_0 + \gamma h \qquad (40)$$

The pressure p_0 is the pressure on the surface of the liquid where $h = 0$. If p_0 is due to atmospheric pressure and the measuring instrument records only the increment above atmospheric pressure,* the measurement gives what is known as "gage pressure" and is $p = \gamma h$.

Gases, on the other hand, are compressible, and here the specific weight varies with vertical distance. For most engineering problems this variation is negligible when considering gas pressures on a structure, since the height of the structure usually represents only a small change in altitude. When necessary, the gas law $p = \gamma KT$, where T is the absolute temperature of the gas and K is a constant, may be used to determine the variation of pressure with altitude. Substitution of $\gamma = p/(KT)$ into Eq. 39 and replacement of the downward measurement h by the upward measurement z, where $dh = -dz$, give $KT\,dp = -p\,dz$. Integration from the conditions of pressure p_0 at zero altitude to pressure p at altitude z yields, under conditions of constant temperature,

$$z = KT \log \frac{p_0}{p}$$

When the temperature of a gas is not constant with altitude, such as with the earth's atmosphere, the effect of changing temperature must be accounted for.

(*b*) *Hydrostatic Pressure on Submerged Rectangular Surfaces.* A surface submerged in a liquid, such as a gate valve in a dam or the wall of a tank, is subjected to fluid pressure normal to its surface and distributed over its area. In problems where fluid forces are appreciable it is necessary to account for the resultant force due to the distribution of pressure on the surface and the position at which this resultant acts. For systems that are open to the earth's atmosphere, the atmospheric pressure p_0 acts over all surfaces and, hence, yields a zero resultant. Thus only the increment above atmospheric pressure, called "gage pressure" or $p = \gamma h$, need be considered.

Consider the special but common case of the action of hydrostatic pressure on a rectangular plate submerged in a liquid. Figure 62a shows such a plate 1-2-3-4 with its top edge horizontal and with the plane of the plate making any particular angle θ with the vertical plane. The horizontal surface of the liquid is represented by the x-y' plane. The fluid pressure (gage) acting

---

† See Table C1, Appendix C, for table of specific weights.

* Atmospheric pressure at sea level may be taken to be 14.7 lb/in.²

normal to the plate at point 2 is represented by the arrow 6-2 and equals the specific weight γ times the vertical depth from the liquid surface to point 2. This same pressure acts at all points along the edge 2-3. At point 1 on the lower edge, the fluid pressure equals γ times the vertical depth to point 1, and this pressure is the same at all points along edge 1-4. The variation of pressure p over the area of the plate is governed by the linear depth relationship and is thereby represented by the altitude of the truncated prism 1-2-3-4-5-6-7-8 with the plate as its base. The resultant force produced by this pressure distribution is represented by R, which acts at some point P known as the *center of pressure.*

It is clear that the conditions that prevail at the vertical section 1-2-6-5 in Fig. 62*a* are identical to those at section 4-3-7-8 and at every other vertical section through the plate. Thus the problem may be analyzed from the two-dimensional view of a vertical section as shown in Fig. 62*b* for section 1-2-6-5. For this section the pressure distribution is trapezoidal. If b is the horizontal width of the plate, an element of plate area over which the pressure $p = \gamma h$ acts is $dA = b\,dy$, and an increment of the resultant force is $dR = p\,dA = bp\,dy$. But $p\,dy$ is merely the shaded increment of trapezoidal area dA', so that $dR = b\,dA'$. Therefore the resultant force acting on the entire plate may

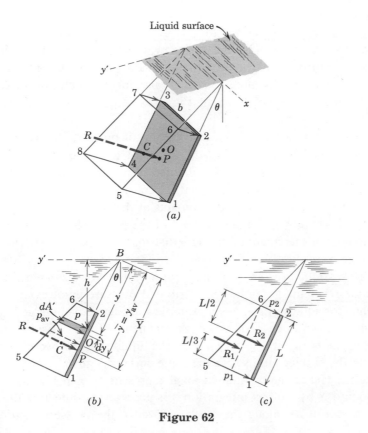

Figure 62

be expressed as

$$R = \int b \, dA' \quad \text{or} \quad R = bA'$$

Care must be taken not to confuse the physical area A of the plate with the geometrical area A' defined by the trapezoidal distribution of pressure.

The trapezoidal area representing the pressure distribution is easily expressed by using its average altitude. Therefore the resultant force R may be written in terms of the average pressure $p_{av} = \frac{1}{2}(p_1 + p_2)$ times the plate area A. The average pressure is also the pressure that exists at the average depth to the centroid O of the plate. Therefore an alternative expression for R is

$$R = p_{av}A = \gamma \bar{h} A$$

where $\bar{h} = \bar{y} \cos \theta$.

The line of action of the resultant force R is obtained from the principle of moments. Using the x-axis (point B in Fig. 62b) as the moment axis yields $R\bar{Y} = \int y(pb \, dy)$. Substituting $p \, dy = dA'$ and cancelling b give

$$\bar{Y} = \frac{\int y \, dA'}{\int dA'}$$

which is simply the expression for the centroidal coordinate of the trapezoidal area A'. In the two-dimensional view, therefore, the resultant R passes through the centroid C of the trapezoidal area defined by the pressure distribution on the vertical section. Clearly $\bar{Y}$ also locates the centroid C of the truncated prism 1-2-3-4-5-6-7-8 in Fig. 62a through which the resultant actually passes.

In dealing with a trapezoidal distribution of pressure, the calculation is usually simplified by considering the resultant as composed of two components, Fig. 62c. The trapezoid is divided into a rectangle and a triangle, and the force represented by each part is considered separately. The force represented by the rectangular portion acts at the center O of the plate and is $R_2 = p_2A$ where A is the area 1-2-3-4 of the plate. The force represented by the triangular increment of pressure distribution is $\frac{1}{2}(p_1 - p_2)A$ and acts through the centroid of the triangular portion as shown.

(*c*) *Hydrostatic Pressure on Cylindrical Surfaces.* For a submerged curved surface the resultant R caused by distributed pressure involves more calculation than for a flat surface. As an example consider the submerged cylindrical surface shown in Fig. 63a where the elements of the curved surface are parallel to the horizontal surface x-y' of the liquid. Vertical sections perpendicular to the surface all disclose the same curve AB and the same pressure distribution. Hence the two-dimensional representation in Fig. 63b may be used. To find R by a direct integration it is necessary to integrate the x- and y-components of dR along the curve AB, since the pressure continuously changes direction. Thus

$$R_x = b \int (p \, dL)_x = b \int p \, dy \qquad \text{and} \qquad R_y = b \int (p \, dL)_y = b \int p \, dx$$

A moment equation would next be required to establish the position of R.

A second method for finding R is often much simpler. The equilibrium of the block of liquid ABC directly above the surface, shown in Fig. 63c, is considered. The resultant R is then disclosed as the equal and opposite reaction of the surface upon the block of liquid. The resultants of the pressures along AC and CB are P_y and P_x, respectively, and are easily obtained. The weight W of the liquid block is calculated from the area ABC of its section multiplied by the constant dimension b and the specific weight. The weight W passes through the centroid of area ABC. The equilibrant R is then determined completely from the equilibrium equations applied to the free-body diagram of the fluid block.

(*d*) *Hydrostatic Pressure on Flat Surfaces of Any Shape.* Figure 64a shows a flat plate of any shape submerged in a liquid. The horizontal surface of the liquid is the plane x-y', and the plane of the plate makes an angle θ with the vertical. The force acting on a differential strip of area dA parallel to the surface of the liquid is $dR = p \, dA = \gamma h \, dA$. The pressure p has the same magnitude throughout the length of the strip, since there is no change of

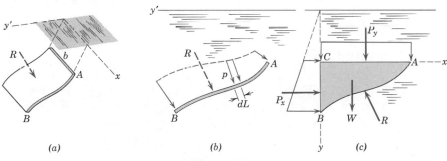

(a)   (b)   (c)

Figure 63

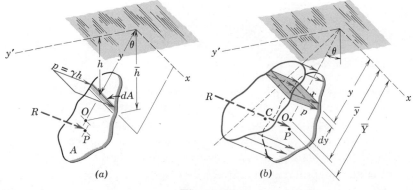

(a)   (b)

Figure 64

depth along the horizontal strip. The total force acting on the exposed area A is obtained by integration and is

$$R = \int dR = \int p \, dA = \gamma \int h \, dA$$

Substituting the centroidal relation $\bar{h}A = \int h \, dA$ gives

▶ $$R = \gamma \bar{h} A \qquad\qquad (41)$$

The quantity $\gamma \bar{h}$ represents the pressure that exists at the depth of the centroid O of the area and is the average pressure over the area.

The resultant R may also be represented geometrically by the volume of the figure shown in Fig. 64b. Here the fluid pressure p is represented as an altitude to the plate considered as a base. The resulting volume is a truncated right cylinder. The force dR acting on the differential area $dA = x \, dy$ yields the elemental volume $dV = p \, dA$ shown by the shaded slice, and the total force is, then, represented by the total volume of the cylinder. Hence

$$R = \int dR = \int dV = V$$

It is seen from Eq. 41 that the average altitude of the truncated cylinder is the average pressure $\gamma \bar{h}$ which exists at a depth corresponding to the centroid O of the area exposed to pressure. For problems where the centroid O or the volume V is not readily apparent a direct integration may be performed to obtain R. Thus

$$R = \int dR = \int p \, dA = \int \gamma h x \, dy$$

where the depth h and the length x of the horizontal strip of differential area must be expressed in terms of y to carry out the integration.

The second requirement of the analysis of fluid pressure is the determination of the position of the resultant force in order to account for the moments of the pressure forces. Using the principle of moments with the x-axis of Fig. 64b as the moment axis gives

$$R\bar{Y} = \int y \, dR \qquad \text{or} \qquad \bar{Y} = \frac{\int y \, dV}{V} \qquad\qquad (42)$$

This second relation satisfies the definition of the coordinate $\bar{Y}$ to the centroid of the volume V, and it is concluded, therefore, that the resultant R passes through the centroid C of the volume described by the plate area as base and the linearly varying pressure as altitude. The point P at which R is applied to the plate is the center of pressure. It should be noted carefully that the center of pressure P and the centroid O of the plate area are *not* the same.

An alternative expression for $\overline{Y}$ involving the so-called *moment of inertia** of the area may be written. Substitution of $R = \gamma \overline{h}A$, $dR = \gamma h\, dA$, $\overline{h} = \overline{y}\cos\theta$, and $h = y\cos\theta$ into the principle of moments $R\overline{Y} = \int y\, dR$ gives

$$\overline{Y} = \frac{\int y^2\, dA}{\overline{y}A} \qquad \text{or} \qquad \overline{Y} = \frac{I_x}{\overline{y}A} \tag{43}$$

where I_x is the moment of inertia $\int y^2\, dA$ of the exposed area about the x-axis at the surface of the liquid. The area moment of inertia may be written as $I_x = k_x^2 A$ where k_x is known as the *radius of gyration* of the area about the x-axis. With this substitution Eq. 43 may be written alternatively as

$$\overline{Y} = \frac{k_x^2}{\overline{y}} \tag{43a}$$

The use of Eqs. 43 requires a knowledge of the properties of moments of inertia of areas. This knowledge is not assumed of the reader at this stage, so further discussion of Eqs. 43 will be omitted.

In accounting for the resultant moment of the pressure forces on the plate about some axis parallel to the plate a direct integration is often just as straightforward as the application of Eqs. 42 or 43. Thus the total moment M of the pressure forces about, say, the x-axis is

$$M = \int y\, dR = \int y\gamma h x\, dy$$

and, again, the depth h and the length x of the horizontal strip of differential area must be expressed in terms of y to carry out the integration.

Sample Problems

'107 A rectangular plate, shown in vertical section AB, is 12 ft high and 18 ft wide (normal to the plane of the paper) and blocks the end of a fresh-water channel 9 ft deep. The plate is hinged about a horizontal axis along its upper edge through A and is restrained from opening the channel by the fixed ridge at B that bears horizontally against the lower edge of the plate. Calculate the force B exerted against the plate by the ridge.

Solution. The free-body diagram of the plate is shown in section and includes the vertical and horizontal components of the force at A, the unspecified weight W of the plate, the unknown horizontal force B, and the resultant R of the triangular distribution of pressure against the vertical face.

The specific weight of fresh water is $\gamma = 62.4$ lb/ft^3 so that the average pressure is

$[p_{av} = \gamma \overline{h}]$ $p_{av} = 62.4(\tfrac{9}{2}) = 280.8$ lb/ft^2

The resultant R of the pressure forces against the plate becomes

$[R = p_{av}A]$ $R = (280.8)(9)(18) = 45{,}490$ lb

---

* Moments of inertia of areas are developed in full in Chapter 8.

This force acts through the centroid of the triangular distribution of pressure, which is 3 ft above the bottom of the plate. A zero moment summation about A establishes the unknown force B. Thus

$$[\Sigma M_A = 0] \qquad\qquad 9(45{,}490) - 12B = 0, \qquad B = 34{,}120 \text{ lb} \qquad\qquad Ans.$$

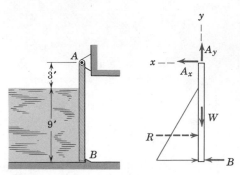

Problem 5/107

5/108 Determine completely the resultant force R exerted on the cylindrical dam surface by the water. The specific weight of fresh water is 62.4 lb/ft³, and the dam has a length, normal to the paper, of $b = 100$ ft.

Solution. The circular block of water $BCDO$ is isolated and its free-body diagram is drawn. The force P_x is

$$P_x = \gamma \bar{h} A = \frac{\gamma r}{2} br = \frac{(62.4)(10)}{2}(100)(10) = 312{,}000 \text{ lb}$$

The weight W of the water is

$$W = \gamma V = (62.4)\frac{\pi(10)^2}{4}(100) = 490{,}000 \text{ lb}$$

Equilibrium of the section of water requires

$$[\Sigma F_x = 0] \qquad\qquad R_x = P_x = 312{,}000 \text{ lb}$$
$$[\Sigma F_y = 0] \qquad\qquad R_y = W = 490{,}000 \text{ lb}$$

The resultant force R exerted by the fluid on the dam is equal and opposite to that shown acting on the fluid and is

$$[R = \sqrt{R_x^2 + R_y^2}] \qquad\qquad R = 581{,}000 \text{ lb} \qquad\qquad Ans.$$

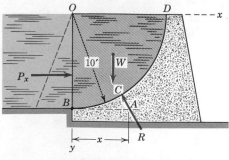

Problem 5/108

The x-coordinate of the point A through which R passes may be found graphically if desired. Algebraically x is found from the principle of moments. Using B as a moment center gives

$$P_x\frac{r}{3} + W\frac{4r}{3\pi} - R_yx = 0, \qquad x = \frac{312{,}000\left(\frac{10}{3}\right) + 490{,}000\left(\frac{40}{3\pi}\right)}{490{,}000} = 6.37 \text{ ft} \qquad Ans.$$

If the point C on the curved surface is desired, it will be necessary to combine the moment relation with the equation of the circle for solution.

5/109　Determine the resultant force R exerted on the semicircular end of the water tank shown in the figure if the tank is filled to capacity. Express the result in terms of the radius r of the circular end and the specific weight γ of water.

　　Solution I. By direct integration R is found from

$$[R = \int dR] \qquad R = \int p \, dA = \int \gamma y(2x \, dy) = 2\gamma \int_0^r y \sqrt{r^2 - y^2} \, dy$$

$$R = \tfrac{2}{3}\gamma r^3 \qquad\qquad Ans.$$

The location of R is determined from the principle of moments, which gives

$$[R\overline{Y} = \int y \, dR] \qquad \tfrac{2}{3}\gamma r^3 \overline{Y} = 2\gamma \int_0^r y^2 \sqrt{r^2 - y^2} \, dy$$

Upon integrating there results

$$\tfrac{2}{3}\gamma r^3 \overline{Y} = \frac{\gamma r^4}{4}\frac{\pi}{2} \qquad \text{and} \qquad \overline{Y} = \frac{3\pi}{16}r \qquad\qquad Ans.$$

　　Solution II. Using Eq. 41 directly to find R gives

$$[R = \gamma\overline{h}A] \qquad R = \gamma\frac{4r}{3\pi}\frac{\pi r^2}{2} = \tfrac{2}{3}\gamma r^3 \qquad\qquad Ans.$$

Application of Eq. 42 for determining the centroid C amounts to the identical integration of *Solution I*. As an alternative Eq. 43 may be used. With the expression $I_x = \tfrac{1}{8}\pi r^4$ for the moment of inertia* of the semicircular area about the x-axis, there results

$$\left[\overline{Y} = \frac{I_x}{\overline{y}A}\right] \qquad \overline{Y} = \frac{\pi r^4/8}{\dfrac{4r}{3\pi}\dfrac{\pi r^2}{2}} = \frac{3\pi}{16}r \qquad\qquad Ans.$$

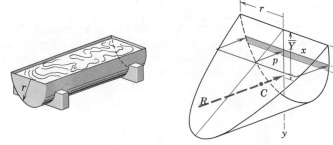

Problem 5/109

* See Table C4, Appendix C.

Problems

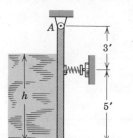

Problem 5/110

5/110 The spring-loaded vertical gate is hinged about a horizontal axis along its upper edge A and closes the end of a rectangular fresh-water channel 4 ft wide (normal to the plane of the paper). Calculate the preset spring force F that will limit the depth of the water to $h = 6$ ft.

Ans. F = 8986 lb

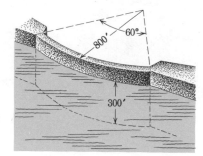

Problem 5/111

5/111 The arched dam has the form of a cylindrical surface of 800-ft radius and subtends an angle of 60 deg. If the water is 300 ft deep, determine the total force P exerted by the water on the face of the dam.

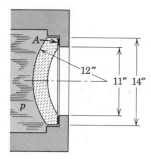

Problem 5/112

5/112 One of the critical problems in the design of deep-submergence vehicles is to provide viewing ports that will withstand tremendous hydrostatic pressures without fracture or leakage. The figure shows the cross section of an experimental acrylic window with spherical surfaces under test in a high-pressure liquid chamber. If the pressure p is raised to a level that simulates the effect of a dive to a depth of 3000 ft in sea water, calculate the average pressure σ supported by the ring gasket A. *Ans. σ = 3484 lb/in.²*

Problem 5/113

5/113 The quonset hut is subjected to a horizontal wind, and the pressure p against the circular roof is approximated by $p_0 \cos \theta$. The pressure is positive on the windward side of the hut and is negative on the leeward side. Determine the total horizontal shear Q on the foundation per unit length of roof measured normal to the paper.

5/114 The sides of a V-shaped fresh-water trough, shown in section, are hinged about their common intersection through O and are held together by a cable and turnbuckle placed every 6 ft along the length of the trough. Calculate the tension T supported by each turnbuckle.

Ans. $T = 1230$ lb

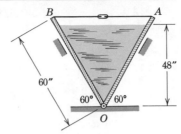

Problem 5/114

5/115 A fresh-water channel 30 ft wide (normal to the plane of the paper) is blocked by a rectangular barrier, shown by its section AEF. Supporting horizontal struts BE are placed every 2 ft along the 30-ft width. Determine the compression C in each strut BE. The weight of the barrier is assumed to be small compared with the other forces acting.

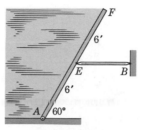

Problem 5/115

5/116 The form for a concrete footing is hinged to its lower edge A and held in place every 4 ft by a horizontal brace BC. Neglect the weight of the form and compute the compression C in each brace resulting from the liquid behavior of wet concrete, which has a density of 150 lb/ft³.

Ans. $C = 1575$ lb

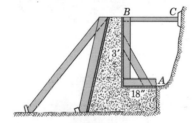

Problem 5/116

5/117 The cover plate for the 8-by-12-in. opening in the tank is bolted in place with negligible initial tensions in the bolts. If the tank is filled with liquid mercury to the level shown, compute the tension induced in each of the bolts at A and at B. Mercury has a specific weight of 847 lb/ft³.

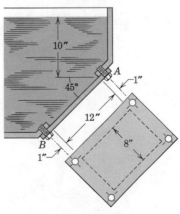

Problem 5/117

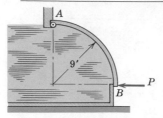

Problem 5/118

5/118 The quarter-circular gate *AB*, shown in section, has a horizontal width of 6 ft (normal to the plane of the paper) and controls the flow of fresh water over the ledge at *B*. The gate has a total weight of 6800 lb and is hinged about its upper edge *A*. Calculate the minimum force *P* required to keep the gate closed. In locating the center of gravity of the gate neglect its thickness compared with the 9-ft radius.

Ans. P = 10,830 lb

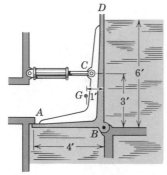

Problem 5/119

5/119 In the figure is shown the cross section of a gate *ABD* which closes an opening 5 ft wide in a salt-water channel. For the water level shown, compute the compressive force *F* in the hydraulic-piston rod needed to maintain a contact force of 200 lb per foot of gate width along the line of contact at *A*. The gate weighs 3400 lb with center of gravity at *G*.

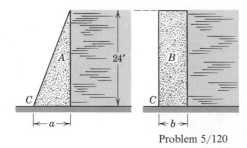

Problem 5/120

5/120 The triangular and rectangular sections are being considered for a small fresh-water concrete dam. From the standpoint of resistance to overturning about *C*, which section will require less concrete, and how much less per foot of dam length? Concrete weighs 150 lb/ft³.

Ans. Section *A* requires 12,480 lb less per foot

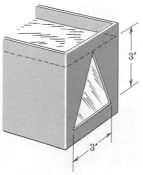

Problem 5/121

5/121 Determine the total force *R* exerted on the triangular window by the fresh water in the tank. The water level is even with the top of the window. Also determine the distance *H* from *R* to the water level.

5/122 A rectangular plate, shown in edge view, is 9 ft high and 8 ft wide (normal to the paper) and separates reservoirs of sea water and oil. The oil has a specific gravity (ratio of density to that of fresh water) of 0.85. Determine the depth h of the water required to make the reaction at B equal to zero.   *Ans. h = 5.64 ft*

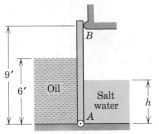

Problem 5/122

5/123 A uniform rectangular plate AB, shown in section, weighs 3200 lb and separates the two bodies of fresh water in a tank which is 10 ft wide (normal to the plane of the figure). Calculate the tension T in the supporting cable.

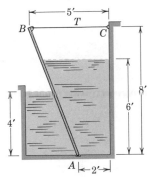

Problem 5/123

5/124 The fresh-water side of a concrete dam has the shape of a vertical parabola with vertex at A. Determine the position b of the base point B through which acts the resultant force of the water against the dam face C.

Ans. b = 93.75 ft

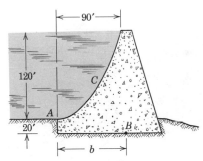

Problem 5/124

5/125 The gate AB is a 580-lb rectangular plate 5 ft high and 3.5 ft wide and is used to close the discharge channel at the bottom of an oil reservoir. As a result of condensation in the tank, fresh water collects at the bottom of the channel. Calculate the moment M applied about the hinge axis B required to close the gate against the hydrostatic forces of the water and oil. The specific gravity of the oil is 0.85.

Ans. M = 12,970 lb-ft

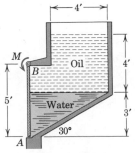

Problem 5/125

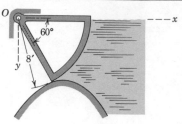

Problem 5/126

5/126 A control gate of cylindrical cross section is used to regulate the flow of water over the spillway of a fresh-water dam as shown. The gate weighs 15,000 lb and has a length normal to the paper of 18 ft. A torque applied to the shaft of the gate at O controls the angular position of the gate. For the position shown determine the horizontal and vertical components F_x and F_y of the total force exerted by the shaft on its bearings at O. Account for the effect of fluid pressure by a direct integration over the cylindrical surface.

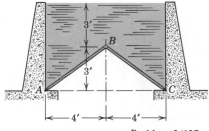

Problem 5/127

5/127 The bottom of a fresh-water channel, shown in section, consists of two uniform rectangular plates AB and BC each weighing 3000 lb and hinged along their common edge B and hinged also to the base of the fixed channel along their lower edges A and C. The length of the channel is 20 ft, measured perpendicular to the plane of the paper. Determine the force P per foot of channel length exerted by each plate on the connecting hinge at B.   *Ans. $P = 1140$ lb/ft*

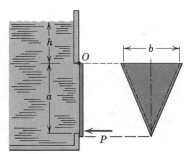

Problem 5/128

5/128 A flat plate seals a triangular opening in the vertical wall of a tank of liquid of specific weight γ. The plate is hinged about the upper edge O of the triangle. Determine the force P required to hold the gate in a closed position against the pressure of the liquid.

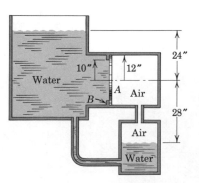

Problem 5/129

◄ **5/129** A circular disk A bears against a gasket around the flange B and seals off the air space from the water space in the connecting circular passage. For the water levels shown, compute the average pressure p on the gasket whose outside and inside diameters are 24 in. and 20 in., respectively. The top of the water tank is open to the atmosphere.   *Ans. $p = 4.18$ lb/in.²*

5/130 A dam consists of the flat-plate barriers *A* and *B* whose weights are small. Supporting struts *C* and *D* are placed every 10 ft of dam section. A mud sample drawn up to the surface weighs 100 lb/ft³. Determine the compression in *C* and *D*. All joints may be assumed to be hinged.

Ans. $C = 111{,}800$ lb, $D = 20{,}800$ lb

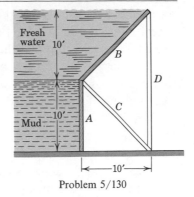

Problem 5/130

5/131 The end of a fresh-water channel with a 60-deg V-section is closed by the slanted triangular plate *A*. Calculate the resultant force *R* exerted on *A* by the water and the height *h* of the point on *A* through which *R* acts.

Ans. $R = 961$ lb, $h = 2$ ft

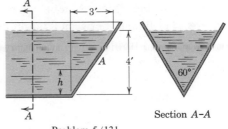

Problem 5/131

Section *A-A*

5/132 The closed tank contains fresh water to the level shown. Air is pumped into the tank until the open U-tube manometer shows a difference of height of $h = 21$ in. between its two columns of water. Determine the total force *R* produced by hydrostatic pressure acting on the cover plate which closes the 16-in.-diameter opening. Also find the distance *b* below the center line of the cover to the center of pressure *P*.

Ans. $R = 240$ lb, $b = 0.484$ in.

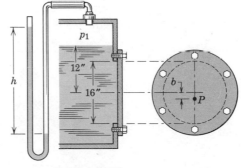

Problem 5/132

30 Buoyancy. The principle of buoyancy, the discovery of which is credited to Archimedes, is easily explained in the following manner for any fluid, gaseous or liquid, in equilibrium. Consider a portion of the fluid defined by an imaginary closed surface, as illustrated by the irregular dotted boundary in Fig. 65*a*. If the body of the fluid could be sucked off from within the closed cavity and replaced simultaneously by the forces that it exerted on the boundary of the cavity, Fig. 65*b*, there would be no disturbance of the equilibrium of the surrounding fluid. Furthermore, a free-body diagram of the fluid portion before removal, Fig. 65*c*, shows that the resultant of the pressure forces distributed over its surface must be equal and opposite to its weight *W* and must pass through the center of gravity of the fluid element. If the element is replaced by a body of the same dimensions, the surface forces acting on the body held in this position will be identical with those acting on the fluid element. Thus the resultant force exerted on the surface of an object immersed in a fluid is equal and opposite to the weight of fluid displaced and passes through the center of gravity of the displaced fluid. This *resultant force* is the force of *buoyancy*. In the case of a liquid whose specific weight is constant the center of gravity of the displaced liquid coincides with the centroid of the displaced volume.

It follows from the foregoing discussion that when the specific weight of an object is less than the specific weight of the fluid in which it is immersed, there will be an unbalance of force in the vertical direction, and the object will rise. When the immersing fluid is a liquid, the object continues to rise until it comes to the surface of the liquid and then comes to rest in an equilibrium position, assuming that the specific weight of the new fluid above the surface is less than the specific weight of the object. In the case of the surface boundary between a liquid and a gas, such as water and air, the effect of the gas pressure on that portion of the floating object above the liquid is balanced by the added pressure in the liquid due to the action of the gas on its surface.

One of the most important problems involving buoyancy is the determination of the stability of a floating object. This situation may be illustrated by considering a ship's hull shown in cross section in an upright position in Fig. 66a. Point B is the centroid of the displaced volume and is known as the *center of buoyancy*. The resultant of the forces exerted on the hull by the water pressure is the force F. Force F passes through B and is equal and opposite to the weight W of the ship. If the ship is caused to list through an angle α, Fig. 66b, the shape of the displaced volume changes, and the center

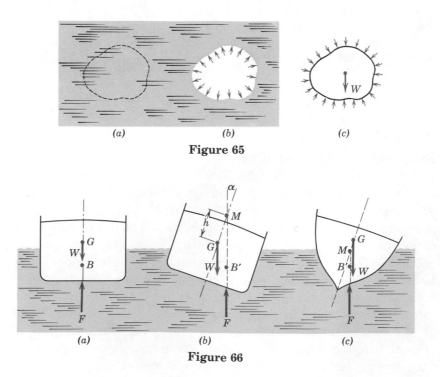

Figure 65

Figure 66

of buoyancy will shift to some new position such as B'. The point of intersection of the vertical line through B' with the centerline of the ship is called the *metacenter M*, and the distance h of M above the center of gravity G is known as the *metacentric height*. For most hull shapes it is found that the metacentric height remains practically constant for angles of list up to about 20 deg. When M is above G, as in Fig. 66b, there is clearly a righting moment which tends to bring the ship back to its original position. The magnitude of this moment for any particular angle of list is a measure of the stability of the ship. If M is below G, as for the hull of Fig. 66c, the moment accompanying any list is in the direction to increase the list. This is clearly a condition of instability and must be avoided in the design of any ship.

Problems

/133   A small balloon for use in recording wind velocity and its direction has an uninflated weight of 14 oz. If the balloon is inflated with 0.60 lb of helium and exerts an upward force of 1.20 lb on its mooring prior to release, determine the diameter d of the spherical balloon. The specific weight of the air is 0.0753 lb/ft³.

Ans. d = 4.08 ft

/134   The submersible diving chamber has a total weight out of water of 14,800 lb including personnel, equipment, and ballast. When the chamber is lowered to a depth of 4000 ft in the ocean, the cable tension is 1800 lb. Compute the total volume V displaced by the chamber.

Problem 5/134

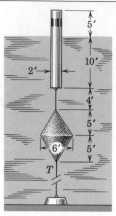

Problem 5/135

5/135 The spar buoy consists of a closed steel cylinder that weighs 1800 lb. It is secured to a submerged stabilizing buoy consisting of two conical shells welded together to form a closed unit weighing 2200 lb. Calculate the tension T in the lower cable, which secures the two buoys in fresh water in the positions shown.

Ans. T = 3840 lb

5/136 The density of ice is 56 lb/ft³. Determine the ratio n of depth below water to height above water for an iceberg of rectangular shape floating in salt water (specific weight 64 lb/ft³).

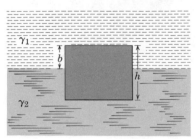

Problem 5/137

5/137 The homogeneous block of specific weight γ is floating between two liquids of specific weights $\gamma_1 < \gamma$ and $\gamma_2 > \gamma$. Determine an expression for the distance b that the block protrudes into the top liquid.

$$Ans.\ b = h\frac{\gamma_2 - \gamma}{\gamma_2 - \gamma_1}$$

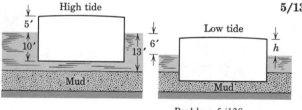

Problem 5/138

5/138 The loaded barge of rectangular shape has a freeboard of 5 ft and a draft of 10 ft when floating in salt water at high tide. As the water level drops 6 ft to the low-tide level, the barge settles in the soft mud that covers the bottom of the harbor. The mud has a specific weight of 90 lb/ft³ and behaves like a liquid. Determine the freeboard h at the low-tide condition.

5/139 Steel chain with a nominal shank diameter of $\frac{1}{4}$ in. weighs 70 lb per 100 feet of length and has a breaking strength of 3400 lb. What length h of chain can be lowered into a deep part of the ocean before the chain breaks under the action of its own weight.   *Ans. h* = 5590 ft

5/140 Show that both the homogeneous solid hemisphere and the hemispherical shell float with stability. If both have the same radius and total weight, which one is the less stable?

Problem 5/140

5/141 The solid floating object is composed of a hemisphere and a circular cylinder of equal radii r. If the object floats with the center of the hemisphere above the water surface, determine the maximum height h which the cylinder may have before the object will no longer float in the upright position illustrated. *Ans.* $h = r/\sqrt{2}$

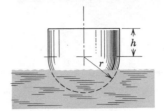

Problem 5/141

5/142 A structure designed for sub-ice observation of sea life in polar waters consists of the cylindrical viewing chamber connected to the surface by the cylindrical shaft open at the top for ingress and egress. Ballast is carried in the rack below the chamber. To ensure a stable condition for the structure, it is necessary that its legs bear upon the ice with a force that is at least 15 per cent of the total buoyancy force of the submerged structure. If the structure less ballast has a weight of 12,600 lb out of water, calculate the required weight w of lead ballast. The specific weight of lead is 710 lb/ft³. *Ans.* $w = 10,320$ lb

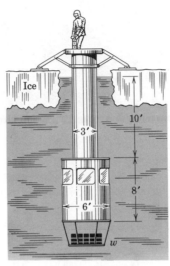

Problem 5/142

5/143 A buoy in the form of a uniform 24-ft pole 8 in. in diameter weighs 400 lb and is secured at its lower end to the bottom of a fresh-water lake with 15 ft of cable. If the depth of the water is 30 ft, calculate the angle θ made by the pole with the horizontal.

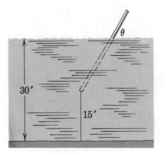

Problem 5/143

5/144 One end of a uniform pole of length l and specific weight γ_1 is hinged about a point a distance h above the surface of a liquid of specific weight γ_2. Find the angle θ assumed by the pole. Assume that $\gamma_2 > \gamma_1$.

$$Ans.\ \theta = \sin^{-1}\left(\frac{h}{l}\sqrt{\frac{\gamma_2}{\gamma_2 - \gamma_1}}\right)$$

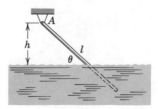

Problem 5/144

5/145 After a Navy tanker moves from a fresh-water anchorage to a berth in salt water, it takes on 100,000 gal of fuel oil (specific gravity 0.88). The draft marks on the hull of the tanker read the same in salt water as they did in fresh water before the transfer. Calculate the final displacement W (total weight) of the ship in long tons. (1 long ton equals 2240 lb.)

Ans. $W = 13,110$ long tons

Problem 5/146

5/146 A homogeneous solid sphere of specific weight γ_s is resting on the bottom of a tank containing a liquid of specific weight γ_l which is greater than γ_s. As the tank is filled, a depth h is reached at which the sphere begins to float. Determine the relation among h, r, and γ_s/γ_l which defines this depth.

$$Ans. \ \frac{\gamma_s}{\gamma_l} = \frac{1}{4}\left(\frac{h}{r}\right)^2\left(3 - \frac{h}{r}\right)$$

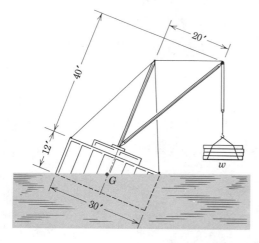

Problem 5/147

◄**5/147** The barge crane of rectangular proportions has a 12- by 30-ft cross section over its entire length of 80 ft. If the maximum permissible submergence and list in sea water are represented by the position shown, determine the corresponding maximum safe load w that the barge can handle at the 20-ft extended position of the boom. Also find the total displacement W in long tons of the unloaded barge (1 long ton equals 2240 lb.) The distribution of machinery and ballast places the center of gravity G of the barge, minus the load w, at the center of the hull.

Ans. $w = 100,800$ lb, $W = 366$ long tons

5/148 The accurate determination of the vertical position of the center of gravity G of a ship is difficult to achieve by calculation. It is more easily obtained by a simple inclining experiment on the loaded ship. With reference to the figure, a known external weight w is placed a distance d from the center line, and the angle of list θ is measured by means of the deflection of a plumb bob. The displacement of the ship and the location of the metacenter M are known. Calculate the metacentric height $\overline{GM}$ for a 12,000-ton ship (long tons) inclined by a 60,000-lb weight placed 26 ft from the center line if a 20-ft plumb line is deflected a distance $a = 8$ in. The weight w is at a distance $b = 6$ ft above M.

Ans. $\overline{GM} = 1.75$ ft

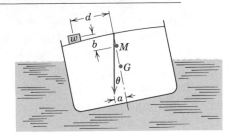

Problem 5/148

5/149 A spherical balloon is made from material that stretches in proportion to the force applied to it. Since the amount of stretch of any element depends directly on the original length and the force acting, the equation $u = Krp$ may be written, where u is the increase in radius of the balloon, r is the radius of the unstretched balloon, p is the internal gas pressure (gage), and K is the constant of proportionality. As the balloon is filled with gas, such as helium, under pressure, the sphere expands and creates greater buoyancy in the air. At the same time the weight of the balloon is increasing as more gas is injected. Find the pressure p for which the balloon has a maximum lift. The specific weight of air is γ_a, and that of the gas is $\gamma_g = k(p + p_0)$ for constant temperature, where k is a constant and where p_0 is the atmospheric pressure.

$$Ans. \ p = \frac{1}{4}\left(\frac{3\gamma_a}{k} - \frac{1}{K} - 3p_0\right)$$

5/150 Determine the minimum ratio of a to b for stability of the rectangular block of specific weight γ_b for very small angles of list. The specific weight of the liquid is γ_l.

$$Ans. \ \frac{a}{b} = \sqrt{6\frac{\gamma_b}{\gamma_l}\left(1 - \frac{\gamma_b}{\gamma_l}\right)}$$

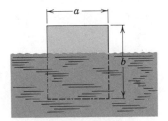

Problem 5/150

31 Equilibrium of Internal Stresses.

Attention has been directed in several of the previous articles in this chapter to the equilibrium of bodies subjected to a continuous variation of forces over their region of application. The variation of tension in a flexible cable, the variation of shear force and bending moment in a beam, and the change in fluid pressure on the surface of an object submerged in a fluid were all described with the same basic procedure, which entailed the isolation of a differential element of the body in question to disclose the forces and their incremental changes. In each of the cases cited, the variation was a function of one dimension only. The tension in the cable varied along the length of the cable; the shear and moment in the beam varied along the length of the beam; and the pressure in the fluid changed with the vertical dimension. In the present article the variation of internal forces in a body with respect to the three spatial coordinates of the body will be introduced. The description of such a variation comes under the heading of *continuum mechanics*. This dependence of a quantity upon more than one independent variable calls for the use of *partial derivatives* and leads to the solution of *partial differential equations*.

Continuum mechanics is an advanced topic in the study of engineering mechanics and is generally beyond the scope of this book. The introduction that follows, however, is designed to illustrate the application of the basic principles of statics to more advanced analytical problems.

Before the equilibrium equations can be written, it is necessary to establish the concept and notation of internal stress. Consider any body, Fig. 67*a*, in equilibrium under the action of an externally applied force system. The external forces will induce internal forces which are transmitted across any arbitrary section, such as section *N*. The isolation of one of the two parts of the body, Fig. 67*b*, discloses on this section a distribution of force whose direction and intensity vary from point to point. The net force acting on a small element of area ΔA located at point O on the section is $\Delta \mathbf{F}$. The *stress* acting on the element is a vector defined as the limiting ratio

$$\boldsymbol{\sigma} = \lim_{\Delta A \to 0} \frac{\Delta \mathbf{F}}{\Delta A} = \frac{d\mathbf{F}}{dA}$$

Thus the stress acting on a given surface at a given point is the intensity of the force and has the same units as pressure, lb/in.^{2} In dealing with stress it is convenient to resolve $\boldsymbol{\sigma}$ into three scalar components in the directions normal and tangent to the surface dA, as shown in Fig. 67*c*. The component σ_3 is known as a *normal stress* and may be tension (away from the surface)

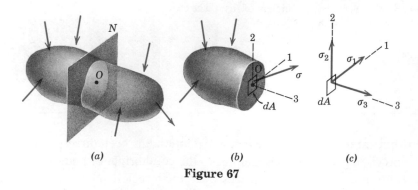

(a)   *(b)*   *(c)*

Figure 67

or compression (toward the surface). Positive normal stress is taken to be tension, and negative normal stress is compression. The components in the tangential directions, σ_1 and σ_2, are known as *shear stresses*. In contrast to the ability of a solid to support shear stress, it should be recalled that the only stress which can be supported by a fluid at rest is a normal stress or pressure.

It is seen from the foregoing description that there are three stress components acting on a given surface at any point in a stressed solid. There are two additional surfaces which are mutually perpendicular that can be exposed at the interior point O of Fig. 67b by taking as the cutting plane N through O plane 2-3 and plane 1-3. Three additional stress components may be designated for each of these other surfaces, thus making a total of nine components of stress associated with the point O. These components, together with their positive directions, are shown in Fig. 68a as they act on the three perpendicular surfaces of an elemental rectangular block of material. In relation to the coordinate directions, these three faces are designated as the positive faces of the elemental block. The positive stresses acting on the three hidden negative faces of the block are opposite in direction to those on the positive visible faces. This condition is required by the principle of action and reaction and is illustrated in Fig. 68b for the case of the normal stresses σ_{xx} and σ_{yy} and the shear stresses σ_{xy} and σ_{yx}. Similar conditions exist for the remaining stress components.*

Special attention is drawn to the use of a right-handed coordinate system and to the subscript notation for stress. The first subscript indicates that the stress acts on the surface which is normal to the coordinate direction, and the second subscript denotes the direction of the stress. Thus σ_{xy} is a shear stress which acts on the face normal to the x-direction and which is pointed in the y-direction for the positive face of the element.

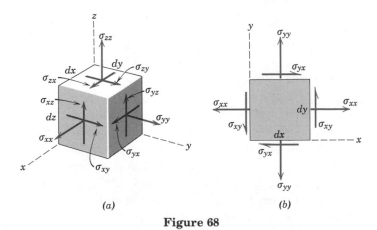

(a)   (b)

Figure 68

* A notation frequently used for the stress components is σ_x, σ_y, σ_z for the normal stresses and τ_{xy}, τ_{xz}, τ_{yz} for the shear stresses.

The state of stress at a point requires the specification of nine components of stress. These stress components may be represented by the array

$$[T] = \begin{bmatrix} \sigma_{xx} & \sigma_{xy} & \sigma_{xz} \\ \sigma_{yx} & \sigma_{yy} & \sigma_{yz} \\ \sigma_{zx} & \sigma_{zy} & \sigma_{zz} \end{bmatrix} \tag{44}$$

which is known as the *stress tensor* $[T]$. It is noted that the first subscript denotes the row, and the second subscript denotes the column. The stress tensor is a *second-order* tensor with each of its components depending on two directions in space. A vector is a *first-order* tensor with each of its components depending on only one direction in space. A scalar quantity is a *zero-order* tensor with no spatial orientation. Other second-order tensor quantities dealt with in mechanics are the *inertia tensor,*[*] which is used in the description of rigid-body motion in three dimensions, and the *strain tensor,* which is used in connection with the stress tensor in the study of deformable bodies.

Stress tensors find use in transforming the components of stress at a point between coordinate systems. A full discussion of the stress tensor is beyond the scope of this book, and only one of its special properties will be developed here. The moment equilibrium of the elemental block of material of Fig. 68*b* about the *z*-axis taken through the lower left-hand corner of the block requires that

$$(\sigma_{xy} \, dy \, dz) \, dx - (\sigma_{yx} \, dx \, dz) \, dy = 0$$

All other stresses on the element contribute zero moment. The moments of any body forces, such as gravitational forces, are of higher order than the moments of the shear forces and therefore disappear. Thus, from the moment equation it is seen that the two shears are equal, and with similar conclusions for the remaining shear components, there result

$$\sigma_{xy} = \sigma_{yx} \qquad \sigma_{xz} = \sigma_{zx} \qquad \sigma_{yz} = \sigma_{zy} \tag{45}$$

With this equality of shear components, the elements on the two sides of the main diagonal $\sigma_{xx}, \sigma_{yy}, \sigma_{zz}$ of the stress tensor of Eq. 44 may be interchanged, and the tensor is said to be *symmetric.*

With the concept and notation of internal stress established, the equations for the equilibrium of stresses at a point may now be written. For the purpose of illustration, a two-dimensional continuum of stress will be examined for a plate of unit thickness acted upon by the stresses σ_{xx}, σ_{yy}, and $\sigma_{xy} = \sigma_{yx}$. All other stress components are taken to be zero, which describes a condition of *plane stress.* An elemental block of dimensions dx and dy is cut from the loaded plate, Fig. 69*a*, and isolated with its free-body diagram in Fig. 69*b* (which shows stress rather than force). Here account must be taken of the change in stress with change in the coordinates, and the diagram must disclose these stress increments. At the location defined by the coordi-

[*] See *Dynamics,* Chapter 8, Art 41.

nates x and y to the negative faces of the element, the stresses σ_{xx}, σ_{yy}, and σ_{xy} are shown in their positive directions. On the positive faces of the element, defined by the coordinates $x + dx$ and $y + dy$, the stresses will have incremental changes which must reflect the rates of change with respect to the coordinates. Therefore the normal stress in the x-direction at $x + dx$ will be σ_{xx} plus the rate at which σ_{xx} changes with respect to x, with y held constant, times the change in x and is given by the expression $\sigma_{xx} + (\partial\sigma_{xx}/\partial x)\,dx$. Here the partial derivative $\partial\sigma_{xx}/\partial x$ means the rate of change of σ_{xx} with respect to x with y held constant. Similar expressions are written for the increments in the other two stress components that are shown on the elemental block in Fig. 69b. In addition to the stresses acting on the surface of the element, there may be body forces present which act throughout the mass of the body. The quantities X and Y represent the intensity of the body forces expressed as force per unit volume of material.

Equilibrium of forces in the x-direction requires

$$\left(\sigma_{xx} + \frac{\partial\sigma_{xx}}{\partial x}\,dx\right)dy - \sigma_{xx}\,dy + \left(\sigma_{xy} + \frac{\partial\sigma_{xy}}{\partial y}\,dy\right)dx - \sigma_{xy}\,dx + X\,dx\,dy = 0$$

This equation and the similar equation written for the y-direction become, upon simplification,

$$\frac{\partial\sigma_{xx}}{\partial x} + \frac{\partial\sigma_{xy}}{\partial y} + X = 0 \quad\text{and}\quad \frac{\partial\sigma_{yy}}{\partial y} + \frac{\partial\sigma_{xy}}{\partial x} + Y = 0 \quad (46)$$

Equations 46 are the partial differential equations of equilibrium for a state of plane stress expressed in x-y coordinates. These equations must be satisfied at every point in the plate. They are necessary conditions for the determination of the stresses, but, in general, they are not sufficient. Sufficiency requires a knowledge of the relationships between the stresses and the internal strains of the material under investigation and the satisfaction of an equation expressing the compatibility between the linear and angular strains. These additional relationships and various solutions to Eqs. 46 will be found in treatments on the theories of elasticity and plasticity.

The stress relationships in this article are developed in rectangular coordi-

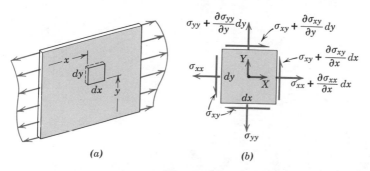

(a)   (b)

Figure 69

nates which are appropriate for problems with rectangular boundaries. It is frequently of advantage to use polar, cylindrical, or spherical coordinates for the description of stresses when one of these systems or some other orthogonal coordinate system more closely fits the boundaries of the stressed body under consideration.

Selected examples of the equilibrium requirements for the stresses in a continuum are given in the problems that follow.

Problems

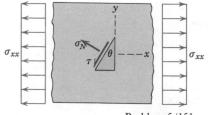

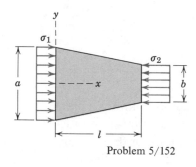

Problem 5/151

5/151 A flat plate is subjected to a uniform field of stress σ_{xx}. Write the expressions for the normal stress σ_N and shear stress τ acting on the face of an element inclined at an angle θ with respect to the y-direction.

$$Ans. \quad \sigma_N = \sigma_{xx} \cos^2\theta, \quad \tau = \tfrac{1}{2}\sigma_{xx} \sin 2\theta$$

Problem 5/152

5/152 The flat plate with parallel ends and tapered sides is in equilibrium under the action of the uniform compressive stress σ_1 acting on its left-hand edge and σ_2 acting on its right-hand edge. An "expert" claims that the stress σ_{xx} at any interior position x varies linearly from σ_1 at $x = 0$ to σ_2 at $x = l$. The absence of shear stress in the plate is also claimed. Set up the assumed expression for σ_{xx} and prove that the expert is wrong.

5/153 A body is subjected to a magnetic field which exerts a torque (couple) given by $\mathbf{M} = \mathbf{i}M_x + \mathbf{j}M_y + \mathbf{k}M_z$ per unit of mass on the body. Reexamine the relationship between the shear components σ_{xy} and σ_{yx}, σ_{xz} and σ_{zx}, and σ_{yz} and σ_{zy}.

5/154 For the case of plane stress derive the expressions for normal stress σ_N and shear stress τ acting on a surface inclined at angle θ with respect to the x-axis. Show that σ_N is a maximum or a minimum when $\tau = 0$. (The similarity between the resulting equations and the equations developed in Art. 44 of Chapter 8 for moments of inertia of areas about inclined axes is to be noted.)

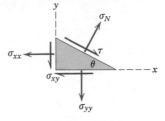

Problem 5/154

$$Ans. \ \sigma_N = \sigma_{xx} \sin^2 \theta + \sigma_{yy} \cos^2 \theta + \sigma_{xy} \sin 2\theta$$

$$\tau = \frac{\sigma_{yy} - \sigma_{xx}}{2} \sin 2\theta - \sigma_{xy} \cos 2\theta$$

$$\tan 2\theta = \frac{2\sigma_{xy}}{\sigma_{yy} - \sigma_{xx}} \ \text{for max. or min. } \sigma_N$$

5/155 By considering the equilibrium of the elemental tetrahedron of stressed material, derive the expression for the normal stress σ_{ss} acting on the inclined surface ABC in terms of the rectangular components of stress at this point. The direction cosines of the normal to ABC are l, m, n. Shear stresses on ABC are omitted from the figure to avoid confusion. (Suggestion: Take the area ABC to be unity.)

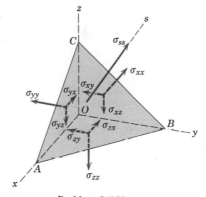

Problem 5/155

5/156 Expand the derivation of the first of Eqs. 46 for equilibrium in the x-direction to include the general case in three dimensions. By inspection write the remaining two equations of equilibrium. Neglect body forces.

$$Ans. \ \frac{\partial \sigma_{xx}}{\partial x} + \frac{\partial \sigma_{xy}}{\partial y} + \frac{\partial \sigma_{xz}}{\partial z} = 0$$

$$\frac{\partial \sigma_{yy}}{\partial y} + \frac{\partial \sigma_{yz}}{\partial z} + \frac{\partial \sigma_{xy}}{\partial x} = 0$$

$$\frac{\partial \sigma_{zz}}{\partial z} + \frac{\partial \sigma_{xz}}{\partial x} + \frac{\partial \sigma_{yz}}{\partial y} = 0$$

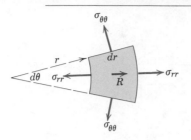

Problem 5/157

◀ **5/157** A circular disk rotates with a constant speed about its center. As a result of symmetry there are no shear stresses in the disk, and both the radial stress σ_{rr} and the tangential stress ("hoop stress") $\sigma_{\theta\theta}$ are functions of r only. The element may be treated as a body in equilibrium by adding a body force $R = \rho r \omega^2$ per unit volume where ρ is the mass density and ω is the angular velocity of the disk. Draw the free-body diagram of the element showing the change in σ_{rr} and derive the differential equation for "equilibrium" of the stresses. Assume unit thickness of the disk.

$$Ans. \quad \frac{d\sigma_{rr}}{dr} + \frac{\sigma_{rr} - \sigma_{\theta\theta}}{r} + \rho r \omega^2 = 0$$

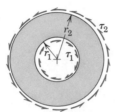

Problem 5/158

◀ **5/158** The flat ring is subjected to a uniform shear stress τ_1 on its inner rim and a uniform shear stress τ_2 on its outer rim to maintain equilibrium. Under this condition of twist, no normal stresses are developed in the ring. Derive the differential equation for the equilibrium of an element of the ring and integrate the equation so as to obtain an expression for the shear stress $\sigma_{r\theta}$ as a function of r and the outer shear τ_2. Compare this result with the relation between τ_1 and τ_2 obtained by satisfying the equilibrium requirement for the entire ring.

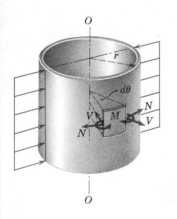

Problem 5/159

◀ **5/159** A long cylindrical shell is loaded uniformly along two opposite elements. The resulting stresses in the shell that affect equilibrium in the cross-sectional plane are functions of θ only and consist of a normal force N, a shear force V, and a moment M all expressed per unit of length of shell section. Draw the free-body diagram of the element showing the increments in the stresses as functions of θ and derive the three differential equations of equilibrium for the element.

$$Ans. \quad \frac{d^2V}{d\theta^2} + V = 0$$

$$\frac{d^2N}{d\theta^2} + N = 0, \quad \frac{dM}{d\theta} + Vr = 0$$

5/160 The thin conical shell is filled with a liquid of specific weight γ and is balanced on its vertex. The shell supports internal forces per unit length of shell section which are designated N_l in the direction of the cone elements and N_θ tangent to the horizontal circular sections. Neglect the weight of the shell and derive the equilibrium equations for the directions along and normal to a cone element.

$$\text{Ans. } \frac{dN_l}{dl} + \frac{N_l - N_\theta}{l} = 0$$

$$N_\theta = \gamma r l \left(1 - \frac{l}{\sqrt{r^2 + h^2}} \right)$$

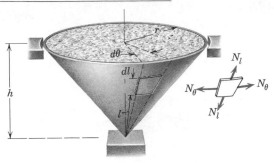

Problem 5/160

5/161 A rectangular differential element of a flat plate supports bending, shear, and normal forces. The positive directions are shown for the normal force N, in-plane shear T, out-of-plane shear V, and bending moment M, all expressed per unit length of section. Additionally, the plate is subjected to a normal loading stress w on its surface. Derive the differential equations for the equilibrium of moments about the x-axis, for the equilibrium of forces in the x-direction, and for the equilibrium of forces in the z-direction.

$$\text{Ans. } \frac{\partial M_x}{\partial y} + V_{yz} = 0$$

$$\frac{\partial N_{xx}}{\partial x} + \frac{\partial T_{xy}}{\partial y} = 0, \quad \frac{\partial V_{yz}}{\partial y} + \frac{\partial V_{xz}}{\partial x} = w$$

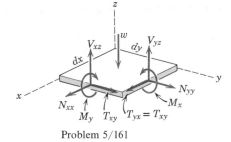

Problem 5/161

5/162 Derive the differential equations of equilibrium in polar coordinates for the element of a plate under a general state of plane stress without body forces. The notation for positive stresses is shown in the figure.

$$\text{Ans. } \frac{\partial \sigma_{rr}}{\partial r} + \frac{1}{r} \frac{\partial \sigma_{r\theta}}{\partial \theta} + \frac{\sigma_{rr} - \sigma_{\theta\theta}}{r} = 0$$

$$\frac{\partial \sigma_{r\theta}}{\partial r} + \frac{1}{r} \frac{\partial \sigma_{\theta\theta}}{\partial \theta} + \frac{2\sigma_{r\theta}}{r} = 0$$

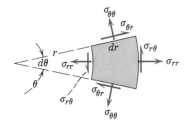

Problem 5/162

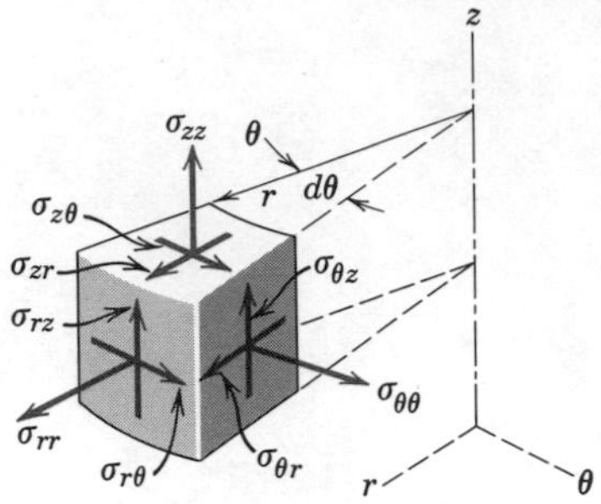

Problem 5/163

◀ **5/163** Derive the differential equations of equilibrium in cylindrical coordinates using the notation shown on the element for positive stresses.

$$\textit{Ans.}\ \frac{\partial \sigma_{rr}}{\partial r} + \frac{\partial \sigma_{rz}}{\partial z} + \frac{1}{r}\frac{\partial \sigma_{r\theta}}{\partial \theta} + \frac{\sigma_{rr} - \sigma_{\theta\theta}}{r} = 0$$

$$\frac{\partial \sigma_{r\theta}}{\partial r} + \frac{\partial \sigma_{\theta z}}{\partial z} + \frac{1}{r}\frac{\partial \sigma_{\theta\theta}}{\partial \theta} + \frac{2\sigma_{r\theta}}{r} = 0$$

$$\frac{\partial \sigma_{zz}}{\partial z} + \frac{\partial \sigma_{rz}}{\partial r} + \frac{1}{r}\frac{\partial \sigma_{\theta z}}{\partial \theta} + \frac{\sigma_{rz}}{r} = 0$$

6 FRICTION

32 Introduction. In the preceding chapters the forces of action and reaction between contacting surfaces have for the most part been assumed to act normal to the surfaces. This assumption characterizes the interaction between smooth surfaces and was illustrated in example 2 of Fig. 27. Although in many instances this ideal assumption involves only a very small error, there are a great many problems wherein the ability of contacting surfaces to support tangential as well as normal forces must be considered. Tangential forces generated between contacting surfaces are known as friction forces and are present to some degree with the interaction between all actual surfaces. Whenever a tendency exists for one contacting surface to slide along another surface, the friction forces developed are always found to be in a direction to oppose this tendency.

In some types of machines and processes it is desirable to minimize the retarding effect of friction forces. Examples are bearings of all types, power screws, gears, the flow of fluids in pipes, and the propulsion of aircraft and missiles through the atmosphere. In other situations effort is made to take advantage of friction, as in brakes, clutches, belt drives, and wedges. Wheeled vehicles depend on friction for both starting and stopping, and ordinary walking depends on friction between the shoe and the ground. Friction forces are present throughout nature and exist to a considerable extent in all machines no matter how accurately constructed or carefully lubricated. A machine or process in which friction is neglected is often referred to as *ideal*. When friction is taken into account, the machine or process is termed *real*. In all real cases where sliding motion between parts occurs, the friction forces result in a loss of energy which is dissipated in the form of heat. In addition to the generation of heat and the accompanying loss of energy, friction between mating parts will cause wear to take place during the period of relative motion between them.

33 Frictional Phenomena. There are a number of separate types of frictional resistance encountered in mechanics, and each of these types will be described briefly in this article prior to a more detailed account of the most common type of friction in the next article.

Dry Friction. Dry friction is encountered when the unlubricated surfaces of two solids are in contact under a condition of sliding or tendency to slide. A friction force tangent to the surfaces of contact is developed both during the interval leading up to impending slippage and while slippage takes place. The direction of the force always opposes the motion or impending motion.

This type of friction is also called *Coulomb* friction. The principles of dry or Coulomb friction were developed largely from the experiments of Coulomb in 1781 and from the work of Morin from 1831 to 1834. Although a comprehensive theory of dry friction is not available, an analytical model sufficient to handle the vast majority of problems in dry friction is available and is described in Art. 34 which follows. This model forms the basis for most of this chapter.

Fluid Friction. Fluid friction is developed when adjacent layers in a fluid (liquid or gas) are moving at different velocities. This motion gives rise to frictional forces between fluid elements, and these forces depend on the relative velocity between layers. When there is no such relative velocity, there is no fluid friction. Fluid friction depends not only on the velocity gradients within the fluid, but also on the viscosity of the fluid, which is a measure of its resistance to shearing action between fluid layers. Frictional forces accompanying viscous action for one-dimensional flow are illustrated in Fig. 70, which shows the cross section AB of a plate moving with a velocity v_0 parallel to a fixed surface C. The two surfaces are separated by a viscous fluid, all particles of which move in the direction of v_0. In this example each lamina or layer of fluid parallel to its bounding surfaces slides over its adjacent layer with no mixing between layers. Such flow is termed *laminar flow*. At $y = 0$ the fluid adheres to surface C and has no velocity. At $y = b$ the fluid adheres to the moving surface AB and its velocity is v_0. Between the two surfaces the velocity v varies linearly with the distance y for this laminar-flow example.

For a layer of fluid of thickness dy, a shear stress τ will be required to maintain a velocity difference dv between the surfaces. This stress is found to be proportional to the velocity gradient, so that the relation

$$\tau = \mu \frac{dv}{dy} \tag{47}$$

may be written. The quantity μ is known as the *absolute viscosity* and represents the shear stress per unit of velocity gradient. The unit of absolute viscosity in the cgs system is the *poise*, which is one (dyne) (sec)/(cm²). More commonly the *centipoise* is used, which is $\frac{1}{100}$ of a poise. In the British gravitational system, one centipoise is equivalent to $1.45(10^{-7})$ (lb)(sec)/(in.²). As a quantitative indication of viscosity several selected values are tabulated below.

Fluid	Temp °F	Absolute viscosity, μ, centipoises
Air	70	0.018
Water	68	1.000
Lubricating oil		
SAE 10	60	100
SAE 30	60	400

The total tangential force acting on the surface of the moving plate due to fluid friction is τA where τ is the shear stress at the surface of the plate and

A is the area in contact with the shearing fluid. In problems where τ varies over the area, the force may be expressed as $\int \tau \, dA$.

When the flow of a fluid is irregular and when mixing occurs across laminar boundaries, the flow is called *turbulent,* and Eq. 47 no longer holds. Turbulence occurs in a large fraction of fluid-flow problems.

Fluid friction plays a large part in the design and operation of bearings of all types. Bearings frequently operate under conditions of partial lubrication where a complete film of lubricating fluid does not separate the surfaces. This case is referred to as *boundary lubrication* and represents a condition somewhere between that of dry friction and that of the fully lubricated fluid bearing. As would be expected a reliable analytical model for this type of friction condition is difficult to construct, and design information must come largely from experimental measurements.

The analysis of fluid friction in fully lubricated journal and thrust bearings, both oil and gas, in centrifugal-pump flow, and in aircraft, missile, and spacecraft performance, to mention a few applications, is basic to the design of these elements and systems. A full treatment of fluid friction is beyond the scope of this book and is best pursued with a direct study of fluid mechanics and lubrication.

Rolling Friction. Rolling friction is a resistance to the rolling of a circular object. The wheel shown in Fig. 71 carries a load L on the axle, and a force P is applied to produce rolling. The deformation of the wheel and supporting surface as shown is greatly exaggerated. The distribution of pressure p over the area of contact is similar to that indicated, and the resultant R of this distribution will act at some point A and will pass through the center of the wheel for equilibrium. The force P necessary to initiate and maintain rolling may be found by equating the moments of all forces about A to zero. This gives

$$P = \frac{a}{r} L = f_r L$$

where the moment arm of P is taken to be r, and $f_r = a/r$ is called the coefficient of rolling friction. The coefficient f_r is the ratio of resisting force to normal load and in this respect is similar to the coefficients of static and

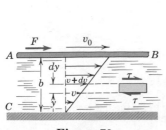

Figure 70

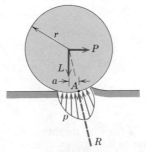

Figure 71

kinetic friction. On the other hand there is no slipping involved in the interpretation of f_r.

The quantity a depends on many factors which are difficult to quantify, so that a comprehensive theory of rolling resistance is not available. This distance a is a function of the elastic and plastic properties of the mating materials, the radius of the wheel, the speed of travel, and the roughness of the surfaces. Some tests indicate only a small variation with wheel radius, and a is often taken to be independent of the rolling radius.

Two typical values of f_r are listed in Table C1, Appendix C. These values are only approximate and may be used merely for rough calculations.

Internal Friction. Internal friction is found in all solid materials that are subjected to cyclical loading. For highly elastic materials the recovery from deformation occurs with very little loss of energy caused by internal friction. For materials which have low limits of elasticity and which undergo appreciable plastic deformations during loading, the amount of internal friction that accompanies this deformation may be considerable. Figure 72 shows a stress-strain* diagram for an element of material that undergoes plastic deformation during one complete cycle. The lack of recovery of both the tensile and compressive deformations gives rise to what is known as a *hysteresis* loop. The magnitude of the area enclosed within the loop is a direct measure of the energy loss per cycle per unit volume due to internal friction. This type of friction attenuates internal vibrations resulting from shock loading, and the attenuation may be accounted for analytically in the study of damped vibrations.† The mechanism of internal friction is associated with the action of shear deformation, and the student should consult a reference on materials science for a detailed description of this shear mechanism.

34   Dry Friction. The mechanism of dry friction will now be explained in some detail with the aid of a very simple experiment. Consider a solid block of weight W resting on a horizontal surface as shown in Fig. 73a. The contacting surfaces possess a certain amount of roughness. The experiment will involve

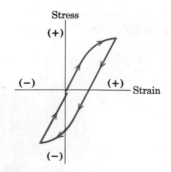

Figure 72

* Strain is defined as the change in length divided by the original length of an element of stressed material. Strain is, therefore, a measure of the intensity of deformation at any point.

† See *Dynamics,* Chapter 9.

the application of a horizontal force P which will vary continuously from zero to a value sufficient to move the block and give it an appreciable velocity. The free-body diagram of the block for any value of P is shown in Fig. 73b, and the tangential friction force exerted by the plane on the block is labeled F. This friction force will *always* be in a direction to oppose motion or the tendency toward motion of the body on which it acts. There is also a normal force N which in this case equals W, and the total force R exerted by the supporting surface on the block is the resultant of N and F. A magnified view of the irregularities of the mating surfaces, Fig. 73c, will aid in visualizing the mechanical action of friction. Support is necessarily intermittent and exists at the mating humps. The direction of each of the reactions on the block R_1, R_2, R_3, etc., will depend not only on the geometric profile of the irregularities but also on the extent of local deformation, as well as the welding that can take place on a minute scale at each contact point. The total normal force N is merely the sum of the n-components of the R's, and the total frictional force F is the sum of the t-components of the R's. When the surfaces are in relative motion, the contacts are more nearly along the tops of the humps, and the t-components of the R's will be smaller than when the surfaces are at rest relative to one another. This consideration helps to explain the well-known fact that the force P necessary to maintain motion is less than that required to start the block when the irregularities are more nearly in mesh.

Assume now that the experiment indicated is performed and the friction force F is measured as a function of P. The resulting experimental relation is indicated in Fig. 73d. When P is zero, equilibrium requires that there be no friction force. As P is increased, the friction force must be equal and opposite to P as long as the block does not slip. During this period the block is in equilibrium, and all forces acting on the block must satisfy the equilibrium equations. Finally a value of P is reached which causes the block to slip and to move in the direction of the applied force. At this same time the friction force drops slightly and rather abruptly to a somewhat lower value. Here it remains

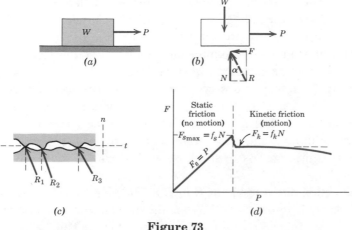

Figure 73

essentially constant for a period but then drops off still more with higher velocities.

The region up to the point of slippage or impending motion is known as the range of *static friction,* and the value of the friction force is determined by the *equations of equilibrium.* This force may have any value from zero up to and including, in the limit, the maximum value. For a given pair of mating surfaces this maximum value of static friction $F_{s_{\max}}$ is found to be proportional to the normal force N. Hence

$$F_{s_{\max}} = f_s N$$

where f_s is the proportionality constant, known as the *coefficient of static friction.* It must be carefully observed that this equation describes only the *limiting* or *maximum* value of the static friction force and not any lesser value. Thus the equation applies *only* to cases where it is known that motion is impending.

After slippage occurs a condition of *kinetic friction* is involved. Kinetic friction force is usually somewhat less than the maximum static friction force. The kinetic friction force F_k is also found to be proportional to the normal force. Hence

$$F_k = f_k N$$

where f_k is the *coefficient of kinetic friction.* It follows that f_k is generally less than f_s. As the velocity of the block increases, the kinetic friction coefficient decreases somewhat, and when high velocities are reached the effect of lubrication by an intervening fluid film may become appreciable. Coefficients of friction depend greatly on the exact condition of the surfaces as well as on the velocity and are subject to a considerable measure of uncertainty.

It is customary to write the two friction force equations merely as

$$F = fN \tag{48}$$

There will be an understanding from the problem whether limiting static friction with its corresponding coefficient of static friction or whether kinetic friction with its corresponding kinetic coefficient is implied. It should be emphasized again that many problems involve a static friction force which is less than the maximum value at impending motion, and therefore under these conditions the friction equation *cannot* be used.

From Fig. 73c it may be observed that for rough surfaces there is a greater possibility for large angles between the reactions and the *n*-direction than for smoother surfaces. Thus a friction coefficient reflects the roughness of a pair of mating surfaces and incorporates a geometric property of these mating contours. It is meaningless to speak of a coefficient of friction for a single surface.

The direction of the resultant R in Fig. 73b measured from the direction of N is specified by $\tan \alpha = F/N$. When the friction force reaches its limiting static value, the angle α reaches a maximum value ϕ_s. Thus

$$\tan \phi_s = f_s$$

When slippage occurs, the angle α will have a value ϕ_k corresponding to the kinetic friction force. In like manner

$$\tan \phi_k = f_k$$

It is customary to write merely

$$\tan \phi = f \tag{49}$$

where application to the limiting static case or to the kinetic case is inferred from the problem at hand. The angle ϕ_s is known as the *angle of static friction*, and the angle ϕ_k is called the *angle of kinetic friction*. This friction angle ϕ for each case clearly defines the limiting position of the total reaction R between two contacting surfaces. If motion is impending, R must be one element of a right circular cone of vertex angle $2\phi_s$, as shown in Fig. 74. If motion is not impending, R will be within the cone. This cone of vertex angle $2\phi_s$ is known as the *cone of static friction* and represents the locus of possible positions for the reaction R at impending motion. If motion occurs, the angle of kinetic friction applies, and the reaction must lie on the surface of a slightly different cone of vertex angle $2\phi_k$. This cone is the *cone of kinetic friction*.

Further experiment shows that the friction force is essentially independent of the apparent or projected area of contact. The true contact area is much smaller than the projected value, since only the peaks of the contacting surface irregularities support the load. Relatively small normal loads result in high stresses at these contact points. As the normal force increases, the true contact area also increases as the material undergoes yielding, crushing, or tearing at the points of contact. A comprehensive theory of dry friction must go beyond the mechanical explanation presented here. For example, there is some evidence to support the theory that molecular attraction may be an important cause of friction under conditions where the mating surfaces are in very intimate contact. Other factors that influence dry friction are the generation of high local temperatures and adhesion at contact points, relative hardness of mating surfaces, and the presence of thin surface films of oxide, oil, dirt, or other substances.

Figure 74

There are three types of problems in dry friction commonly encountered in mechanics.

(1) In the *first* type the condition of impending motion is to be investigated. It will be clear from the wording of the problem that the requirement of limiting static friction should be used.

(2) In the *second* type of problem impending motion need not exist, and therefore the friction force may be smaller than that given by Eq. 48 with the static coefficient. In this event the friction force will be determined only by the equations of equilibrium. In such a problem it may be asked whether or not the existing friction is sufficient to maintain the body at rest. To answer the question, equilibrium may be assumed, and the corresponding friction force necessary to maintain this state can be calculated from the equations of equilibrium. This friction force may then be compared with the maximum static friction that the surfaces can support as calculated from Eq. 48 with $f = f_s$. If F is less than that given by Eq. 48, it follows that the assumed friction force can be supported and therefore the body is at rest. If the calculated value of F is greater than the limiting value, it follows that the given surfaces cannot support that much friction force, and therefore motion exists and the friction becomes kinetic.

(3) The *third* type of problem involves relative motion between the contacting surfaces, and here the kinetic coefficient of friction applies. For this case Eq. 48 with $f = f_k$ will always give the kinetic friction force directly.

The foregoing discussion applies to all dry contacting surfaces and, to a limited extent, to moving surfaces which are partially lubricated. Some typical values of the coefficients of friction are given in Table C1, Appendix C. These values are only approximate and are subject to considerable variation, depending on the exact conditions prevailing. They may be used, however, as typical examples of the magnitudes of frictional effects. When a reliable calculation involving friction is required, it is often desirable to determine the appropriate friction coefficient by experiment wherein the surface conditions of the problem are duplicated as closely as possible.

Sample Problems

6/1 Determine the range of values which the weight W may have so that the 100-lb block shown in the figure will neither start moving up the plane nor slip down the plane. The coefficient of static friction for the contact surfaces is 0.30.

Solution. The maximum value of W will be given by the requirement for motion impending up the plane. The friction force on the block therefore acts down the plane as shown in the free-body diagram of the block for Case I in the figure. Applying the equations of equilibrium gives

$$[\Sigma F_y = 0] \qquad N - 100 \cos 20° = 0, \qquad N = 94 \text{ lb}$$

$$[F = fN] \qquad F = 0.30(94.0) = 28.2 \text{ lb}$$

$$[\Sigma F_x = 0] \qquad W - 28.2 - 100 \sin 20° = 0, \qquad W = 62.4 \text{ lb} \qquad\qquad Ans.$$

The minimum value of W is determined when motion is impending down the plane. The friction force on the block will act up the plane to oppose the tendency to move as shown in the free-body diagram for Case II. Equilibrium in the x-direction requires

$[\Sigma F_x = 0]$ $W + 28.2 - 100 \sin 20° = 0,$ $W = 6.0$ lb   *Ans.*

Thus W may have any value from 6.0 lb to 62.4 lb, and the block will remain at rest.

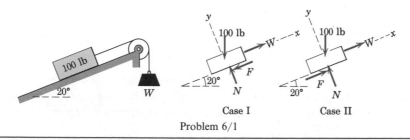

Case I

Case II

Problem 6/1

6/2 Determine the amount and direction of the friction force acting on the 100-lb block shown if, first, $P = 50$ lb, and, second, $P = 10$ lb. The coefficient of static friction is 0.20, and the coefficient of kinetic friction is 0.17. The forces are applied with the block initially at rest.

Solution. There is no way of telling from the statement of the problem whether the block will remain in equilibrium or whether it will begin to slip following the application of P. It is therefore necessary to make an assumption. Assume the friction force to be up the plane, as shown by the solid arrow. A balance of forces in both x- and y-directions gives

$[\Sigma F_x = 0]$ $P \cos 20° + F - 100 \sin 20° = 0$

$[\Sigma F_y = 0]$ $N - P \sin 20° - 100 \cos 20° = 0$

Case I. P $= 50$ lb

Substitution into the first of the two equations gives

$$F = -12.8 \text{ lb}$$

The negative sign means that *if* the block is in equilibrium, the friction force acting on it is in the direction opposite to that assumed and therefore is down the plane as represented by the dotted arrow. Conclusion on the magnitude of F cannot be reached, however, until it is verified that the surfaces are capable of supporting 12.8 lb of friction force. This may be done by substituting $P = 50$ lb into the second equation, which gives

$$N = 111.1 \text{ lb}$$

The maximum static friction force that the surfaces can support is then

$[F = fN]$ $F = 0.20(111.1) = 22.2$ lb

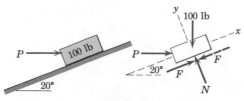

Problem 6/2

Since this force is greater than that required for equilibrium, it follows that the assumption of equilibrium was correct. The answer is, then,

$$F = 12.8 \text{ lb down the plane} \qquad\qquad Ans.$$

Case II. P = 10 lb
Substitution into the two equilibrium equations gives

$$F = 24.8 \text{ lb}, \qquad N = 97.4 \text{ lb}$$

But the maximum possible static friction force is

$$[F = fN] \qquad\qquad F = 0.20(97.4) = 19.5 \text{ lb}$$

It follows that 24.8 lb of friction cannot be supported. Therefore equilibrium cannot exist, and the correct value of the friction force is obtained by using the kinetic coefficient of friction accompanying the motion down the plane. Hence the answer is

$$[F = fN] \qquad\qquad F = 0.17(97.4) = 16.6 \text{ lb up the plane} \qquad\qquad Ans.$$

It should be noted that even though ΣF_x is no longer equal to zero, equilibrium does exist in the *y*-direction, so that $\Sigma F_y = 0$.

6/3   A homogeneous rectangular block of weight W rests on a horizontal plane and is subjected to the horizontal force P as shown. If the coefficient of friction is f, determine the greatest value which h may have so that the block will slide without tipping.

Solution. If the block is on the verge of tipping, the entire reaction between the plane and the block will be at A. The free-body diagram of the block for this condition is shown in the right side of the figure. If P is sufficient to cause slipping, the friction force is the limiting value fN, and the angle θ becomes $\theta = \tan^{-1}f$. The resultant of F and N passes through a point B through which P must also pass, since three coplanar forces in equilibrium are concurrent. Hence from the geometry of the block

$$\tan \theta = f = \frac{b/2}{h}, \qquad h = \frac{b}{2f} \qquad\qquad Ans.$$

If h were greater than this value, moment equilibrium about A would not be satisfied. For h less than $b/2f$ the resultant of F and N would be concurrent with P and W at a point below B. Thus this resultant would act not at A but at some point to the right of A.

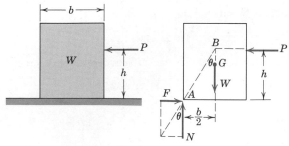

Problem 6/3

Problems

6/4 Determine the maximum angle θ that an inclined plane may have with the horizontal and not cause a block resting upon it to slide down. The coefficient of static friction is f. (This angle is known as the *angle of repose*.)

6/5 The figure shows a device that secures a rope or cable under tension by reason of large friction forces developed. For the position shown determine the total reaction R on each of the cam bearings. The coefficient of friction is 0.40.

Ans. R = 1471 lb

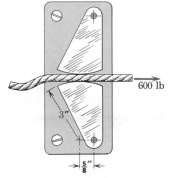

Problem 6/5

6/6 The tongs are used to handle hot steel tubes that are being heat treated in an oil bath. For a 20-deg jaw opening, what is the minimum coefficient of friction f between the jaws and the tube that will enable the tongs to grip the tube without slipping?

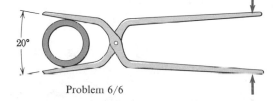

Problem 6/6

6/7 Determine the maximum value of the distance d at which the lower end of the prop of negligible weight can be set and still support the heavy hinged plank without slipping. The coefficient of friction is 0.40.   *Ans. d* = 22.3 in.

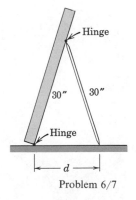

Problem 6/7

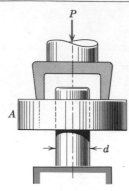

Problem 6/8

6/8 A force P is required to slide the collar A onto its fixed shaft with a force fit. Show that the torque M required on the collar to turn it on the fixed shaft against friction after P is removed is $M = Pd/2$, where d is the shaft diameter.

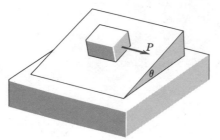

Problem 6/9

6/9 A uniform block of weight W is at rest on an incline θ. Determine the maximum force P that can be applied to the block in the direction shown before slipping begins. The coefficient of friction between the block and the incline is f.

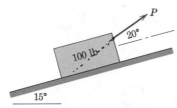

Problem 6/10

6/10 The coefficients of static and kinetic friction between the 100-lb block and the inclined plane are 0.30 and 0.20, respectively. Determine (*a*) the friction force F acting on the block when P with a magnitude of 20 lb is applied to the block at rest, (*b*) the force P required to initiate motion up the incline from rest, and (*c*) the friction force F acting on the block if $P = 60$ lb.

Ans. (*a*) $F = 7.09$ lb
(*b*) $P = 52.6$ lb
(*c*) $F = 15.2$ lb

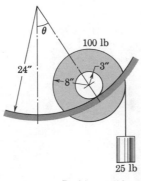

Problem 6/11

6/11 The 100-lb wheel rolls on its hub up the circular incline under the action of the 25-lb weight attached to a cord around the rim. Determine the angle θ at which the wheel comes to rest, assuming that friction is sufficient to prevent slippage. What is the minimum coefficient of friction f that will permit this position to be reached with no slipping?

6/12 The 20-in.-diameter wheel weighs 80 lb. Determine the couple M required to roll the wheel over the 2-in. step on the 5-deg incline and specify the minimum coefficient of friction f that must exist to prevent the wheel from slipping.

$Ans.$ $M = 534$ lb-in., $f = 0.896$

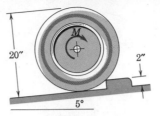

Problem 6/12

6/13 The roll of paper is to be rolled slowly up the incline by a tension T applied to the paper as it is pulled horizontally off the roll. The coefficient of friction between the roll and the incline is 0.20. Prove whether the paper rolls without slipping or whether it slips.

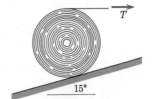

Problem 6/13

6/14 The 75-lb block with center of gravity at G is resting in the horizontal position on the support at C and the light vertical strut AB. If the coefficients of friction for all pairs of contacting surfaces are as shown, compute the lowest value of P which would initiate slippage of some part of the system. Where does slippage occur first?

$Ans.$ $P = 22.5$ lb
Slippage occurs first at C

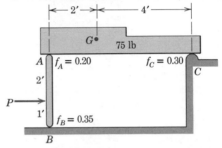

Problem 6/14

6/15 The homogeneous rectangular block of weight W rests on the inclined plane that is hinged about a horizontal axis through O. If the coefficient of static friction between the block and the plane is f, specify the conditions that determine whether the block tips before it slips or slips before it tips as the angle θ is gradually increased.

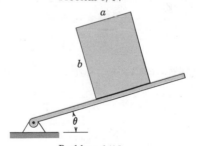

Problem 6/15

6/16 The circular collar A shown in section is mated with part B with a shrink fit which sets up a pressure or compressive stress p between the parts. The pressure has the values shown at the extremities of the overlap and is closely approximated by the relation $p = p_0 + kx^2$ in between. If a torque of 3000 lb-ft is required to turn collar A inside of part B, calculate the effective coefficient of friction f between the two parts.

$Ans.$ $f = 0.758$

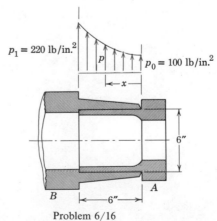

Problem 6/16

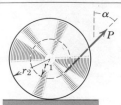

Problem 6/17

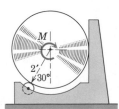

Problem 6/18

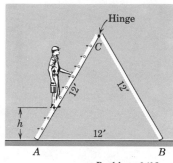

Problem 6/19

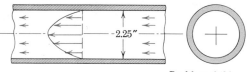

Problem 6/20

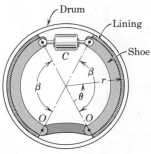

Problem 6/21

6/17 The wheel shown will roll to the left when the angle α of the cord is small. When α is large the wheel rolls to the right. Determine by inspection from the geometry of the free-body diagram the angle α for which the wheel will not roll in either direction. If the coefficient of friction is f and the weight of the wheel is W, determine the value of P for which the wheel will slip for the critical value of α.

6/18 The uniform cylinder weighs 400 lb and is supported by the roller, which turns with negligible friction. If the coefficient of friction between the cylinder and the vertical surface is 0.6, calculate the torque M required to turn the cylinder. Also find the reaction R on the bearing of the roller as M is applied.
  Ans. $M = 206$ lb-ft, $R = 343$ lb

6/19 Find the greatest height h of the step which the 200-lb man can reach without causing the hinged stepladder to collapse. The coefficient of friction at A and B is 0.50, and each of the two uniform sections of the hinged ladder weighs 25 lb.

6/20 Air is flowing in a straight pipe of $2\frac{1}{4}$-in. inside diameter under laminar flow conditions with a maximum velocity in the center of the pipe of 2 ft/sec. If the velocity distribution as a function of radius is parabolic, compute the friction force F per foot of pipe length caused by the air flow against the walls of the pipe. Air temperature is 70°F. *Ans.* $F = 9.45(10^{-6})$ lb/ft

6/21 The two brake shoes and their lining pivot about the points O and are expanded against the brake drum through the action of the hydraulic cylinder C. The pressure p between the drum and the lining may be shown to vary directly as the sine of the angle θ measured from the pin O for each shoe and has a value p_0 at $\theta = \beta$. The width of the lining in contact with the drum is b. Write the expression for the braking torque M_f on the wheel if the coefficient of friction between the drum and the lining is f.

6/22 In the figure are shown the elements of a rolling mill. Determine the maximum thickness b which the slab may have and still enter the rolls by means of the friction between the slab and the rolls. Assume that the coefficient of friction is f and that $b - a$ is small compared with d.

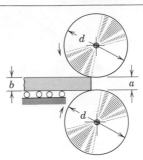
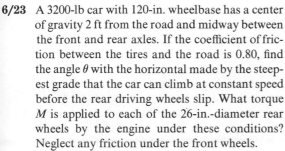

Problem 6/22

6/23 A 3200-lb car with 120-in. wheelbase has a center of gravity 2 ft from the road and midway between the front and rear axles. If the coefficient of friction between the tires and the road is 0.80, find the angle θ with the horizontal made by the steepest grade that the car can climb at constant speed before the rear driving wheels slip. What torque M is applied to each of the 26-in.-diameter rear wheels by the engine under these conditions? Neglect any friction under the front wheels.

Ans. $\theta = 25°28'$, $M = 745$ lb-ft

6/24 Calculate the torque M required to spin the uniform 60-lb wheel in its position against the vertical wall. The coefficient of friction for each pair of contacting surfaces is 0.40.

Ans. $M = 348$ lb-in.

Problem 6/24

6/25 Determine the force P that will begin to rotate the cylinder of weight W against the action of friction. The coefficient of friction for both pairs of contacting surfaces is f.

Problem 6/25

6/26 Calculate the force T required to rotate the 400-lb reel of telephone cable that rests on its hubs and bears against a vertical wall. The coefficient of friction for each pair of contacting surfaces is 0.60.

Ans. $T = 148.2$ lb

Problem 6/26

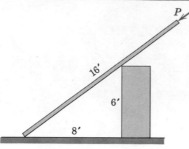

Problem 6/27

6/27 Determine the force P required to move the uniform 100-lb plank from its rest position shown if the coefficient of friction at both contact locations is 0.50.

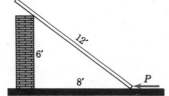

Problem 6/28

6/28 Calculate the force P required to start the uniform 80-lb plank sliding over the 6-ft wall from the position shown if the coefficient of friction for each pair of contacting surfaces is 0.60.

Ans. $P = 79.3$ lb

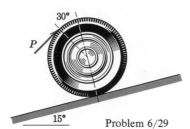

Problem 6/29

6/29 The force P is applied tangentially to the rim of the 50-lb wheel at the position shown to prevent rolling down the incline. The coefficient of friction between the wheel and the incline is 0.30. Determine the friction force F that the incline exerts on the wheel.

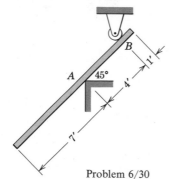

Problem 6/30

6/30 The uniform 12-ft plank weighs 50 lb and is placed at rest on the fixed corner at A and bears against the roller at B. The plank makes an angle of 45 deg with respect to the horizontal. If the coefficients of static and kinetic friction between the plank and the corner are 0.90 and 0.75, respectively, calculate the friction force F acting at A. *Ans.* $F = 35.4$ lb

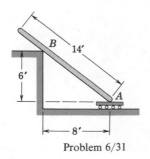

Problem 6/31

6/31 The lower end A of the uniform 100-lb plank rests on rollers that are free to move on the horizontal surface. If the coefficients of static and kinetic friction between the plank and the corner B are 0.80 and 0.70, respectively, compute the friction force F acting at B if the plank is released from rest in the position shown.

6/32 A force of 200 lb is developed in the hydraulic cylinder C to activate the block brake. If the coefficient of static friction between the blocks and the rim of the wheel is 0.60, compute the maximum torque M which can be applied to the wheel without causing rotation. The wheel is mounted in a fixed bearing at its center. Assume that the forces between the blocks and the wheel act at the centers of the contact faces of the blocks.
Ans. $M = 4470$ lb-in.

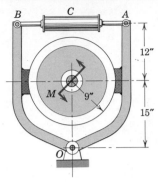

Problem 6/32

6/33 The uniform slender rod of weight W and length l is on the verge of slipping when placed against the vertical wall in the position shown. Find the expression for the coefficient of friction f, which is the same for both pairs of contacting surfaces.

Problem 6/33

6/34 The uniform bar with center of gravity at G is supported by the pegs A and B, which are fixed in the wheel. If the coefficient of friction between the bar and the pegs is f, determine the angle θ through which the wheel may be turned about its horizontal axis through O before the bar begins to slip.

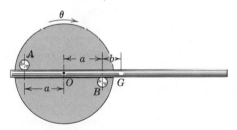

Problem 6/34

6/35 The semicylindrical shell of weight W and radius r is rolled through an angle θ by the horizontal force P applied to its rim. If the coefficient of friction is 0.20, calculate the angle θ at which the shell slips on the horizontal surface as P is gradually increased.
Ans. $\theta = 27°16'$

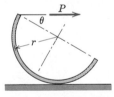

Problem 6/35

6/36 The left-hand jaw of the C-clamp can be slid along the frame to increase the capacity of the clamp. To prevent slipping of the jaw on the frame when the clamp is under load, the dimension x must exceed a certain minimum value. Find this value corresponding to given dimensions a and b and a coefficient of friction f between the frame and the loose-fitting jaw.

$$Ans. \ x = \frac{a - bf}{2f}$$

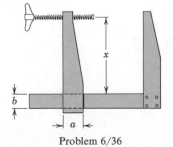

Problem 6/36

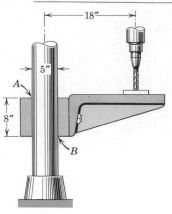

Problem 6/37

6/37 The coefficient of friction between the collar of the drill-press table and the vertical column is 0.30. Will the collar and table slide down the column under the action of the drill thrust if the operator forgets to secure the clamp, or will friction be sufficient to hold it in place? Neglect the weight of the table and collar compared with the drill thrust and assume that contact occurs at the points A and B.

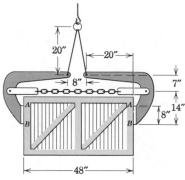

Problem 6/38

6/38 The friction tongs are designed to lift 1000-lb crates with a nominal width of 48 in. From the configuration of the links determine whether slippage is more likely for crates that are slightly wider (with contact at A) or slightly narrower (with contact at B) than the nominal size. Determine the minimum coefficient of friction f between the tongs and the crate that will prevent slippage for the case where slippage is more likely, and calculate the corresponding tension T in the horizontal chain that connects the jaws of the tongs.

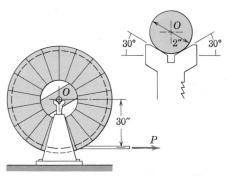

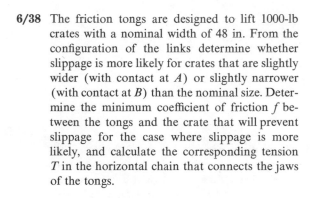

Problem 6/39

6/39 The reel of telephone cable weighs 6000 lb and is supported on its shaft in the V-notched blocks on both sides of the reel. The reel is raised off the ground by jacking up the supports so that cable may be pulled off in the horizontal direction as shown. The shaft is fastened to the reel and turns with it. If the coefficient of friction between the shaft and the V-surfaces is 0.30, calculate the pull P in the cable required to turn the reel.   *Ans.* $P = 63.8$ lb

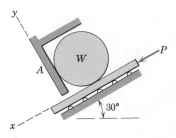

Problem 6/40

6/40 The roller of weight W rests on a plate of negligible weight that slides without resistance on its under side. The roller bears against the fixed perpendicular barrier A. If the coefficient of friction between the roller and each of its contacting surfaces is 0.30, determine the force P required to move the plate.

6/41 A bulldozer rolls the 1500-lb log up the 20-deg incline by pushing with the blade, which is normal to the incline. If the coefficient of friction between the blade and the log is 0.50 and that between the log and the ground is 0.80, calculate the force component *P*, normal to the blade, that must be exerted against the log.

Ans. P = 1026 lb

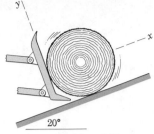

Problem 6/41

6/42 The 10-ft oak plank 1 ft wide and 4 in. thick with a specific weight of 50 lb/ft³ is placed on the concrete rim of a freshwater tank as shown. If the coefficient of friction between the plank and the rim is 0.80, calculate the friction force *F* acting at *A*.　　*Ans. F* = 23.7 lb

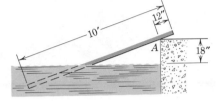

Problem 6/42

6/43 The uniform square crate weighs 400 lb and is to be moved down the incline by a horizontal pull *P* on the rope attached to the upper edge of the crate. The coefficient of friction between the crate and the incline is 0.40. Determine the force *P* required to initiate motion, either tipping or sliding.

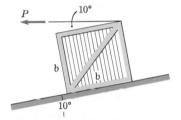

Problem 6/43

6/44 Determine the maximum distance *d* at which the lower corner of the uniform 200-lb metal block may be placed from the hinge at *C* without causing slippage to occur at *A*. The coefficient of friction at *A* is 0.30.　　*Ans. d* = 32.4 in.

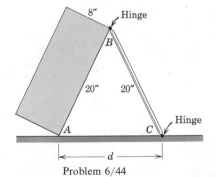

Problem 6/44

6/45 A uniform bar of weight *W* and length *l* rests on a horizontal surface with its weight evenly distributed along its length. If the coefficient of friction between the bar and the supporting surface is *f*, write expressions for the horizontal force *P*, applied at the end of the bar, required to move the bar and the distance *a* to the axis *O* about which the bar is observed to rotate.

Ans. P = 0.414 *fW*, *a* = 0.293 *l*

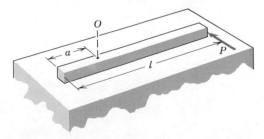

Problem 6/45

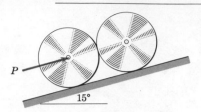

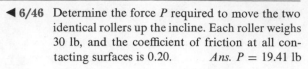

◄ 6/46 Determine the force P required to move the two identical rollers up the incline. Each roller weighs 30 lb, and the coefficient of friction at all contacting surfaces is 0.20.   *Ans. $P = 19.41$ lb*

Problem 6/46

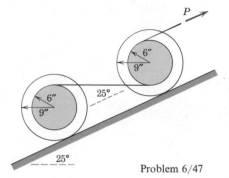

◄ 6/47 Each of the two identical wheels with attached hubs weighs 40 lb. The wheels are placed on the 25-deg incline with their connecting horizontal cord wrapped securely around their hubs. The wheels are prevented from rolling by the force P applied parallel to the incline. The coefficient of static friction between the wheels and the incline is 0.50. Determine the friction force F acting on the upper wheel.   *Ans. $F = 14.96$ lb*

Problem 6/47

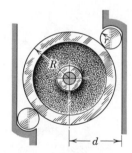

◄ 6/48 The device shown prevents clockwise rotation in the horizontal plane of the central wheel by means of frictional locking of the two small rollers. For given values of R and r and for a common coefficient of friction f at all contact surfaces, determine the range of values of d for which the device will operate as described.

$$Ans. \quad \frac{2r + (1 - f^2)R}{1 + f^2} < d < (R + 2r)$$

Problem 6/48

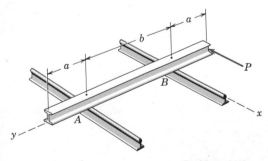

◄ 6/49 An I-beam of weight W is supported by two fixed horizontal rails as shown. Compute the applied load P that is just sufficient to cause the beam to slip and determine the corresponding friction force at A as slippage begins. The coefficient of friction between the beam and the rails is f.

$$Ans. \quad P = \frac{fWb}{2(a + b)}, \quad F_A = \frac{fWa}{2(a + b)}$$

Problem 6/49

◀ **6/50** The coefficient of friction between the uniform 20-ft pole and the horizontal surface is limited to 0.40. When placed in the vertical plane shown, the pole will not slip provided that θ is either less than a certain critical value or greater than a certain larger critical value. Find the range of θ for which the pole is unstable.

Ans. Unstable for $24.0° < \theta < 57.2°$

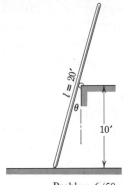

Problem 6/50

◀ **6/51** The uniform circular shaft, which weighs 100 lb and has a diameter of 4 in., rests on the horizontal and parallel rails and bears against the vertical post at C as shown. A horizontal force P is applied to the cord wrapped around the shaft and is increased until slippage occurs. If the coefficient of friction at all contacting surfaces is 0.40, determine the mode of slippage and find the friction force at the point of contact A between the shaft and the rail as slippage impends.

Ans. $F_A = 4.93$ lb

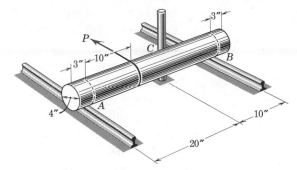

Problem 6/51

35 Friction in Machines. There are a number of important and interesting applications of the principles of dry friction in the design and operation of various machines. Several of these applications are discussed in the following sections, and the problems at the end of the article contain examples illustrative of all of the sections.

(*a*) *Wedges.* A wedge is one of the simplest and most useful of machines and is used as a means of producing small adjustments in the position of a body or as a means of applying large forces. Wedges are largely dependent on friction. When sliding of a wedge is impending, the resultant force on each sliding surface of the wedge will be inclined from the normal to the surface by an amount equal to the friction angle. The component of the resultant along the surface is the friction force, which is always in the direction to oppose the motion of the wedge.

Figure 75a shows a wedge that is used to position or lift a heavy load W. The coefficient of friction for each pair of surfaces is $f = \tan \phi$. The force P required to start the wedge is found from the equilibrium triangles of the forces on the load and on the wedge. The free-body diagrams are shown in Fig. 75b, where the reactions are inclined at an angle ϕ from their respective normals and are in the direction to oppose the motion. The weight of the wedge is neglected. From these diagrams the equilibrium equations

$$\mathbf{W} + \mathbf{R}_2 + \mathbf{R}_3 = 0 \quad \text{and} \quad \mathbf{R}_1 + \mathbf{R}_2 + \mathbf{P} = 0$$

may be written. The solutions of these equations are shown in the *c*-part of the figure, where R_2 is found first in the upper diagram using the known value of *W*. The force *P* is then found from the lower triangle once the value of R_2 has been established.

If *P* is removed, the wedge will remain in place as long as α is less than ϕ, in which case the wedge is said to be self-locking. If the wedge is self-locking and is to be withdrawn, a pull *P'* would be required. In this event the reactions R_1 and R_2 would act on the opposite sides of their normals to oppose the new impending motion, and the solution would proceed along lines similar to those described for the case of raising the load.

Wedge problems lend themselves to graphical solutions as indicated in Fig. 75*c*. The accuracy of a graphical solution is easily held to within tolerances consistent with the uncertainty of friction coefficients. Algebraic solutions may also be obtained from the trigonometry of the equilibrium polygons.

(*b*) *Screws.* Screws are used for fastenings and for transmitting power or motion. In each case the friction developed in the threads largely determines the action of the screw. For transmitting power or motion the square thread is more efficient than the V-thread, and the analysis illustrated here is confined to the square thread.

Consider the square-threaded jack, Fig. 76, under the action of the axial load *W* and a moment *M* applied about the axis of the screw. The force *R* exerted by the thread of the jack frame on a small representative portion of the thread of the screw is shown on the free-body diagram of the screw. Similar reactions exist on all segments of the screw thread where contact occurs with the thread of the base. If *M* is just sufficient to turn the screw, the thread of the screw will slide around and up on the fixed thread of the frame. The angle ϕ made by *R* with the normal to the thread will be the angle of friction, so that $\tan \phi = f$. The moment of *R* about the vertical axis of the screw is $Rr \sin (\alpha + \phi)$, and the total moment due to all reactions on the threads

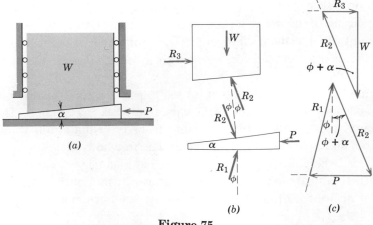

(*a*)   (*b*)   (*c*)

Figure 75

is $\Sigma R r \sin (\alpha + \phi)$. Since $r \sin (\alpha + \phi)$ appears in each term, it may be factored out. The moment equilibrium equation for the screw becomes

$$M = [r \sin (\alpha + \phi)]\Sigma R$$

Equilibrium of forces in the axial direction further requires

$$W = \Sigma R \cos (\alpha + \phi) = [\cos (\alpha + \phi)]\Sigma R$$

Dividing M by W gives

$$M = Wr \tan (\alpha + \phi) \qquad (50)$$

where $\phi = \tan^{-1} f$. The angle α is determined from the lead L or advancement per revolution of the screw. Thus $\alpha = \tan^{-1} (L/2\pi r)$. This relation is easily seen by unwrapping the thread for one turn of the screw and forming the right triangle whose base is the mean circumference $2\pi r$ and whose altitude is the lead L.

Equation 50 gives the moment required to start the screw upward or to maintain upward movement, depending on whether the static or kinetic coefficient of friction is used. If the moment M is removed, the friction force changes direction so that ϕ is measured to the other side of the normal. Thus the screw will remain in place and be self-locking provided that $\phi > \alpha$ and will be on the verge of unwinding if $\phi = \alpha$. A moment M' must be applied in the direction opposite to that shown in Fig. 76 to lower the screw when $\phi > \alpha$ and is given by

$$M' = Wr \tan (\phi - \alpha)$$

If $\phi < \alpha$ the screw will unwind by itself and would require a moment to keep it from unwinding equal to

$$M = Wr \tan (\alpha - \phi)$$

(*c*) *Journal Bearings.* A journal bearing is a bearing that gives lateral support to a shaft in contrast to axial or thrust support. For dry bearings and for

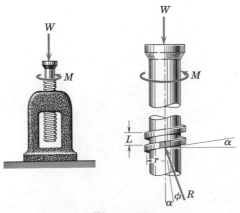

Figure 76

many partially lubricated bearings an analysis by the principles of dry friction gives a satisfactory approximation for design purposes. A dry or partially lubricated journal bearing with contact or near contact between the shaft and the bearing is shown in Fig. 77, where the clearance between the shaft and bearing is greatly exaggerated. As the shaft begins to turn in the direction shown, it rolls up the inner surface of the bearing until slippage occurs. Here it remains in a more or less fixed position during rotation. The torque M required to maintain rotation and the radial load L on the shaft will cause a reaction R at the contact point A. For equilibrium in the vertical direction R must equal L but will not be collinear with it. The force R will be tangent to a small circle of radius r_f called the *friction circle*. The angle between R and its normal component N is the friction angle ϕ. Equating the sum of the moments about O to zero gives

$$M = Rr_f = Rr \sin \phi \qquad (51)$$

For a small coefficient of friction the angle ϕ is small, and the sine and tangent may be interchanged with only small error. Since $f = \tan \phi$, a good approximation to the torque is

$$M = fRr \qquad (51a)$$

This relation gives the amount of torque or moment applied to the shaft which is necessary to overcome friction for a dry or partially lubricated journal bearing.

The behavior of a fully lubricated bearing where the lubricant is able to form a complete fluid wedge under the shaft is quite different from that of the dry or partially lubricated bearing. An analysis of the fully lubricated bearing is beyond the scope of this treatment but may be found in references on lubrication, machine design, and fluid mechanics. If the lateral loads on a fully lubricated bearing are small, a first approximation to the frictional moment may be developed by assuming the shaft to be centered in the bearing as shown in Fig. 78, where the radial clearance c is greatly exaggerated. A linear distribution of velocity may be taken for the fluid layers in the small

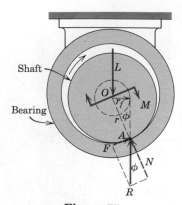

Figure 77

clearance space with a change from zero at the fixed inner surface of the bearing to the peripheral velocity v of the shaft at its outer surface. For the radial clearance c, the velocity gradient has the magnitude $|dv/dr| = v/c = r\omega/c$, where ω is the angular velocity of the shaft in radians per second. The shear stress on the surface of the shaft from Eq. 47 is

$$\tau = \mu \left| \frac{dv}{dr} \right| = \frac{\mu r \omega}{c}$$

and the frictional moment for a bearing of length l with surface area $A = 2\pi rl$ becomes

$$M = \tau Ar = \frac{2\pi \mu r^3 l \omega}{c} \tag{52}$$

where μ is the absolute viscosity of the lubricant.

(*d*) *Disk and Pivot Friction.* Friction between circular surfaces under normal pressure is encountered in pivot bearings, clutch plates, and disk brakes. Consider the two flat circular disks of Fig. 79 whose shafts are mounted in bearings (not shown) so that they can be brought into contact under the axial force P. The maximum torque that this clutch can transmit will be equal to the torque M required to slip one disk against the other. If p is the normal pressure at any location between the plates, the frictional force acting on an elemental area is $fp\,dA$, where f is the friction coefficient and dA is the area $r\,dr\,d\theta$ of the element. The moment of this elemental friction force about the shaft axis is $fpr\,dA$, and the total moment is

$$M = \int fpr\,dA$$

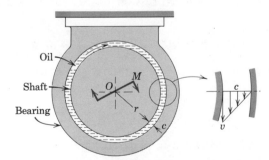

Figure 78

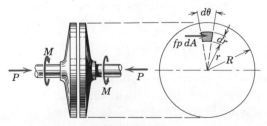

Figure 79

where the integral is evaluated over the area of the disk. To carry out this integral the variation of f and p with r must be known.

In the following examples f is assumed to be constant. Furthermore, if the surfaces are new, flat, and well supported, it is reasonable to assume that the pressure p is constant and uniformly distributed so that $\pi R^2 p = P$. Substituting this constant value of p in the expression for M gives

$$M = \frac{fP}{\pi R^2} \int_0^{2\pi} \int_0^R r^2 \, dr \, d\theta = \tfrac{2}{3} fPR \tag{53}$$

This result may be interpreted as being equivalent to the moment due to a friction force fP acting at a distance $2R/3$ from the center of the shaft.

If the friction disks are rings, the limits of integration are the inside and outside radii R_i and R_o, respectively, and the frictional torque becomes

$$M = \tfrac{2}{3} fP \, \frac{R_o^3 - R_i^3}{R_o^2 - R_i^2} \tag{53a}$$

After some wear of the surfaces has taken place, it is found that the frictional moment decreases somewhat. When the wearing-in period is over, the surfaces retain their new relative shape and further wear is therefore constant over the surface. This wear depends on the circumferential distance traveled and the pressure p. Since the distance traveled is proportional to r, the expression $rp = K$ may be written, where K is a constant. The value of K is determined by equating the axial forces to zero, or

$$P = \int p \, dA = K \int_0^{2\pi} \int_0^R dr \, d\theta = 2\pi KR$$

With $pr = K = P/(2\pi R)$, the expression for M may be written

$$M = \int fpr \, dA = \frac{fP}{2\pi R} \int_0^{2\pi} \int_0^R r \, dr \, d\theta$$

which becomes

$$M = \tfrac{1}{2} fPR \tag{54}$$

The frictional moment for worn-in plates is, therefore, only $(\tfrac{1}{2})/(\tfrac{2}{3})$, or $\tfrac{3}{4}$ as much as for new surfaces.

If the friction disks are rings of inside radius R_i and outside radius R_o, substitution of these limits in the integrations shows that the frictional torque for worn-in surfaces is

$$M = \tfrac{1}{2} fP(R_o + R_i) \tag{54a}$$

(*e*) *Belt Friction.* The impending slippage of flexible members such as belts and ropes over sheaves and drums is of importance in the design of belt drives of all types, band brakes, and hoisting rigs. In Fig. 80 is shown a drum subjected to the two belt tensions T_1 and T_2, the torque M necessary to prevent rotation, and a bearing reaction R. With M in the direction shown, T_2

is greater than T_1. The free-body diagram of an element of the belt of length $r\,d\theta$ is also shown in the figure. The force analysis of this element proceeds in a manner similar to that which has been illustrated for other variable-force problems where the equilibrium of a differential element is established. The tension increases from T at the angle θ to $T + dT$ at the angle $\theta + d\theta$. The normal force is a differential dN, since it acts on a differential element of area. Likewise the friction force, which must act on the belt in a direction to oppose slipping, is a differential and is $f\,dN$ for impending motion. Equilibrium in the t-direction gives

$$T \cos \frac{d\theta}{2} + f\,dN = (T + dT) \cos \frac{d\theta}{2}$$

or

$$f\,dN = dT$$

since the cosine of a differential quantity is unity. Equilibrium in the n-direction requires that

$$dN = (T + dT) \sin \frac{d\theta}{2} + T \sin \frac{d\theta}{2}$$

or

$$dN = T\,d\theta$$

In this reduction it must be remembered that the sine of a differential angle equals the angle and that the product of two differentials must be neglected in the limit compared with the first-order differentials remaining. Combining the two equilibrium relations gives

$$\frac{dT}{T} = f\,d\theta$$

Integrating between corresponding limits yields

$$\int_{T_1}^{T_2} \frac{dT}{T} = \int_0^{\beta} f\,d\theta$$

or

$$\log \frac{T_2}{T_1} = f\beta$$

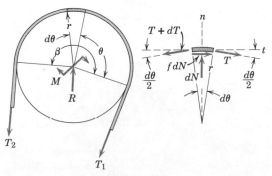

Figure 80

where the log (T_2/T_1) is a natural logarithm to the base e. Solving for T_2 gives

$$T_2 = T_1 e^{f\beta} \qquad (55)$$

It should be noted that β is the total angle of belt contact and is expressed in radians. If a rope were wrapped around a drum n times, the angle β would be $2\pi n$ radians. Equation 55 holds equally well for a noncircular section where the total angle of contact is β. This conclusion is evident from the fact that the radius r of the circular drum of Fig. 80 does not enter into the equations for the equilibrium of a differential element of the belt.

The relation expressed by Eq. 55 also applies to belt drives where both the belt and the pulley are rotating at constant speed. In this case the equation describes the ratio of belt tensions for slipping or impending slipping. When the speed of rotation becomes large, there is a tendency for the belt to leave the rim, so that Eq. 55 will involve some error.

Problems

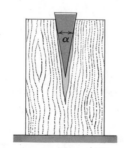

Problem 6/52

6/52 A wedge will be self-locking provided that its angle α is less than a critical value. If the coefficient of friction between the wedge and the material being split is f, what is the critical value of α?

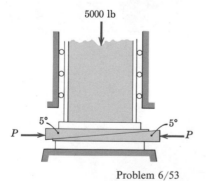

5000 lb

$P \longrightarrow$ 5°   5° $\longleftarrow P$

Problem 6/53

6/53 The two 5-deg wedges shown are used to adjust the position of the column under a vertical load of 5000 lb. Determine the magnitude of the forces P required to raise the column if the coefficient of friction for all surfaces is 0.40.

Ans. $P = 4520$ lb

6/54 If the loaded column of Prob. 6/53 is to be lowered, calculate the horizontal forces P' required to withdraw the wedges.

6/55 A screw jack with square threads having a mean radius of 1 in. supports a load of 1000 lb. If the coefficient of friction is 0.25, what is the greatest lead L (advancement per turn) of the screw for which the screw will not unwind by itself? For this condition what torque M applied to the screw would be required to raise the load?

Ans. $L = 1.57$ in., $M = 533$ lb-in.

6/56 The turnbuckle supports a tension T of 15,000 lb. Each of the screws has a mean diameter of 1.354 in. and has a single thread with a lead (advancement per turn) of 1/3 inch, one being right-handed and the other left-handed. If a torque of 270 lb-ft is required to loosen the turnbuckle by turning the body with both screws prevented from rotating, calculate the effective coefficient of friction f in the threads.

Problem 6/56

6/57 A new and flat collar bearing supports the thrust P in the shaft. If a constant moment M of 30 lb-in. is applied to the shaft, (*a*) what value of P is required to maintain a constant speed of rotation? The coefficient of friction is known to be 0.15. (*b*) For the condition after the bearing has worn in, compute the increase in thrust ΔP which is necessary in order to maintain constant speed with the same applied moment.

       Ans. (*a*) $P = 85.7$ lb,   (*b*) $\Delta P = 3.2$ lb

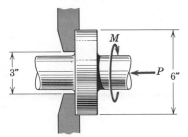

Problem 6/57

6/58 A force $P = 5W$ is required to raise the load W with the cord making $1\frac{1}{4}$ turns around the fixed shaft. Calculate the coefficient of friction f between the cord and the shaft.

6/59 For a given coefficient of friction and a given number of turns around the shaft for the configuration of Prob. 6/58, a force P of 300 lb is required to raise W and a force P of 48 lb is required to lower W. Find W.   *Ans.* $W = 120$ lb

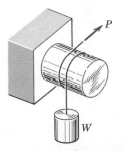

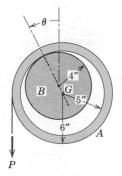

Problem 6/58

6/60 The steel ring A weighs 40 lb with inside and outside radii of 5 in. and 6 in., respectively, and rests on a fixed horizontal shaft of 4-in. radius. If a downward force $P = 30$ lb applied to the periphery of the ring is just sufficient to cause the ring to slip, calculate the coefficient of friction f and the angle θ.

       Ans. $f = 0.600$, $\theta = 30°57'$

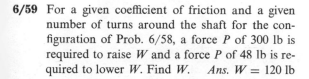

Problem 6/60

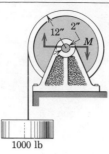

Problem 6/61

6/61 A torque M of 12,300 lb-in. must be applied to the 2-in.-diameter shaft of the hoisting drum to raise the 1000-lb load at constant speed. The drum and shaft together weigh 200 lb. Calculate the coefficient of friction f for the bearing.

Ans. $f = 0.258$

6/62 Calculate the torque M on the shaft of the hoisting drum of Prob. 6/61 that is required to lower the 1000-lb load at constant speed. Use the value $f = 0.258$ calculated in Prob. 6/61 for the coefficient of friction.

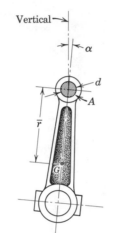

Problem 6/63

6/63 The shaft A fits loosely in the wrist-pin bearing of the connecting rod with center of gravity at G as shown. With the rod initially in the vertical position the shaft is rotated slowly until the rod slips at the angle α. Write an exact expression for the coefficient of friction.

$$Ans. \; f = \frac{1}{\sqrt{\left(\dfrac{d/2}{\bar{r} \sin \alpha}\right)^2 - 1}}$$

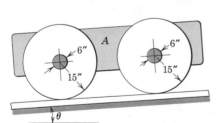

Problem 6/64

6/64 Each of the four wheels of the vehicle weighs 150 lb and is made of hardened steel. The diameter of the central hole in each wheel is 6.004 in., so that it fits easily over the 6-in. shaft fixed to the 800-lb frame A. The weight of the frame is supported equally by the four wheels. If the vehicle, initially at rest, is on the verge of rolling when on an incline θ of 2 deg, compute the coefficient of static friction f in the 6-in.-diameter wheel bearings. Assume that all resistance to rotation of the wheels is in the form of journal friction.

6/65 A device for lowering a person in a sling down a rope at a constant controlled rate is shown in the figure. The rope passes around a central shaft fixed to the frame and leads freely out of the lower collar. The number of turns is adjusted by turning the lower collar, which winds or unwinds the rope around the shaft. Entrance of the rope into the upper collar at A is equivalent to $\frac{1}{4}$ of a turn, and passage around the corner at B is also equivalent to $\frac{1}{4}$ of a turn. Friction of the rope through the straight portions of the collars averages 2 lb for each collar. If three complete turns around the shaft, in addition to the corner turns, are required for a 150-lb man to lower himself at a constant rate without exerting a pull on the free end of the rope, calculate the coefficient of friction f between the rope and the contact surfaces of the device. Neglect the small helix angle of the rope around the shaft.   *Ans.* $f = 0.196$

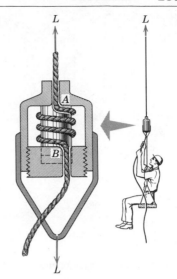

Problem 6/65

6/66 Determine the horizontal force P, applied to the 10-deg wedge, necessary to start moving the 1000-lb block upward. The coefficient of friction between all contacting surfaces is 0.30, and the weight of the wedge is negligible.

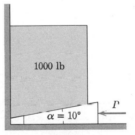

Problem 6/66

6/67 A 5-deg steel wedge is forced under the end of the 4200-lb machine with a force $P = 1100$ lb. If the coefficient of friction between the wedge and both the machine and the horizontal floor is 0.30, determine the position x of the center of gravity G of the machine. The machine is prevented from sliding horizontally by a rigid ledge at A.   *Ans.* $x = 30.8$ in.

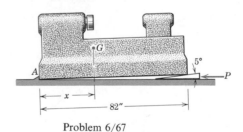

Problem 6/67

6/68 The horizontal position of the 1800-lb concrete block is to be adjusted by the 5-deg wedge. If the coefficient of friction between all mating surfaces is 0.70, what vertical force P is required to move the block? Neglect the weight of the wedge.

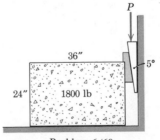

Problem 6/68

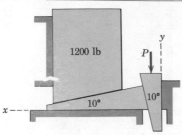

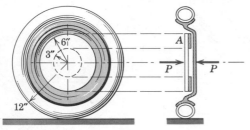

Problem 6/69

6/69 The two 10-deg wedges are positioned so that a downward force P on the one wedge will result in an elevation of the 1200-lb load. The coefficient of friction for all sliding surfaces is 0.20, and the weights of the wedges are negligible. Determine P. *Ans.* $P = 492$ lb

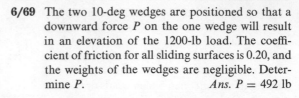

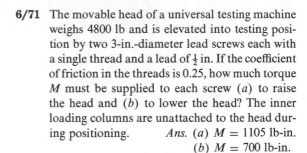

Problem 6/70

6/70 The front wheels of an experimental rear-drive vehicle have a radius of 12 in. and are equipped with disk-type brakes consisting of a ring A with outside and inside radii of 6 in. and 3 in., respectively. The ring, which does not turn with the wheel, is forced against the wheel disk with a force P. If the pressure between the ring and the disk wheel is uniform over the mating surfaces, compute the friction force F between each front tire and the horizontal road for an axial force $P = 200$ lb when the vehicle is powered at constant speed with the wheels turning. The coefficient of friction between the disk and ring is 0.35.

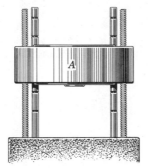

Problem 6/71

6/71 The movable head of a universal testing machine weighs 4800 lb and is elevated into testing position by two 3-in.-diameter lead screws each with a single thread and a lead of $\frac{1}{2}$ in. If the coefficient of friction in the threads is 0.25, how much torque M must be supplied to each screw (*a*) to raise the head and (*b*) to lower the head? The inner loading columns are unattached to the head during positioning. *Ans.* (*a*) $M = 1105$ lb-in.
(*b*) $M = 700$ lb-in.

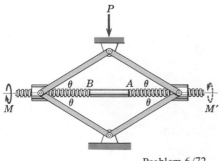

Problem 6/72

6/72 The device shown is used as a jack. The screw has a double square thread with a mean diameter of $\frac{7}{8}$ in. and a lead of $\frac{1}{3}$ in. Section A of the screw has right-hand threads, and section B has left-hand threads. For $\theta = 30$ deg determine (*a*) the torque M which must be applied to the screw to raise a load $P = 1500$ lb and (*b*) the torque M' needed to lower the load. The coefficient of friction in the threads is 0.20.

6/73 The screw of the small press has a mean diameter of $1\frac{1}{4}$ in. and has a double square thread with a lead of 0.4 in. The flat thrust bearing at A is shown in the enlarged view and has surfaces which are well worn. If the coefficient of friction for both the threads and the bearing at A is 0.25, calculate the torque M on the handwheel required (a) to produce a compressive force of 1000 lb and (b) to loosen the press from the 1000-lb compression.   *Ans.* (a) $M = 300$ lb-in.
(b) $M = 165$ lb-in.

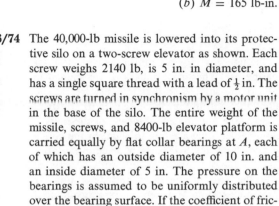
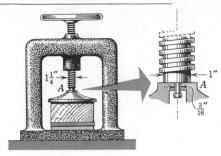

Problem 6/73

6/74 The 40,000-lb missile is lowered into its protective silo on a two-screw elevator as shown. Each screw weighs 2140 lb, is 5 in. in diameter, and has a single square thread with a lead of $\frac{1}{2}$ in. The screws are turned in synchronism by a motor unit in the base of the silo. The entire weight of the missile, screws, and 8400-lb elevator platform is carried equally by flat collar bearings at A, each of which has an outside diameter of 10 in. and an inside diameter of 5 in. The pressure on the bearings is assumed to be uniformly distributed over the bearing surface. If the coefficient of friction for the collar bearing and the screws at B is 0.15, calculate the torque M which must be applied to each screw (a) to raise the elevator and (b) to lower the elevator.
Ans. (a) $M = 26{,}410$ lb-in.
(b) $M = 22{,}490$ lb-in.

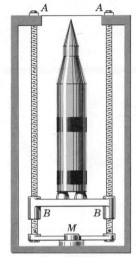

Problem 6/74

6/75 The uniform drum A with center of gravity at midlength is suspended by a rope which passes over the fixed cylindrical surface B. The coefficient of static friction between the rope and the surface over which it passes is f. Determine the maximum value that the dimension a may have before the drum tips out of its horizontal position.

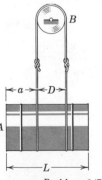
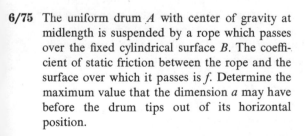

Problem 6/75

6/76 The linkage is initially at rest under the action of the torques M_1 and M_2. If M_1 is increased gradually until the linkage moves, write an exact expression for the angle α between the resultant compressive force in link AB and its center line as motion impends. The coefficient of friction for each bearing is f.

$$Ans.\ \alpha = \sin^{-1}\frac{fd}{l\sqrt{1+f^2}}$$

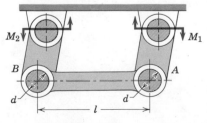

Problem 6/76

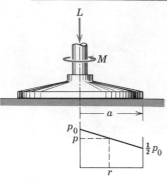

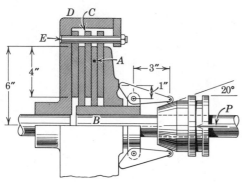

Problem 6/77

6/77 For the flat sanding disk of radius *a* the pressure *p* developed between the disk and the sanded surface decreases linearly with *r* from a value p_0 at the center to $p_0/2$ at $r = a$. If the coefficient of friction is *f*, derive the expression for the torque *M* required to turn the shaft under an axial force *L*.

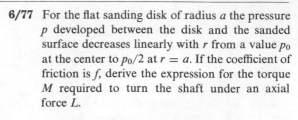

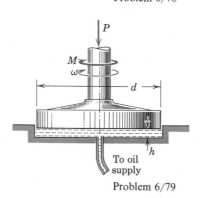

Problem 6/78

6/78 In the figure is shown a multiple-disk clutch for marine use. The driving disks *A* are splined to the driving shaft *B* so that they are free to slip along the shaft but must rotate with it. The disks *C* drive the housing *D* by means of the bolts *E* along which they are free to slide. In the clutch shown there are five pairs of friction surfaces. Assume the pressure to be uniformly distributed over the area of the disks and determine the maximum torque *M* which can be transmitted if the coefficient of friction is 0.15 and $P = 100$ lb.
$$Ans.\ M = 2680\ \text{lb-in.}$$

Problem 6/79

6/79 A flat disk bearing supports a load *P* with a fully lubricated oil space. The pressure in the oil is maintained through a regulated supply line at the center of the bearing. The oil has a viscosity *μ* and at any radial distance has a tangential velocity which varies linearly from zero at the stationary bottom of the oil space to the corresponding velocity of the rotating bearing surface directly above. Derive an expression for the torque *M* necessary to overcome the frictional resistance of the oil for a constant angular velocity *ω* of the shaft and for a given vertical clearance *h* in the bearing.

6/80 A 3.000-in.-diameter shaft 20 in. long is protected by a fixed outer tube of the same length which has an inside diameter of 3.030 in. If the shaft is rotating at the constant speed of 3000 rev/min, compute the resisting torque M acting on the shaft as a result of the viscosity of the air film between the shaft and the tube. Neglect the effect of air leakage at the ends of the tube and assume a linear velocity gradient between the shaft and tube surfaces as indicated. The pressure is atmospheric and the temperature is 70°F. If the clearance between the shaft and tube were cut in half, what would be the resisting torque?

Ans. $M = 0.0232$ lb-in.

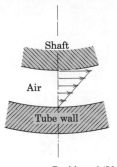

Problem 6/80

6/81 The tape slides around the two fixed pegs as shown and is under the action of the horizontal tensions $T_1 = 4$ lb and $T_2 = 16$ lb. Determine the coefficient of friction f between the tape and the pegs.   *Ans. $f = 0.313$*

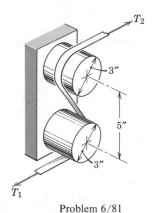

Problem 6/81

6/82 The assembly shown weighs 100 lb with center of gravity at G and is supported by the two wires, which pass around the fixed pegs and are kept under equal tensions by the equalizer plate A and the adjustable spring S. Calculate the minimum tension T in the spring which will ensure that the assembly remains suspended as shown. The coefficient of friction between the wires and the pegs is 0.30.   *Ans. $T = 56.6$ lb*

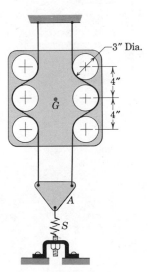

Problem 6/82

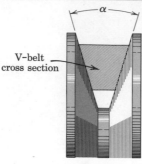

V-belt
cross section

Problem 6/83

6/83 Replace the flat belt and pulley of Fig. 80 by a V-belt and matching grooved pulley as indicated by the cross-sectional view accompanying this problem. Derive the relation between the belt tensions, the angular contact, and the coefficient of friction for the V-belt when slipping impends. Use of a V-belt with $\alpha = 35$ deg would be equivalent to increasing the coefficient of friction for a flat belt of the same material by what factor n?

Ans. $T_2 = T_1 e^{f\beta'}$ where $\beta' = \dfrac{\beta}{\sin(\alpha/2)}$

$n = 3.33$

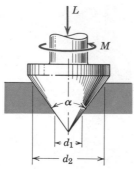

Problem 6/84

6/84 Determine the expression for the torque M required to turn the shaft whose thrust L is supported by a conical pivot bearing. The coefficient of friction is f, and the bearing pressure is constant.

Ans. $M = \dfrac{fL}{3\sin\dfrac{\alpha}{2}}\dfrac{d_2{}^3 - d_1{}^3}{d_2{}^2 - d_1{}^2}$

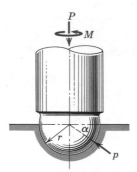

Problem 6/85

6/85 The spherical thrust bearing on the end of the shaft supports an axial load P. Determine the expression for the moment M required to turn the shaft against friction in the bearing. Assume that the pressure p is directly proportional to $\sin\alpha$ and that the coefficient of friction is f.

Ans. $M = fPr$

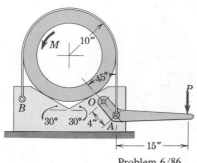

Problem 6/86

6/86 Find the couple M required to turn the pipe in the V-block against the action of the flexible band. A force $P = 25$ lb is applied to the lever which is pivoted about O. The coefficient of friction between the band and the pipe is 0.30, and that between the pipe and the block is 0.40. The weights of the parts are negligible.

Ans. $M = 1834$ lb-in.

6/87 In the figure is shown a differential band brake. Determine the force P required to brake the wheel under a clockwise torque M if the coefficient of friction is f. What happens if $a_1/a_2 < e^{f\beta}$?

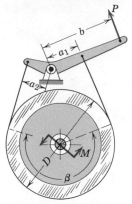

Problem 6/87

6/88 The tapered pin is forced into a mating tapered hole in the fixed block with a force $P = 400$ lb. If the force required to remove the pin is $P' = 300$ lb, calculate the coefficient of friction between the pin and the surface of the hole.

Ans. $f = 0.122$

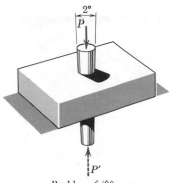

Problem 6/88

6/89 The roller chain is used as a pipe grip. If the coefficient of friction between the chain and the fixed pipe is 0.25, determine the minimum value of h which will ensure that the grip will not slip on the pipe regardless of P. Neglect the weights of the chain and handle, and neglect any friction between the left end of the handle and the pipe.

Ans. $h = 3.63$ in.

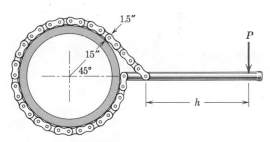

Problem 6/89

6/90 A 5-deg wedge is used to lift the 1000-lb cylinder as shown. If the coefficient of friction is $\frac{1}{4}$ for all surfaces, determine the force P required to move the wedge.

Ans. $P = 378$ lb

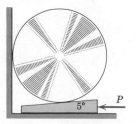

Problem 6/90

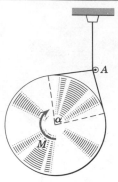

Problem 6/91

◀ **6/91** A light flexible cord is passed around the circular disk of weight W and ends in a small pulley that is free to find its equilibrium position on the cord. If the coefficient of friction between the cord and the disk is 0.50, compute the angle α between the normals to the cord at the tangency points for the position where the disk is on the verge of turning under the action of a couple M applied to the disk. (*Suggestion:* Solve the resulting equation for α graphically.)

Ans. $\alpha = 87°20'$

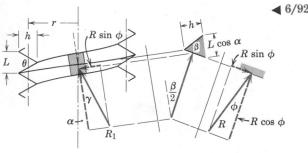

Problem 6/92

◀ **6/92** Replace the square thread of the screw jack in Fig. 76 by a V-thread as indicated in the figure accompanying this problem and determine the moment M on the screw required to raise the load W. The force R acting on a representative small section of the thread is shown with its relevant projections. The vector R_1 is the projection of R in the plane of the figure containing the axis of the screw. The analysis is begun with an axial force and a moment summation and includes substitutions for the angles γ and β in terms of θ, α, and the friction angle $\phi = \tan^{-1} f$. The helix angle of the single thread is exaggerated for clarity.

Ans. $M = Wr \dfrac{\tan\alpha + f\sqrt{1 + \tan^2\dfrac{\theta}{2}\cos^2\alpha}}{1 - f\tan\alpha\sqrt{1 + \tan^2\dfrac{\theta}{2}\cos^2\alpha}}$

where $\alpha = \tan^{-1}\dfrac{L}{2\pi r}$

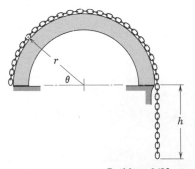

Problem 6/93

◀ **6/93** The chain has a weight μ per unit length. Determine the overhang h below the fixed cylindrical guide for which the chain will be on the verge of slipping. The coefficient of friction is f. (*Hint:* The resulting differential equation involving the variable chain tension T at the corresponding angle θ is of the form $dT/d\theta + KT = f(\theta)$, a first-order, linear, nonhomogeneous equation with constant coefficient. The solution is

$$T = Ce^{-K\theta} + e^{-K\theta}\int e^{K\theta}f(\theta)\,d\theta$$

where C and K are constants.)

Ans. $h = \dfrac{2fr}{1 + f^2}(1 + e^{f\pi})$

7 VIRTUAL WORK

36 Introduction. In the previous chapters the equilibrium of a body has been analyzed by isolating it with a free-body diagram and writing the zero-force and zero-moment summation equations. For the most part this approach has been employed for a body whose equilibrium position was known or specified and where one or more of the external forces was an unknown to be determined.

There is a separate class of problems in which bodies are composed of interconnected members that allow relative motion between the parts, thus permitting various possible equilibrium configurations to be examined. For problems of this type, the force- and moment-equilibrium equations, although valid and adequate, are generally not the most direct and convenient approach. Here a method based on the concept of the work done by a force is found to be more useful and direct. Also, the method provides a deeper insight into the behavior of mechanical systems and permits an examination into the question of the stability of systems in equilibrium. A development of this method follows.

37 Work. The term *work* is used in a quantitative sense as contrasted to its common nontechnical usage. The work done by a force $\mathbf{F}$ during a differential displacement $d\mathbf{s}$ of its point of application O, Fig. 81a, is by definition the scalar quantity

$$dU = \mathbf{F} \cdot d\mathbf{s}$$

The magnitude of this dot product of the vector force and the vector displacement is $dU = F\,ds\,\cos\alpha$, where α is the angle between $\mathbf{F}$ and $d\mathbf{s}$. This expression may be interpreted as the displacement multiplied by the force component $F\cos\alpha$ in the direction of the displacement, as represented by the dotted lines in Fig. 81b. Alternatively the work dU may be interpreted as the force multiplied by the displacement component $ds\cos\alpha$ in the direction of the force, as represented by the full lines in Fig. 81b. With this definition of work, it should be noted that the component $F\sin\alpha$ normal to the displace-

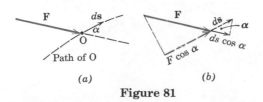

(a)   (b)

Figure 81

ment does no work. Work is positive if the working component $F \cos \alpha$ is in the direction of the displacement and negative if it is in the opposite direction. Work is a scalar quantity with dimensions of (distance) (force) and is usually expressed in foot-pound units. Dimensionally, work and moment are the same. In order to distinguish between the two quantities it is recommended that work be expressed as foot pounds (ft-lb) and moment as pound feet (lb-ft). It should be noted that work is a scalar as given by the dot product and involves the product of a force and a distance, both measured along the same line. Moment, on the other hand, is a vector as given by the cross product and involves the product of force and distance measured at right angles to the force.

During a finite displacement s of the point of application of a force the force does an amount of work equal to

$$U = \int \mathbf{F} \cdot d\mathbf{s} = \int (F_x \, dx + F_y \, dy + F_z \, dz)$$

or

$$U = \int F \cos \alpha \, ds$$

In order to carry out this integration it is necessary to know the relation between the force components and their respective coordinates or the relations between F and s and between $\cos \alpha$ and s.

In the case of concurrent forces acting on a body the work done by their resultant equals the total work done by the several forces. This may be seen from the fact that the component of the resultant in the direction of the displacement equals the sum of the components of the several forces in the same direction.

The work done by a couple $\mathbf{M}$ acting on a body obeys a vector relation analogous to that for the work of a force. Consider the couple $\mathbf{M} = 2\mathbf{r} \times \mathbf{F}$ formed by the equal and opposite forces $\mathbf{F}$ acting on the disk, Fig. 82. If the disk has an angular displacement $d\theta_z$ about the z-axis, each force does work in the amount $Fr \, d\theta_z$, where $r \, d\theta_z$ is the movement of the point of application of $\mathbf{F}$ along the arc. The work of the couple is, then, $dU = 2Fr \, d\theta_z = M \, d\theta_z$. For infinitesimal rotations of the disk $d\theta_x$ about the x-axis and $d\theta_y$ about the

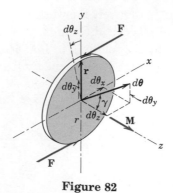

Figure 82

y-axis, the points of application of the forces have no displacement in the direction of the forces, so no work is done. The only component of the resultant infinitesimal rotation $d\boldsymbol{\theta} = \mathbf{i}\, d\theta_x + \mathbf{j}\, d\theta_y + \mathbf{k}\, d\theta_z$ of the disk which results in work by the couple is the *z*-component $d\theta_z = d\theta \cos \gamma$. Therefore, the work of the couple may be expressed as the dot product

$$dU = \mathbf{M} \cdot d\boldsymbol{\theta}$$

which is analogous to the expression for the work of a force. The angle θ is expressed in radians, and the work has the units of ft-lb. If the component of $d\boldsymbol{\theta}$ along the axis of the couple is in the direction opposite to the couple vector, negative work is done. If the disk of Fig. 82 should have translation in any direction (movement without rotation), the work done by one of the forces cancels the work done by the other force, so that the only work done by a couple occurs during rotation.

The total work done by a couple during a finite rotation of the body on which it acts is

$$U = \int \mathbf{M} \cdot d\boldsymbol{\theta} = \int (M_x\, d\theta_x + M_y\, d\theta_y + M_z\, d\theta_z)$$

or

$$U = \int M \cos \gamma \, d\theta$$

where γ represents the angle measured between the two vectors $\mathbf{M}$ and $d\boldsymbol{\theta}$. Caution is needed when dealing with rotation vectors, because it may be shown* that, whereas infinitesimal rotations may be added vectorially as in Fig. 82, finite rotations do not obey the commutative law for vector addition. Therefore a finite rotation cannot be multiplied by a couple to obtain work unless the rotation has been about a fixed axis where the direction of the rotation vector remains unchanged.

Consider now a particle whose equilibrium position is determined by the forces which act upon it. Any assumed and arbitrary small displacement $\delta \mathbf{s}$ away from this natural position is called a *virtual displacement*. The term *virtual* is used to indicate that the displacement does not exist in reality but is only assumed for the purpose of comparing various possible equilibrium positions in the process of selecting the correct one. The work done by any force $\mathbf{F}$ acting on the particle during the virtual displacement is called *virtual work* and is

$$\delta U = \mathbf{F} \cdot \delta \mathbf{s} \qquad \text{or} \qquad \delta U = F\, \delta s \cos \alpha$$

where α is the angle between $\mathbf{F}$ and $\delta \mathbf{s}$. The difference between $d\mathbf{s}$ and $\delta \mathbf{s}$ is that $d\mathbf{s}$ refers to an infinitesimal change in an actual movement, whereas $\delta \mathbf{s}$ refers to an infinitesimal virtual or assumed movement. Mathematically both quantities are first-order differentials.

A virtual displacement may also be a rotation $\delta \boldsymbol{\theta}$ of a body. The virtual work done by a couple $\mathbf{M}$ during a virtual angular displacement $\delta \boldsymbol{\theta}$ is, then, $\delta U = \mathbf{M} \cdot \delta \boldsymbol{\theta}$.

* See *Dynamics*, Art. 37, Chapter 7.

The force **F** or couple **M** may be regarded as remaining constant during any infinitesimal virtual displacement $\delta\mathbf{s}$ or $\delta\boldsymbol{\theta}$. If account is taken of the change in magnitude or direction of **F** or **M** during the infinitesimal motion, higher-order terms will result which disappear in the limit. This consideration is identical with that which permits the writing of an element of area under the curve $y = f(x)$ as $dA = y\,dx$.

38 Equilibrium of a Rigid Body. A rigid body may be considered as composed of a set of joined particles with distances between them remaining fixed. The equilibrium conditions for the body in terms of the virtual work of the forces that act on it are established first by considering the equilibrium of a single particle.

If a particle is in equilibrium, the virtual work done by all forces acting on it during any arbitrary virtual displacement $\delta\mathbf{s}$ away from the equilibrium position is

$$\delta U = \mathbf{F}_1 \cdot \delta\mathbf{s} + \mathbf{F}_2 \cdot \delta\mathbf{s} + \mathbf{F}_3 \cdot \delta\mathbf{s} + \cdots$$
$$= \delta\mathbf{s} \cdot \Sigma\mathbf{F} = \Sigma F_x\,\delta x + \Sigma F_y\,\delta y + \Sigma F_z\,\delta z = 0$$

The sum is zero, since $\Sigma\mathbf{F} = \mathbf{0}$ and also $\Sigma F_x = 0$, $\Sigma F_y = 0$, and $\Sigma F_z = 0$. The equation $\delta U = 0$ is, then, an alternative statement of the equilibrium conditions for a particle. This condition of zero virtual work for equilibrium is both necessary and sufficient, since it may be applied to virtual displacements taken one at a time in each of the three mutually perpendicular directions and is therefore equivalent to the three known scalar requirements for equilibrium.

The principle of zero virtual work for the equilibrium of a single particle usually does not simplify this already simple problem, since the virtual displacement appears as a factor in each term and hence cancels. The concept of virtual work for a particle is introduced so that it may be applied to systems of particles in the development that follows.

The extension of the principle of virtual work from a particle to a system of rigidly connected particles which form a rigid body is easily done. For a rigid body the distances between its particles remain constant, and therefore no part of the work done by forces external to the body can be absorbed by internal friction or by internal elastic deformations which are assumed absent in a rigid body.

Since the virtual work done on each particle of the body in equilibrium is zero, it follows that the virtual work done on the entire rigid body is zero. Only the virtual work done by *external* forces appears in the evaluation of $\delta U = 0$ for the entire body, since the internal forces occur in pairs of equal, opposite, and collinear forces and the net work done by these forces during any movement is zero. This net work is zero because the displacement components of the two particles along the lines of action of the forces are identical for a rigid body.

Again, as in the case of a particle, the principle of virtual work offers no real advantage to the solution for a single rigid body in equilibrium. Any assumed virtual displacement defined by a linear or angular movement will appear in each term in $\delta U = 0$ and when canceled will leave the same expression as would have been obtained by using one of the force or moment equations of equilibrium directly. This condition is illustrated in Fig. 83, where it is required to determine the reaction R under the roller for the hinged plate of negligible weight under the action of a given force P. A small assumed rotation $\delta\theta$ of the plate about O is consistent with the constraining hinge at O and is taken as the virtual displacement. The work done by P is $-Pa\ \delta\theta$, and the work done by R is $+Rb\ \delta\theta$. Therefore the principle $\delta U = 0$ gives

$$-Pa\ \delta\theta + Rb\ \delta\theta = 0$$

Canceling out $\delta\theta$ leaves

$$Pa - Rb = 0$$

which is simply the equation of moment equilibrium about O. Therefore nothing is gained by the use of the virtual-work principle for a single rigid body. Use of the principle will have a decided advantage for interconnected bodies as described in the next article.

39   Systems of Rigid Bodies. The principle of virtual work will now be extended to describe an interconnected system of rigid bodies for which the advantage of the principle will become apparent. First it is necessary to make certain observations concerning mechanical systems.

   Two or more rigid bodies fastened together by mechanical connections which are frictionless and which cannot absorb energy by elongation or compression constitute an *ideal* mechanical system. Several examples of ideal systems are shown in Fig. 84. In each system there is assumed to be no internal friction in the connections. For the system of Fig. 84a it is permissible to have friction on the sides of the piston, since such friction forces are *external* to the system composed of the piston, connecting rod, and crank and, if known, can be treated like any other external force. The rope-and-pulley arrangement of Fig. 84b is an ideal mechanical system as long as the stretch in the rope is negligible and as long as the rope does not slip on the pulleys, thereby inducing kinetic friction and heat loss. By isolating an ideal system with a free-body diagram showing all forces external to the system, as indicated with the three links in Fig. 84c, it is seen that during any and all possible

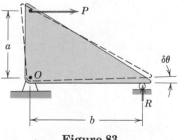

Figure 83

movements of the system or its parts the *net work done by the internal forces at the connections is zero.* This is so because the internal forces exist in pairs of equal and opposite forces, as shown for joint B in Fig. 84c, and the work of one force will cancel the work of the other force, since their displacements are identical. Hence the principle of virtual work may be applied to the entire system where only the external forces can do work during a virtual displacement. The principle may now be restated as follows: *For any ideally connected system in equilibrium the net work done by all forces externally applied to the system is zero for any and all possible virtual displacements consistent with the constraints of the connections.*

In order to apply the virtual work principle it is necessary to observe both the freedoms and the constraints for possible displacements of a mechanical system. In the mechanisms of Figs. 84a, b, and c only one coordinate is required in each case to specify the position of all parts of the system. Thus the angle of the crank determines the position of the piston. The movement of the free end of the rope completely specifies the position of the weight W and the angular positions of the pulleys. And the angular position of link AB uniquely determines the positions of the remaining two links. These mechanisms and others where only one coordinate is required to specify the system configuration are said to have *one degree of freedom.* In Fig. 84d the position of W will

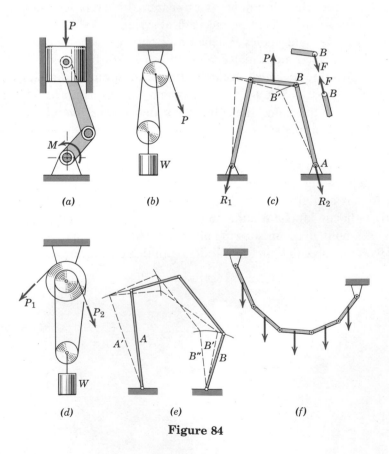

Figure 84

depend on the independent displacements of the two free ends of the rope, and therefore this device has two degrees of freedom. The five-bar linkage of Fig. 84*e* (one link is the ground) has two degrees of freedom since a rotation of *A* to *A'* could be accompanied by a rotation of *B* to *B'* or *B''* or to any number of other possible positions. Thus the positions of both *A* and *B* must be specified before the positions of the remaining two links are uniquely determined. The six-bar linkage of Fig. 84*f* has three degrees of freedom. A rigid body free to move in space has six degrees of freedom, three components of translation and three components of rotation. The *number of degrees of freedom*, then, equals the *number of independent coordinates* required to determine uniquely the configuration of the system. The number of possible independent virtual displacements of a system will equal the number of degrees of freedom.

External reactions at fixed supports do no work during any virtual displacement consistent with the constraints at these supports. These external forces, such as R_1 and R_2 in Fig. 84*c*, are called *reactive forces* as distinguished from the externally applied *active forces*, such as *P* in this same figure, which do work during a possible virtual displacement. A body force such as weight may be treated in the same way as any other active force. With the observation that only the active forces do work, the principle of virtual work may now be restated in a somewhat restricted but more useful form. *The virtual work done by external active forces on an ideal mechanical system in equilibrium is zero for any and all virtual displacements consistent with the constraints.* In this form the principle finds its greatest use for ideal systems. The principle may be stated symbolically by the equation

$$\blacktriangleright \qquad \delta U = 0 \qquad\qquad (56)$$

where δU stands for the total virtual work done by all external active forces during a virtual displacement. For a system where work is done by the forces or moments in one or more connections, such as with hinged links connected by a torsion spring at the hinge, the works do not cancel and must be included in the $\delta U = 0$ equation.

The real advantages of the method of virtual work can only now be seen. There are essentially two. First, it is not necessary to dismember ideal systems to establish the relations between the active forces, as is generally the case with the equilibrium method based on force and moment summations. Second, the relations between the active forces may be determined directly without reference to the reactive forces. These advantages make the method of virtual work particularly useful in determining the position of equilibrium of a system under known loads. This type of problem is in contrast with the problem of determining the forces acting on a body whose equilibrium position is fixed.

In the method of virtual work a diagram which isolates the system under consideration must be drawn. Unlike the free-body diagram, where all forces are shown, the diagram for the method of work need show only the *active* forces, since the reactive forces do not enter into the application of $\delta U = 0$

in the restricted sense for constrained systems. Such a drawing may be termed an *active-force diagram.*

Most problems involve a single degree of freedom, and, since the number of possible virtual displacements equals the number of degrees of freedom, such problems will require one application of Eq. 56 for the one virtual displacement. For systems of n degrees of freedom it is necessary to solve n equations, each expressing zero virtual work of all active forces due to each of the n possible virtual displacements considered separately while the remaining ones are held zero.

Sample Problems

7/1 Each of the two uniform hinged bars has a weight W and a length l, and is supported and loaded as shown. For a given force P determine the angle θ for equilibrium.

Solution. The active-force diagram for the system composed of the two members is shown separately and includes the two weights in addition to the force P. All other forces acting externally on the system are reactive forces which do no work during a virtual movement δx and are not shown.

The principle of virtual work requires that the total work of all external active forces be zero for any virtual displacement consistent with the constraints. Thus for a movement δx the virtual work becomes

$$[\delta U = 0] \qquad\qquad P\,\delta x + 2W\,\delta h = 0$$

It should be evident from the geometry that δh will be negative for a positive δx. Each of these virtual displacements will now be expressed in terms of the variable θ, the required quantity. Hence

$$h = \frac{l}{2}\cos\frac{\theta}{2} \qquad \text{and} \qquad \delta h = -\frac{l}{4}\sin\frac{\theta}{2}\,\delta\theta$$

Similarly,

$$x = 2l\sin\frac{\theta}{2} \qquad \text{and} \qquad \delta x = l\cos\frac{\theta}{2}\,\delta\theta$$

Substitution into the equation of virtual work gives

$$Pl\cos\frac{\theta}{2}\,\delta\theta - 2W\frac{l}{4}\sin\frac{\theta}{2}\,\delta\theta = 0$$

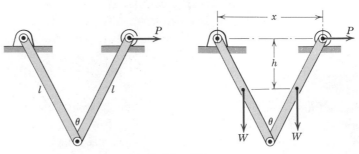

Problem 7/1

from which

$$\tan \frac{\theta}{2} = \frac{2P}{W} \quad \text{or} \quad \theta = 2 \tan^{-1} \frac{2P}{W} \qquad Ans.$$

To obtain this result by the principles of force and moment summation, it would be necessary to dismember the frame and take into account all forces acting on each member. Solution by the method of virtual work involves a simpler operation.

7/2 The weight W is brought to an equilibrium position by the application of the couple M to the end of one of the two parallel links that are hinged as shown. The links have negligible weight, and all friction is assumed to be absent. Determine the expression for the equilibrium angle θ assumed by the links with the vertical for a given value of M.

Solution. The given sketch constitutes the active-force diagram for the complete mechanism, since W and M are the only external forces and moments that do work on the system during a change in θ.

The vertical position of the center of gravity G is designated by the distance h below the fixed horizontal reference line and is $h = b \cos \theta + c$. The work done by W during a movement δh in the direction of W is

$$+ W \, \delta h = W \, \delta(b \cos \theta + c)$$
$$= W(-b \sin \theta \, \delta\theta + 0)$$
$$= - Wb \sin \theta \, \delta\theta$$

The minus sign shows that the work is negative for a positive value of $\delta\theta$. The constant c drops out since its derivative is zero.

With θ measured positive in the clockwise sense, $\delta\theta$ is also positive clockwise. Thus the work done by the clockwise couple M is $+M \, \delta\theta$. Substitution into the virtual work equation gives

$$[\delta U = 0] \qquad\qquad M \, \delta\theta + W \, \delta h = 0$$

which yields

$$M \, \delta\theta = Wb \sin \theta \, \delta\theta$$

$$\theta = \sin^{-1} \frac{M}{Wb} \qquad Ans.$$

Inasmuch as $\sin \theta$ cannot exceed unity, M is limited to equilibrium values that do not exceed Wb.

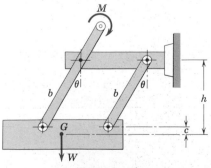

Problem 7/2

The advantage of the virtual-work solution for this problem is readily seen when a solution by force- and moment-equilibrium is attempted. For the latter approach, separate free-body diagrams of all of the three moving parts would have to be drawn and the internal reactions at the pin connections accounted for. To carry out these steps, it would be necessary to include in the analysis the horizontal position of G with respect to the attachment points of the two links, even though reference to this position would finally drop out of the equations when solved. It is evident, then, that the virtual-work method in this problem deals directly with cause and effect and avoids reference to irrelevant quantities.

7/3 A horizontal force P is applied to the end of one of three identical hinged links each of which has a weight W and a length l. Determine the equilibrium configuration.

Solution. The equilibrium configuration may be specified by determining the angles θ_1, θ_2, and θ_3. Thus three independent coordinates or measurements are required to describe uniquely the positions of the bars. The problem, consequently, is one of *three degrees of freedom*. It will be necessary to apply the principle $\delta U = 0$ three times, where each application is made by varying *one* of the coordinates at a time with the remaining two held constant.

The active-force diagram shows the three weights and the applied load P. The geometry of the figure gives

$$h_1 = \frac{l}{2}\cos\theta_1 \qquad h_2 = l\left(\cos\theta_1 + \tfrac{1}{2}\cos\theta_2\right)$$

$$h_3 = l\left(\cos\theta_1 + \cos\theta_2 + \tfrac{1}{2}\cos\theta_3\right)$$

$$x = l\left(\sin\theta_1 + \sin\theta_2 + \sin\theta_3\right)$$

If θ_3 is allowed to vary first and θ_1 and θ_2 are held constant, the principle of virtual work requires

$$[\delta U = 0]_{\theta_3} \qquad P\,\delta x + W\,\delta h_3 = 0 \qquad Pl\cos\theta_3\,\delta\theta_3 - W\frac{l}{2}\sin\theta_3\,\delta\theta_3 = 0$$

$$\tan\theta_3 = \frac{2P}{W} \qquad\qquad\qquad Ans.$$

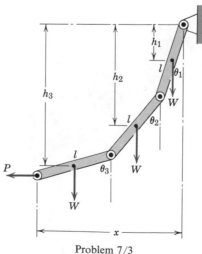

Problem 7/3

Next, θ_2 is allowed to vary and θ_1 and θ_3 are held constant, so that

$$[\delta U = 0]_{\theta_2} \qquad P\,\delta x + W\,\delta h_3 + W\,\delta h_2 = 0$$

$$Pl\cos\theta_2\,\delta\theta_2 - Wl\sin\theta_2\,\delta\theta_2 - W\frac{l}{2}\sin\theta_2\,\delta\theta_2 = 0$$

$$\tan\theta_2 = \frac{2P}{3W} \qquad\qquad\qquad Ans.$$

Lastly, θ_1 is allowed to vary while θ_2 and θ_3 are held constant. Thus

$$[\delta U = 0]_{\theta_1} \qquad P\,\delta x + W\,\delta h_3 + W\,\delta h_2 + W\,\delta h_1 = 0$$

$$Pl\cos\theta_1\,\delta\theta_1 - Wl\sin\theta_1\,\delta\theta_1 - Wl\sin\theta_1\,\delta\theta_1 - W\frac{l}{2}\sin\theta_1\,\delta\theta_1 = 0$$

$$\tan\theta_1 = \frac{2P}{5W} \qquad\qquad\qquad Ans.$$

7/4   The angular position of the telescoping link OA about the z-axis is controlled by the torque $\mathbf{M}$ applied to the shaft OB. End A is confined to slide along a fixed shaft CD. The linkage has a weight $\mathbf{W}$ with center of gravity G which is always one third of the distance from O to A. In the absence of friction, determine the torque $\mathbf{M}$ required to maintain equilibrium for a given value of h.

    Solution. The active-force diagram for the system contains $\mathbf{M}$ and $\mathbf{W}$ only, since all other external forces are reactions which do no work. The system has only one degree of freedom, which can be chosen as h, γ, or the distance $\overline{OA}$. The virtual work done by $\mathbf{M}$ is

$$\mathbf{k}M \cdot \mathbf{k}\,\delta\gamma = M\,\delta\gamma$$

The virtual work done by $\mathbf{W}$ is

$$\mathbf{W} \cdot \delta\mathbf{r} = -\mathbf{k}W \cdot \tfrac{1}{3}\,\delta(\mathbf{i}a + \mathbf{j}h\,\text{ctn}\,\beta + \mathbf{k}h) = -\tfrac{1}{3}W\,\delta h$$

But $h\,\text{ctn}\,\beta = a\tan\gamma$, so that

$$\delta h\,\text{ctn}\,\beta = a\sec^2\gamma\,\delta\gamma = \frac{1}{a}\,(h^2\,\text{ctn}^2\,\beta + a^2)\,\delta\gamma$$

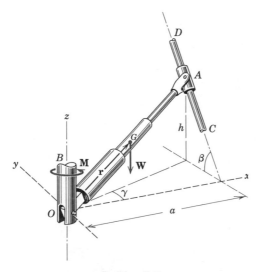

Problem 7/4

Thus the virtual work equation for the system becomes

$$[\delta U = 0] \qquad\qquad M\,\delta\gamma - \tfrac{1}{3}W\,\delta h = 0$$

$$\left[M - \frac{W}{3a\,\mathrm{ctn}\,\beta}\,(h^2\,\mathrm{ctn}^2\,\beta + a^2) \right]\delta\gamma = 0$$

Hence

$$M = \frac{W}{3a}\,(h^2\,\mathrm{ctn}\,\beta + a^2\tan\beta) \qquad\qquad Ans.$$

Problems

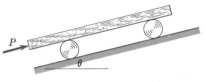

Problem 7/5

7/5 The heavy plank of weight W is supported by the two rollers, each of weight w. Determine the force P, applied to the plank parallel to the incline, necessary to start moving the plank up the incline. Assume that the rollers do not slip. Note that the plank moves twice as far as the centers of the rollers. $Ans.\ P = (W + w)\sin\theta$

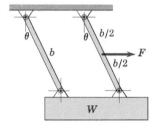

Problem 7/6

7/6 The weight W is suspended by the parallel hinged links of negligible weight. Determine the equilibrium angle θ reached by the links under the action of a horizontal force F applied at the midpoint of one of the links.

$$Ans.\ \theta = \tan^{-1}\frac{F}{2W}$$

7/7 Find the force Q exerted by the paper punch of Prob. 4/56 repeated here.

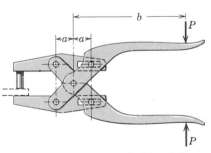

Problem 7/7

7/8 The vertical position of the sliding weight W is controlled through the link AB, whose lower end is attached to the horizontal piston rod of the fixed hydraulic cylinder. Write the expression for the force F that must act on the piston in the cylinder to maintain the weight at some particular position specified by the angle θ. Neglect friction.   *Ans.* $F = W \operatorname{ctn} \theta$

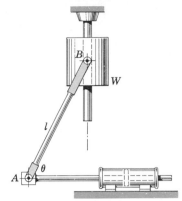

Problem 7/8

7/9 Determine the torque M on the activating lever of the dump truck necessary to balance the load W with center of gravity at G when the dump angle is θ. The polygon $ABDC$ is a parallelogram.

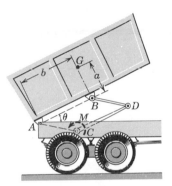

Problem 7/9

7/10 The toggle press of Prob. 4/61 is repeated here. Determine the required force F on the handle to produce a compression R on the roller for any given value of θ.   *Ans.* $F = 0.8R \cos \theta$

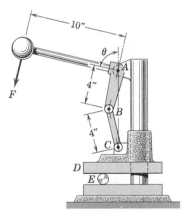

Problem 7/10

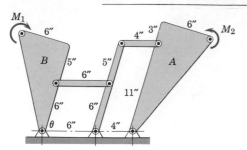

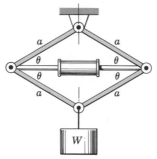

Problem 7/11

7/11 The two hinged triangular plates *A* and *B* are connected by the parallelogram linkage shown. The axes of all hinge pins are in the vertical direction, so that any possible movements of the linkage must occur in the horizontal plane. From a knowledge of the principle of virtual work and from observation of the geometry of possible virtual displacements, determine by inspection the value of the couple M_2 applied to plate *A* to provide equilibrium of the system when the couple M_1 is applied to plate *B*.

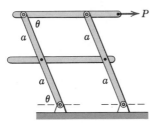

Problem 7/12

7/12 The hydraulic cylinder is used to spread the linkage and elevate the load *W*. For the position shown determine the compression *C* in the cylinder. Neglect the weights of all parts other than *W*. *Ans.* $C = W \operatorname{ctn} \theta$

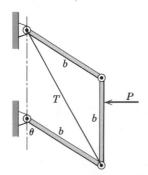

Problem 7/13

7/13 Each of the four uniform links has a weight *W*. Determine the force *P* required to hold them in place in the vertical plane shown.

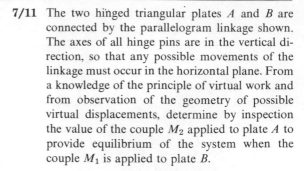

Problem 7/14

7/14 Determine the tension *T* in the diagonal wire which keeps the parallelogram linkage from collapsing under the action of the horizontal force *P*. Neglect the weights of the links.

$$Ans.\ T = P \cos \theta \Big/ \sin \frac{\theta}{2}$$

7/15 The toggle mechanism is used to position the weight W in the smooth vertical guides. Determine the expression for the horizontal force P required to support W for any value of h. Would the action be any more effective if P were applied in a direction other than horizontal?

$$Ans. \ P = W \bigg/ \sqrt{\left(\frac{2b}{h}\right)^2 - 1}$$

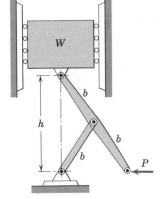

Problem 7/15

7/16 The hydraulic cylinder OA and link OB are arranged to control the tilt of the load which has a weight W and a center of gravity at G. The lower corner C is free to roll horizontally as the cylinder linkage elongates. Determine the force P in the cylinder necessary to maintain equilibrium at a given angle θ.

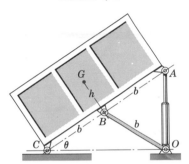

Problem 7/16

7/17 The figure shows a sectional view of a mechanism for engaging a marine clutch. The position of the conical collar on the shaft is controlled by the engaging force P, and a slight movement to the left causes the two levers to bear against the back side of the clutch plate, which generates a uniform pressure p over the contact area of the plate. This area is that of a circular ring with 6-in. outside and 3-in. inside radii. Determine by the method of virtual work the force P required to generate a clutch-plate pressure of 30 lb/in.² The clutch plate is free to slide along the collar to which the levers are pivoted.

$$Ans. \ P = 309 \ \text{lb}$$

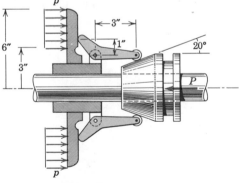

Problem 7/17

7/18 A power-operated loading platform for the back of a truck is shown in the figure. The position of the platform is controlled by the hydraulic cylinder, which applies force at C. The links are pivoted to the truck frame at A, B, and F. Determine the force P supplied by the cylinder in order to support the platform in the position shown. The weight of the platform and links may be neglected compared with that of the 500-lb crate with center of gravity at G.

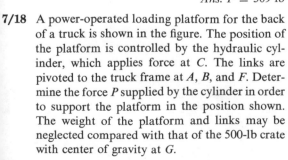

Problem 7/18

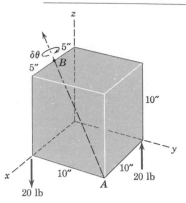

Problem 7/19

7/19 The 10-in. cube undergoes an infinitesimal rotation $\delta\theta$ about the axis AB in the sense shown. Determine the virtual work δU done by the couple composed of the two 20-lb forces.

7/20 The propeller shaft for an automobile with rear-wheel drive turns 3.27 revolutions for every revolution of the 26-in.-diameter wheels. If the car weighs 3400 lb and is moving with constant speed up a 10-percent grade, compute the torque M which the engine must supply to the propeller shaft. Neglect air resistance and all mechanical friction.   *Ans.* $M = 112.1$ lb-ft

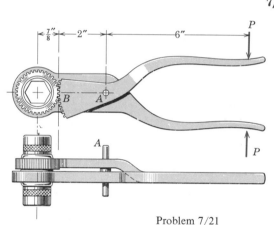

Problem 7/21

7/21 Determine the torque M exerted by the spaceman's anti-torque wrench described with Prob. 4/67 and repeated here. The gripping force on the handles is $P = 30$ lb.

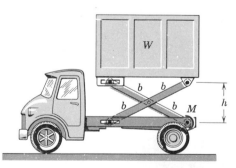

Problem 7/22

7/22 The cargo box of the food-delivery truck for aircraft servicing has a loaded weight W and is elevated by the application of a torque M on the lower end of the link which is hinged to the truck frame. The horizontal slots allow the linkage to unfold as the cargo box is elevated. Express M as a function of h.

$$Ans.\ M = 2Wb\sqrt{1 - \left(\frac{h}{2b}\right)^2}$$

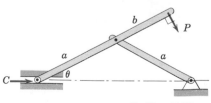

Problem 7/23

7/23 For a given force P applied normal to the link as shown, determine the force C required to maintain equilibrium of the linkage at the angle θ.

$$Ans.\ C = \frac{P}{2\sin\theta}\left(\frac{b}{a} + \cos 2\theta\right)$$

7/24 The postal scale consists of a sector of weight w_0 hinged at O and with center of gravity at G. The pan and vertical link AB have a weight w_1 and are hinged to the sector at B. End A is hinged to the uniform link AC of weight w_2, which in turn is hinged to the fixed frame. The figure $OBAC$ forms a parallelogram, and the angle GOB is a right angle. Determine the relation between the weight W to be measured and the angle θ assuming that $\theta = \theta_0$ when $W = 0$.

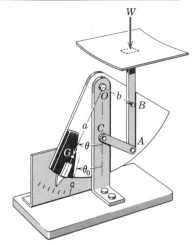

Problem 7/24

7/25 A device for counting the body radiation of a patient is shown. The radiation counter A has a weight W and is positioned by turning the screw of lead L with a torque M which controls the distance BC. Relate the torque M to the load W for given values of b and θ. Neglect all friction and the weight of the linkage compared with W.

$$Ans. \ M = \frac{5WL}{4\pi} \tan \frac{\theta}{2}$$

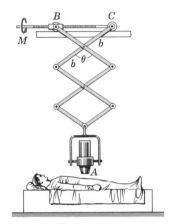

Problem 7/25

7/26 Determine the expression for the oil pressure p required in the hydraulic cylinder C in order that its piston rod may support the legs of the heavy table of weight W for a given value of θ. The piston area is A. Neglect the weights of all parts other than W.

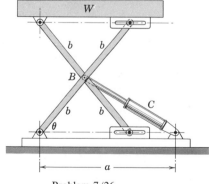

Problem 7/26

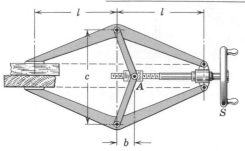

Problem 7/27

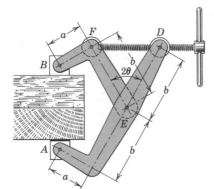

Problem 7/28

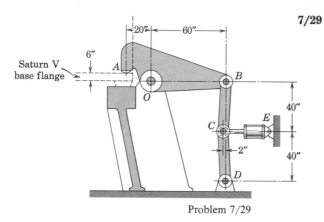

Problem 7/29

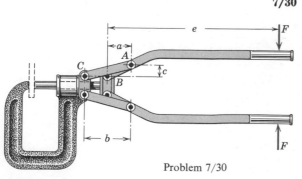

Problem 7/30

7/27 The special-purpose clamp is operated by turning the screw in a right-handed sense to force the threaded block A to the right, thereby tightening the jaws through the action of the toggle links. Express the clamping force C between the jaws in terms of the torque M applied to the hand-wheel of the screw. Neglect all friction. One turn of the screw advances the threads a distance L (termed the lead of the screw).

$$Ans.\ C = \frac{\pi M c}{4Lb}$$

7/28 Determine the force F between the jaws of the clamp in terms of a torque M exerted on the handle of the adjusting screw. The screw has a lead (advancement per revolution) L, and friction is to be neglected.

7/29 The sketch shows the approximate configuration of one of the four toggle-action hold-down assemblies that clamp the base flange of the Saturn V rocket vehicle to the pedestal of its platform prior to launching. Calculate the preset clamping force F at A if the link CE is under tension produced by a fluid pressure of 2000 lb/in.^{2} acting on the left side of the piston in the hydraulic cylinder. The piston has a net area of 16 in.^{2} The weight of the assembly is considerable, but it is small compared with the clamping force produced and is therefore neglected here.

$$Ans.\ F = 960{,}000\ \text{lb}$$

7/30 Determine the force P developed at the jaws of the rivet squeezer of Prob. 4/70, repeated here.

7/31　The toggle pliers of Prob. 4/60 are repeated here with symbolic dimensions. Determine the clamping force C as a function of α for a given handle gripping force P.

$$Ans. \quad C = P\frac{e}{c}\left(\frac{a}{b}\,\text{ctn}\,\alpha - 1\right)$$

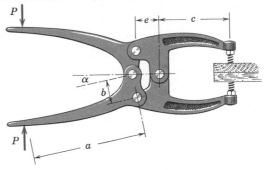

Problem 7/31

7/32　In the screw-activated clamp, a torque M applied to the handle of the screw tightens the clamp by increasing the distance BD, thereby producing a clamping force C. The screw is threaded through the pivoted collar at D and has a lead (advancement per revolution) of 1/6 in. Assume no friction and express the clamping force C in pounds in terms of the torque M measured in pound-inches for the position $\theta = 30$ deg.

$$Ans. \quad C = 26.9\,M\ \text{lb}$$

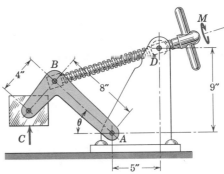

Problem 7/32

7/33　Each of the uniform links has a weight W and is subjected to a couple M at its lower end applied in the direction shown. Determine the angles θ_1 and θ_2 for equilibrium.

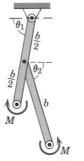

Problem 7/33

7/34　Each of the uniform hinged links has a weight w per unit length. Determine the angles θ_1 and θ_2 assumed by the linkage under the action of the horizontal force P.

$$Ans. \quad \theta_1 = \tan^{-1}\frac{9wa}{P}, \quad \theta_2 = \tan^{-1}\frac{3wa}{P}$$

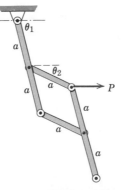

Problem 7/34

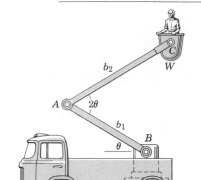

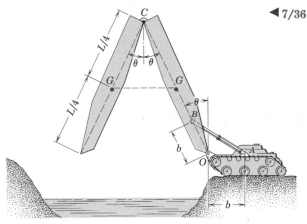

Problem 7/35

◀ **7/35** For the high-access maintenance vehicle, a motor-and-gear unit mounted in the joint of the arms at A supplies a torque internal to the joint and maintains angle CAB at exactly twice θ as the load W is being raised. If $b_2 > b_1$, determine the external moment M_1 which must be applied at B to BA in order to begin raising the load. Neglect the weights of the two members. Discuss the work done on the system and justify the inclusion of the internal moment at A in applying the virtual-work equation.

Problem 7/36

◀ **7/36** The scissors-action military bridge of span L is launched from a tank through the action of the hydraulic linkage AB. If each of the two identical bridge sections has a weight W with center of gravity at G, relate the force F in the hydraulic cylinder to the angle θ for equilibrium. An internal mechanism at C maintains the angle 2θ between the two sections. Why must the work of the internal moment at C be evaluated? Discuss the choice of the method of virtual work for this problem.   *Ans.* $F = \dfrac{WL\sqrt{2}}{b}\tan\theta\ \sqrt{1+\sin\theta}$

Problem 7/37

◀ **7/37** The figure shows a mechanism for measuring the depth of an underwater trench being formed on the ocean floor by a water jet for the laying of telephone cables. The carriage A moves horizontally at constant speed behind the water jet and may be assumed to be a stable platform. Each of the uniform links BC and DE has a weight W_1 and the link and probe EFC has a weight W_2. The specific weight of the link is γ, and that of the sea water is γ_w. The effective coefficient of friction between the probe and the bottom of the trench is f. Determine the vertical component N of the force exerted on EFC at F by the bottom of the trench.

$$Ans.\ N = (W_1 + W_2)\left(1 - \frac{\gamma_w}{\gamma}\right)\frac{1}{1 + f\tan\theta}$$

▌**7/38** A force $\mathbf{F} = \mathbf{i}F_x + \mathbf{j}F_y + \mathbf{k}F_z$ acts on a particle whose position vector is $\mathbf{r} = \mathbf{i}x + \mathbf{j}y + \mathbf{k}z$. The particle is confined to move along a line whose direction cosines are l, m, and n. Write an expression for the virtual work done by $\mathbf{F}$ in terms of a virtual change δx in the coordinate of the particle.

▌**7/39** Each of the two uniform links has a length b and a weight W. The connections at A and O are ball-and-socket joints, and the connection at B permits sliding along the y-axis and rotation of link AB about this axis. For given values of a and h, determine the necessary value of the force F, which is parallel to the x-axis, and the force P.

$$Ans.\ F = \frac{W\sqrt{b^2 - h^2 - a^2}}{h}, \quad P = \frac{Wa}{2h}$$

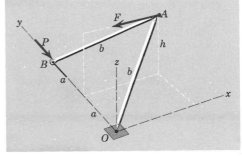

Problem 7/39

▌**7/40** Determine the force Q at the jaw of the shear of Prob. 4/69 repeated here. (*Hint:* Replace the 40-lb force by a force and a couple at the center of the small gear. The absolute angular displacement of the gear must be carefully determined.)

$$Ans.\ Q = 1318\ \text{lb}$$

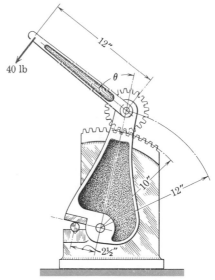

Problem 7/40

40 Systems with Elastic Members.

When a system of interconnected bodies containing elastic members is in equilibrium, some of the work done on the system by the active forces is absorbed in the elastic members during their compressions or extensions. The principle of virtual work as stated for ideal systems without elastic members must be extended to cover the effect of work done on or by the elastic members. Before extending the principle, the work of deformation of an elastic body by a force will be discussed.

The work done on an elastic member is stored in the member and is called *elastic potential energy* V_e. This energy is potentially usable and may be re-covered by allowing the member to do work on some body during the relief of its compression or extension. Consider a spring, Fig. 85, which is being compressed by a force F. The spring is assumed to be elastic and linear, i.e., the force F is taken to be directly proportional to the deflection x. This rela-tion is written as $F = kx$, where k is the *spring constant* and is a measure of the stiffness of the spring. The work done by F during a movement dx is $dU = F\,dx$, so that the elastic potential energy of the spring for a compres-sion x is

$$V_e = \int_0^x F\,dx = \tfrac{1}{2}kx^2$$

Thus the potential energy of the spring equals the triangular area in the dia-gram of F versus x. The virtual work done by F during a displacement δx is the virtual change in potential energy of the spring and is

$$\delta V_e = F\,\delta x = kx\,\delta x$$

If the spring is stretched instead of compressed, a tensile force is required which also does positive work on the spring, since the movement would like-wise be in the direction of the force. Any body that exhibits a linear relation between applied force and resulting deformation is said to be elastic and may be analyzed in this same manner.

The principle of conservation of energy requires that the work done on a mechanical system by external forces be equal to the change in energy of the system. This principle must hold for infinitesimal virtual displacements as well as for finite real displacements. Thus the principle of virtual work may now be extended as follows: *when a system of bodies with elastic members is in equilibrium, the net work done by all active forces external to the system dur-ing any virtual displacement consistent with the constraints equals the change in the elastic energy of the system.* This principle may be stated symbolically

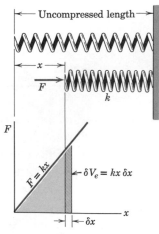

Figure 85

by the equation

▶ $$\delta U = \delta V_e \tag{57}$$

where δU is the total virtual work of all external active forces including the weights of the members and where δV_e is the net change in the total elastic energy of the system during the virtual displacement. If δU is positive, δV_e will be positive, or if δU is negative, δV_e will also be a negative quantity.

For problems with a single degree of freedom Eq. 57 need be applied only once. If there are n degrees of freedom, it will be necessary to apply the principle n times. For each application the movements of the system resulting from one virtual displacement at a time are considered while the remaining $n - 1$ coordinates are held constant.

Equation 57 finds important use in the analysis of both statically determinate and statically indeterminate systems. The elasticity of the system may come from the inclusion of manufactured springs, such as the coiled spring of Fig. 85, or may be due to the elastic deformation of the loaded structural members. In the case of a slender two-force member of uniform cross-sectional area A and length l the action may be treated as that of a one-dimensional spring. If E is the modulus of elasticity and e is the unit deformation or strain, then the stress by Hooke's law is $\sigma = eE$, tensile or compressive. This relation is converted to the form $F = kx$ by writing $\sigma A = (EA/l)(el)$, so that EA/l is the equivalent spring constant for an elastic bar. The strain energy of the bar, therefore, is

$$V_e = \frac{1}{2} kx^2 = \frac{1}{2}\left(\frac{EA}{l}\right)(el)^2 = \frac{\sigma^2 A l}{2E}$$

For an elastic body which contains a two- or three-dimensional continuum of stress, generalized Hooke's law relationships permit writing expressions for the strain energy V_e of the body in terms of the stresses or the strains. These expressions will be found in references on elasticity and stress analysis and are not developed further here.

A variation of the principle of virtual work as presented in Eq. 57 is called the unit-load method and is commonly used in structural analysis for finding the deflection of any part of a loaded structure. The work done by a fictitious or virtual unit load during the application of the real loads is equated to the increase of elastic energy which results when the internal forces due to the fictitious load act over the displacements caused by the real loads. The deflection Δ of any point on a structure may be found by placing the unit load at that point and in the direction of Δ, so that the work of the virtual load is $(1)(\Delta)$. The work-energy equation is then solved for Δ.

Sample Problem

7/41   The collar A is free to slide on the fixed vertical shaft and compresses the spring under the action of the weight W. If the spring is uncompressed for the position equivalent to

$\theta = 0$, determine the angle θ for equilibrium. The spring stiffness is k, there is negligible friction in the joints, and the weights of the arms, collar, and spring are negligible.

Solution. The spring compression is

$$x = 2a \sin \frac{\theta}{2}$$

The variation in elastic potential energy due to a virtual change in θ is

$$\delta V_e = kx \, \delta x = k\left(2a \sin \frac{\theta}{2}\right) \delta\left(2a \sin \frac{\theta}{2}\right) = 2ka^2 \sin \frac{\theta}{2} \cos \frac{\theta}{2} \, \delta\theta$$

The work done by the weight is

$$\delta U = W \, \delta h = W \, \delta\left(3a \sin \frac{\theta}{2}\right) = \frac{3Wa}{2} \cos \frac{\theta}{2} \, \delta\theta$$

For the equilibrium position

$$[\delta U = \delta V_e] \qquad \frac{3Wa}{2} \cos \frac{\theta}{2} \, \delta\theta = 2ka^2 \sin \frac{\theta}{2} \cos \frac{\theta}{2} \, \delta\theta$$

$$\cos \frac{\theta}{2} = 0 \text{ and } \sin \frac{\theta}{2} = \frac{3W}{4ka}$$

The first solution gives the physically impossible value $\theta = \pi$. The second solution gives

$$\theta = 2 \sin^{-1} \frac{3W}{4ka} \qquad\qquad Ans.$$

which has the limitation that $k > 3W/(4a)$. If k is less than this value, the weight would move the mechanism to the position $\theta = \pi$ if it were physically possible.

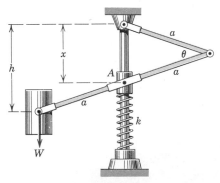

Problem 7/41

Problems

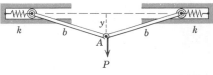

Problem 7/42

7/42 Each of the two springs has a stiffness k and is unstretched when $y = 0$. Neglect the weights of the links and determine the force P required to produce a given displacement y of point A.

$$Ans. \ P = 2ky\left(\frac{b}{\sqrt{b^2 - y^2}} - 1\right)$$

7/43 The uniform wheel of weight W is supported in the vertical plane by the light band ABC and the spring of stiffness k. If the wheel is released initially from the position where the force in the spring is zero, determine the clockwise angle θ through which the wheel rotates from the initial position to the final equilibrium position.

$$Ans. \; \theta = \frac{W}{4kr}$$

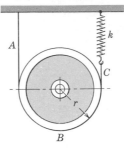

Problem 7/43

7/44 Determine the equilibrium value of x for the spring-supported bar. The spring has a stiffness k and is unstretched when $x = 0$. The force F acts in the direction of the bar, and the weight of the bar is negligible.

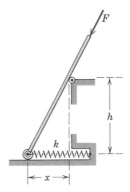

Problem 7/44

7/45 When $u = 0$, the spring of stiffness k is uncompressed. As u increases, the rod slides through the pivoted collar at A and compresses the spring between the collar and the end of the rod. Determine the force P required to produce a given displacement u. Assume the absence of friction and neglect the weight of the rod.

$$Ans. \; P = \left(1 - \frac{b}{\sqrt{b^2 + u^2}}\right)ku$$

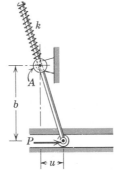

Problem 7/45

7/46 Determine the equilibrium value of the coordinate y for the mechanism under the action of the vertical load P. The spring of stiffness k is unstretched when $y = 0$, and the weight of the link is negligible compared with P.

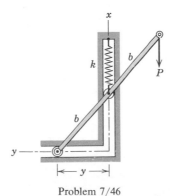

Problem 7/46

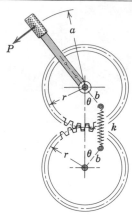

Problem 7/47

7/47 The handle is fastened to one of the spring-connected gears, which are mounted in fixed bearings. The spring of stiffness k connects two pins mounted in the faces of the gears. When the handle is in the vertical position, $\theta = 0$ and the spring force is zero. Determine the force P required to maintain equilibrium at an angle θ.

$$Ans.\ P = \frac{4kb^2}{a} \sin \theta \,(1 - \cos \theta)$$

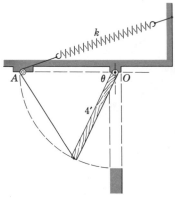

Problem 7/48

7/48 The figure shows the cross section of a uniform 100-lb ventilator door hinged along its upper horizontal edge at O. The door is controlled by the spring-loaded cable which passes over the small pulley at A. The spring has a stiffness of 12 lb per foot of stretch and is undeformed when $\theta = 0$. Determine the angle θ for equilibrium.

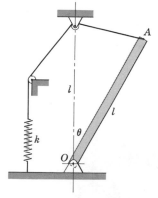

Problem 7/49

7/49 The figure shows the edge view of a uniform skylight door of weight W with center of gravity midway between A and O. The door is counterbalanced by the action of the spring, which has a stiffness k and which is unstretched when the door is in the vertical plane with $\theta = 0$. Determine the necessary stiffness k to balance the door in an equilibrium position.

$$Ans.\ k = \frac{W}{2l},\ \text{independent of } \theta$$

7/50 Specify the stiffness k of the spring to ensure equilibrium for the hoisting frame. When $\theta = 0$, the spring is unstretched. Links AB and CD are uniform members each weighing 200 lb, and the member BD with its load weighs 300 lb.

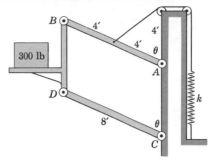

Problem 7/50

7/51 For a horizontal force F of 50 lb, determine the angle θ for equilibrium of the spring-loaded linkage. The rod DG passes through the swivel at E and compresses the spring, which has a stiffness of 25 lb/in. and an uncompressed condition corresponding to $\theta = 0$. Neglect the weights of the members.   *Ans.* $\theta = 19°28'$

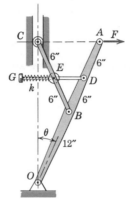

Problem 7/51

7/52 The cross section of a trap door hinged at A and having a weight W and a center of gravity at G is shown in the figure. The spring is compressed by the rod which is pinned to the lower end of the door and which passes through the swivel block at B. When $\theta = 0$, the spring is undeformed. Show that with the proper stiffness k of the spring, the door will be in equilibrium for any angle θ.

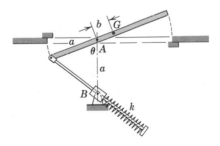

Problem 7/52

7/53 The spring-loaded roller maintains tension in the tape that runs across the rollers A and B. The spring has a stiffness k and is uncompressed when $x = x_0$. Determine the equilibrium value of x for a tape tension T. Neglect the radii of the rollers compared with the other dimensions and solve for the case where x is very small compared with b.

$$Ans.\ x = \frac{x_0}{1 + \dfrac{2T}{kb}}$$

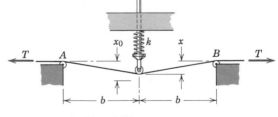

Problem 7/53

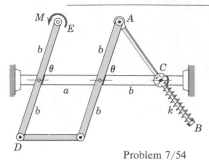

Problem 7/54

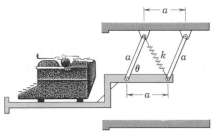

Problem 7/55

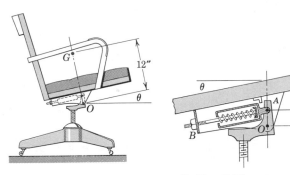

Problem 7/56

7/54 In the mechanism shown, the rod AB slides through the pivoted collar at C and compresses the spring when a couple M is applied to link DE. The spring has a stiffness k and is uncompressed for the position equivalent to $\theta = 0$. Determine the angle θ for equilibrium. The weights of the parts are negligible.

$$Ans. \ \theta = \sin^{-1}\frac{M}{kb^2}$$

7/55 A simplified desk-typewriter lift is shown in the figure. The spring has a stiffness k and has an unstretched length of $a/2$. If the weight of the typewriter and shelf is W, determine the stiffness k of the spring required to maintain equilibrium for a specified angle θ.

7/56 The figure shows a tilting desk chair together with a detail of the spring-loaded tilting mechanism. The frame of the seat is pivoted about the fixed point O on the base. The increase in distance between A and B as the chair tilts back about O is the increase in compression of the spring. The spring, which has a stiffness of 550 lb/in., is uncompressed when $\theta = 0$. For small angles of tilt it may be assumed with negligible error that the axis of the spring remains parallel to the seat. The center of gravity of a 175-lb person who sits in the chair is at G on a line through O perpendicular to the seat. Determine the angle of tilt θ for equilibrium. (*Hint:* The deformation of the spring may be visualized by allowing the base to tilt through the required angle θ about O while the seat is held in a fixed position.)

$$Ans. \ \theta = 17°21'$$

7/57 An exploration device, which unfolds from the body *A* of an unmanned space vehicle resting on the moon's surface, consists of a spring-loaded pantograph with detector head *B*. It is desired to select a spring that will limit the vertical contact force *P* to 20 lb in the position for which $\theta = 120$ deg. If the mass of the arms and head is negligible, specify the necessary spring stiffness *k*. The spring is uncompressed when $\theta = 30$ deg.

Ans. $k = 8.32$ lb/in.

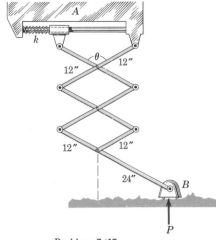

Problem 7/57

7/58 Determine the torque *M*, applied to link *OA* at *O*, that is required for equilibrium of the mechanism for a given value of the angle θ. Each of the uniform links has a weight *W*, and the spring of stiffness *k* is uncompressed in the position equivalent to $\theta = 0$. The weight of the freely sliding collar at *B* is negligible.

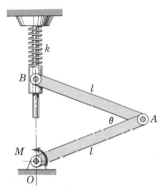

Problem 7/58

7/59 The front-end suspension of Prob. 4/62 is repeated here. The frame *F* must be jacked up so that $h = 14$ in. in order to relieve the compression in the coil springs. Determine the value of *h* when the jack is removed. Each spring has a stiffness of 640 lb/in. The load *L* is 2400 lb, and the central frame *F* weighs 80 lb. Each wheel and attached link has a weight of 70 lb with a center of gravity 27 in. from the vertical center line.

Ans. $h = 10.83$ in.

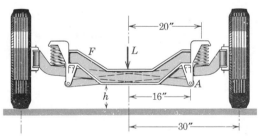

Problem 7/59

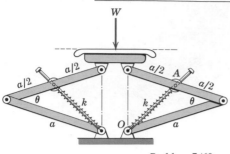

Problem 7/60

7/60 A spring-loaded pantograph provides electrical contact through the upper shoe and is designed to support a load W at an angle θ between the equal arms. Each of the springs has a stiffness k and is uncompressed when $OA = a$. Determine the stiffness k of each spring which is necessary to support the load W for a given angle θ. The weights of the parts are small compared with W.

$$Ans. \; k = \frac{W}{2a \sin \frac{\theta}{2}} \; \frac{\sqrt{1 + 8 \sin^2 \frac{\theta}{2}}}{2 - \sqrt{1 + 8 \sin^2 \frac{\theta}{2}}}$$

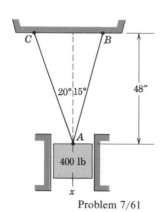

Problem 7/61

◄**7/61** The 400-lb weight is suspended by the two 1/8-in.-diameter steel wires as shown. Determine the vertical deflection x of point A as a result of the application of the 400-lb weight. Also calculate the stresses σ_{AB} and σ_{AC} in the wires. Use $30(10^6)$ lb/in.^{2} for the modulus of elasticity of steel.

$$Ans. \; x = 0.0301 \text{ in.}$$
$$\sigma_{AB} = 17{,}570 \text{ lb/in.}^2$$
$$\sigma_{AC} = 16{,}630 \text{ lb/in.}^2$$

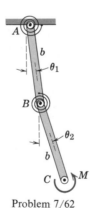

Problem 7/62

◄**7/62** Each of the two uniform bars AB and BC of length b has a weight W. The torsion springs resist rotation with respect to their attachment with a large torque of C lb-in. per radian of spring twist for each spring. The springs are untwisted when $\theta_1 = \theta_2 = 0$. Determine θ_1 and θ_2 assuming the angular deflections to be small.

$$Ans. \; \theta_1 = \frac{M/C}{1 + \frac{5}{2} \frac{Wb}{C} + \frac{3}{4}\left(\frac{Wb}{C}\right)^2}$$

$$\theta_2 = \frac{\left(2 + \frac{3}{2} \frac{Wb}{C}\right)\frac{M}{C}}{1 + \frac{5}{2} \frac{Wb}{C} + \frac{3}{4}\left(\frac{Wb}{C}\right)^2}$$

7/63 Three wires of identical diameter and material meet at a point A under a condition of no load. Determine the tension in each of the wires resulting from the load L. Assume that the deflections are small compared with the length of each wire.

$$\text{Ans. } T_1 = 0.388L$$
$$T_2 = 0.406L$$
$$T_3 = 0.265L$$

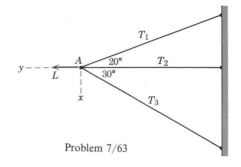

Problem 7/63

7/64 The telescoping link OA weighs 24 lb with center of gravity always one third of the distance from O to A. End O is fixed with a ball-and-socket joint, and end A is free to slide along the fixed 45-deg shaft under the action of the force $P = 18$ lb. The coiled spring inside the telescoping tubes is secured to each end of the tubes and has a stiffness of 1.5 lb/in. The spring is unstretched when the distance s between A and C equals zero. Calculate the equilibrium value of s.

$$\text{Ans. } s = 29.1 \text{ in.}$$

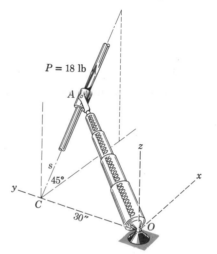

Problem 7/64

41 Systems with Friction; Mechanical Efficiency. When sliding friction is present to any appreciable degree in a system of bodies, some of the work done on the system by the external active forces is dissipated in the form of heat generated by the negative work of the friction forces during movement of the system. When there is sliding of one contact surface relative to a fixed surface the friction force present will always do negative work. The work is negative since the direction of the force is always opposite to the direction of the movement of the body on which it acts. Thus the friction force fW for the sliding block in Fig. 86a does work on the block during the displacement x in an amount equal to $-fWx$. During a virtual displacement δx the friction force does work equal to $-fW\,\delta x$. The friction force acting on the rolling wheel in Fig. 86b, on the other hand, does no work if the wheel does not slip as it rolls. In Fig. 86c the moment M_f about the center of the pinned joint or bearing of all frictional forces that act at the contacting surfaces will do negative work during any relative angular movement between the two parts. Thus, if a virtual angular displacement $\delta\theta$ occurs, the negative work done is $-M_f\,\delta\theta_1 - M_f\,\delta\theta_2 = -M_f(\delta\theta_1 + \delta\theta_2)$, or merely $-M_f\,\delta\theta$. For each part, M_f is in the direction to oppose the relative motion. The negative work done by kinetic friction forces cannot be regained, since all of it is dissipated in the form of heat. Negative frictional work can be caused by kinetic friction forces which are either external or internal to the system.

The principle of virtual work as expressed by Eq. 57 for frictionless systems with elastic potential energy must be modified to account for energy loss when kinetic friction is present. The principle of conservation of energy requires that the work done on the system by external active forces during any movement must equal the change in the energy stored in the elastic members of the system plus that lost by frictional work. Thus, if δQ stands for the total negative work done on a system in equilibrium by all kinetic friction forces during a virtual displacement consistent with the constraints, then

$$\delta U = \delta V_e + \delta Q \tag{58}$$

It has been seen in the previous two articles that a major advantage of the method of virtual work is the analysis of an entire system of connected members without taking them apart. If there is appreciable friction internal to a mechanical system, however, it is usually necessary to dismember the system in order to compute the friction forces. In such cases a prime advantage of the virtual-work method is lost. Consequently the method finds limited use for determining the equilibrium configuration for systems where kinetic friction is appreciable.

Because of energy loss by friction the ratio of useful work derived from a machine to the work input to the machine during the same interval is always less than unity. This ratio is a measure of the mechanical efficiency e of a machine. Thus

$$e = \frac{\text{useful work done by machine}}{\text{work input to machine}}$$

The mechanical efficiency of simple machines which have one degree of freedom and which operate in a uniform manner may be determined by the principle of work by evaluating the numerator and denominator of the expression for e during an infinitesimal displacement. If a machine operates in a uniform manner with time, any elastic changes due to elongations or compressions that occurred during the initial application of the loads will remain constant during the continuous operation and will not be involved in the expressions.

As an example consider the work done by the tension T in the cable,

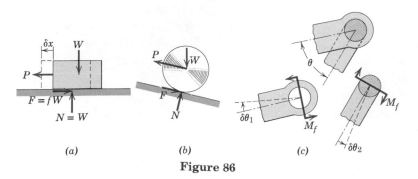

(a)   (b)   (c)

Figure 86

Fig. 87, in causing the body to slide up the inclined plane at constant speed. The active-force diagram discloses only three forces that do work on the system composed of the body and cable. During a differential movement ds up the plane the work input to the body is

$$T\,ds = (W \sin \theta + fW \cos \theta)\,ds$$

The useful work done by the machine consists of the work in raising the weight W the vertical distance $ds \sin \theta$ and is

$$W\,ds \sin \theta$$

The efficiency of the inclined plane is then

$$e = \frac{W\,ds \sin \theta}{(W \sin \theta + fW \cos \theta)\,ds} = \frac{1}{1 + f/\tan \theta}$$

As a second example of mechanical efficiency consider the screw jack described in Art. 35 and shown in Fig. 76. Equation 50 gives the moment M applied to the screw necessary to raise the weight W, where the screw has a mean radius r and a helix angle α, and where the friction angle is ϕ. This relation is

$$M = Wr \tan (\alpha + \phi)$$

During a small rotation $d\theta$ of the screw in a direction to raise the load the moment M does work in the amount of

$$M\,d\theta = Wr\,d\theta \tan (\alpha + \phi)$$

The work done by the machine is the work done in raising the load through the vertical distance $(d\theta/2\pi)L = r\,d\theta \tan \alpha$, where L is the lead or advancement per revolution of the screw. Thus the efficiency of the screw is

$$e = \frac{Wr\,d\theta \tan \alpha}{Wr\,d\theta \tan (\alpha + \phi)} = \frac{\tan \alpha}{\tan (\alpha + \phi)}$$

As the friction in the threads is decreased, the friction angle ϕ becomes smaller, and the efficiency approaches unity.

42 Energy Criterion for Equilibrium. Energy represents the capacity to do work. A body or system of bodies is said to have *potential energy* when, by virtue of its elastic state or its position in a field of force, it is capable of doing work on some other body. There are these two types of mechanical potential

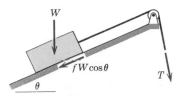

Figure 87

energy. Potential energy V_e associated with the elastic state has been discussed in Art. 40 and for a simple spring is $V_e = \frac{1}{2}kx^2$, where k is the stiffness and equals the constant ratio of applied force to resulting deformation x. This expression represents recoverable energy provided that the limit of the elasticity of the spring material has not been exceeded. The second type of potential energy V_g is that associated with the position of a body in a field of force. More exactly it is the work done on a body in changing its position in the force field to which it is subjected. The most common force field is, of course, the earth's gravitational field, and for experiments near the surface of the earth the intensity may be assumed to be constant.

In the previous examples of virtual work in Arts. 39 and 40 the effect of position in the earth's gravitational field was accounted for by treating the weight or gravitational attraction as an external force which did work during any vertical virtual displacement δh of the center of gravity. Thus the weight W of the body in Fig. 88 does work in the amount $\delta U = -W\,\delta h$, where the minus sign accounts for the fact that the displacement and the force are opposite in direction. If, on the other hand, the body moves downward, the weight W does positive work, since the displacement would be in the same direction as the force. Viewed in this way, the weight is treated like any other active force, and its work appears on the δU-side of the virtual-work equation, as has been demonstrated in Arts. 39 and 40.

An alternative to the foregoing treatment may be adopted where the work done by the gravity forces is expressed by a change in potential energy of the body. This alternative treatment is a useful representation when describing a mechanical system in terms of its total energy. The gravitational potential energy V_g of a body is defined simply by the work done on the body to bring it to the position under consideration from some arbitrary datum plane where the potential energy is defined as zero. When the body is raised, for example, this work is converted into energy that is potentially available, since the body is capable of doing work on some other body as it returns to its original lower position. If V_g is taken to be zero at $h = 0$, Fig. 88, then at a height h above the datum plane the gravitational potential energy of the body is $V_g = +Wh$.

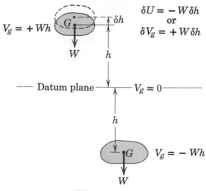

Figure 88

If the body is a distance h below the datum plane, its gravitational potential energy is $-Wh$. Inasmuch as the work done by the weight W is equal in magnitude but opposite in sign to the change in the potential energy, the effect of a vertical change of position may be accounted for on the energy side of the equation rather than on the work side of the equation. It is important to note that the datum plane for zero potential energy is perfectly arbitrary inasmuch as it is only the *change* in energy which is of concern, and this change is independent of the position of the datum plane. Also, the gravitational potential energy is independent of the path followed in arriving at the particular level h under consideration.

In applying the work-energy equation, care must be exercised not to include the effect of the weight W both as a work term and as an energy term in the same problem. Either alternative may be adopted and should be applied consistently within any given problem.

With the introduction of the gravitational potential energy, the total mechanical potential energy of a body or system of bodies is the sum of the recoverable elastic energy V_e and the energy of position V_g in the earth's gravitational field and is

$$V = V_e + V_g$$

For a mechanical system with gravitational forces and elastic members the work-energy relation of Eq. 57 may now be rewritten alternatively as

$$\blacktriangleright \qquad \delta U = \delta V \qquad \text{or} \qquad \delta U = \delta V_e + \delta V_g \qquad \textbf{(59)}$$

Here δU is the virtual work of all external active forces exclusive of the weights of the members which are accounted for in the δV_g term.

In addition to the replacement of the work of gravity forces by the corresponding changes in potential energy, the work of other active forces applied externally to the system may be converted to potential energy changes as shown in Fig. 89. Here the work done by **F** during a virtual displacement $\delta \mathbf{s}$ of its point of application is equivalent to the change $-W \delta s \cos \alpha$ in potential energy of the weight W for the equivalent system. Hence the work done by external active forces on a mechanical system may be replaced by the corresponding potential energy changes with opposite sign for the equivalent system.

With the replacement of the work terms by energy terms, the principle

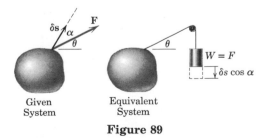

Given System   Equivalent System

Figure 89

of virtual work for a mechanical system without internal kinetic friction, expressed by Eq. 57 or 59, may now be written as

$$\blacktriangleright \qquad \delta(V_e + V_g) = 0 \qquad \text{or} \qquad \delta V = 0 \qquad \textbf{(60)}$$

Equation 60 expresses the requirement that the equilibrium configuration of a mechanical system be one for which the total potential energy V of the system has a stationary value. Thus for a mechanical system whose possible position is described by the variable x, application of Eq. 60 is mathematically equivalent to the condition $dV/dx = 0$, which gives a stationary value of V, either a maximum, a minimum, or a point of inflection in the curve of V versus x. It is well known that the natural equilibrium position for a mechanical system without friction is one for which the potential energy is a minimum, which is a stationary value of V as long as V and its derivatives are continuous functions.

   The principle of zero variation in potential energy for equilibrium may be applied to *conservative systems* only, i.e., systems free from kinetic friction forces that do negative work. Depending on the amount of kinetic friction present, a system will approach the equilibrium position of minimum potential energy but will never quite reach it.

   As an example of the principle of minimum energy, consider the simple system in equilibrium shown in Fig. 90, consisting of the spring that is deflected an amount x by the action of the attached weight W. If $x = 0$ is taken arbitrarily as the position for which $V = 0$, the potential energy for the displacement x is

$$V = V_e + V_g = \tfrac{1}{2}kx^2 - Wx$$

The condition for equilibrium given by Eq. 60 requires

$$\delta V = kx\,\delta x - W\,\delta x = 0, \qquad x = W/k$$

This value agrees with the result that can be obtained by inspection. The contribution of V_e and V_g to the total potential energy V for various possible values of x is also shown in Fig. 90, and the value $x = W/k$ is seen to give a minimum value for V.

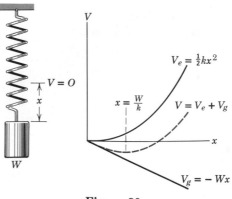

Figure 90

▽   ▽   ▽   ▽   ▽

Situations arise where the configuration of a mechanical system depends on the specification of more than one coordinate. Several examples of such situations were illustrated in parts *d*, *e*, and *f* of Fig. 84. The number of independent coordinates needed to specify the configuration of the system gives the number of degrees of freedom of the system. To determine the conditions of equilibrium for a system with *n* degrees of freedom, the variation of *V* with respect to each of the *n* independent coordinates taken one at a time, with the remaining ones held constant, must be evaluated and set equal to zero. This procedure will give *n* equations in the *n* unknown coordinate values. If x_i represents one of the *n* independent coordinates, the condition

$$\frac{\partial V}{\partial x_i} = 0 \tag{61}$$

must be satisfied for each of the *n* coordinates. The independent coordinates are called *generalized coordinates* and are not restricted to the components in orthogonal coordinate systems such as *x*, *y*, *z* or *r*, *θ*, *φ*.

Up to this point the discussion of potential energy V_g has been limited to the case where the gravitational field of force was parallel and of constant intensity. In the more general case, consider a force $\mathbf{F} = \mathbf{i}F_x + \mathbf{j}F_y + \mathbf{k}F_z$ whose components are functions of the coordinates in a field or region of force, such as a gravitational or magnetic field whose intensity varies with position. The work done by $\mathbf{F}$ during a virtual displacement of its point of application, located by the position vector $\mathbf{r} - \mathbf{i}x + \mathbf{j}y + \mathbf{k}z$, is

$$\delta U = \mathbf{F} \cdot \delta \mathbf{r} = F_x \, \delta x + F_y \, \delta y + F_z \, \delta z$$

If this work is an exact differential $-\delta V$ then, since

$$\delta V = \frac{\partial V}{\partial x} \, \delta x + \frac{\partial V}{\partial y} \, \delta y + \frac{\partial V}{\partial z} \, \delta z$$

the force components become*

$$F_x = -\frac{\partial V}{\partial x} \qquad F_y = -\frac{\partial V}{\partial y} \qquad F_z = -\frac{\partial V}{\partial z} \tag{62}$$

The quantity *V* is known as the potential energy or merely the potential function. The negative sign is arbitrary but is chosen to be consistent with the usual designation of the sign of potential energy change in the earth's gravitational field.

A force field whose components form an exact differential is known as a *conservative* field. It follows that the work done is

$$\int \mathbf{F} \cdot \delta \mathbf{r} = - \int_{V_1}^{V_2} \delta V = -(V_2 - V_1)$$

---

* Recall that the quantity $d\phi = P \, dx + Q \, dy + R \, dz$ is an exact differential if $\frac{\partial P}{\partial y} = \frac{\partial Q}{\partial x}, \frac{\partial P}{\partial z} = \frac{\partial R}{\partial x}, \frac{\partial Q}{\partial z} = \frac{\partial R}{\partial y}$.

which depends *only* on the end points of the motion and *not* on the particular path followed. The work done by a nonconservative force field depends on the particular path between the end points. The introduction of kinetic friction forces into the system, for example, results in nonconservative forces.

The two examples of potential energy discussed previously agree with the more general mathematical formulation of potential energy. Hence, for the force field of a simple spring whose energy is $V_e = \frac{1}{2}kx^2$, the force exerted *by* the spring on the body which stretches it a distance x is

$$F_x = -\frac{\partial V_e}{\partial x} = -\frac{d}{dx}\left(\tfrac{1}{2}kx^2\right) = -kx$$

which is in the direction opposite to the force $+kx$ that stretches the spring. For a weight W in the gravitational field of the earth at its surface, $V_g = Wh$, and the force component in the upward direction of positive h is

$$F_h = -\frac{\partial V_g}{\partial h} = -\frac{d(Wh)}{dh} = -W$$

which agrees with the known downward direction of the weight. The more general formulation of force-field potentials finds important use when the intensity and the direction of the field vary with position in the field.

43   **Stability of Equilibrium.** The requirement of zero variation of the potential energy for equilibrium, as pointed out in the previous article and as stated by Eq. 60, is equivalent to the requirement

$$\frac{dV}{dx} = 0$$

for a system with a single degree of freedom, where x is the coordinate defining the position of the system. This equation defines the condition for a stationary value of V, that is, a minimum, a maximum, a stationary point of inflection, or a constant value, in the relation of V versus x. Although the natural equilibrium configuration usually represents a minimum value of V, there are the other possibilities for stationary values which also represent equilibrium positions. Figure 91 shows an example of each type of equilibrium for the simple case of a cylinder supported by surfaces of four different curvatures. The equilibrium is said to be *stable,* Fig. 91*a,* when a slight displacement away from the equilibrium position results in an increase of potential energy and a corresponding tendency to return to the original position. Thus a minimum value of V always defines a stable position. The

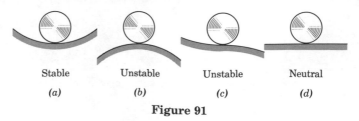

Stable   Unstable   Unstable   Neutral

(*a*)   (*b*)   (*c*)   (*d*)

Figure 91

equilibrium is said to be *unstable*, Fig. 91*b*, when a slight displacement away from the equilibrium position results in a decrease of potential energy and a corresponding tendency to move farther away from the original position. A maximum value of V always defines an unstable position. The third possibility for a stationary value of V occurs at a point of inflection, Fig. 91*c*. Here the equilibrium is classified as unstable, since a slight displacement toward the low side would result in a further tendency to move away from the inflection point. Physically, this equilibrium condition is often quite insensitive to small displacements, and the behavior approaches that of the fourth case, Fig. 91*d*, which represents *neutral equilibrium*. Here there is no tendency to move away from or return to an original rest position.

The stability conditions that are illustrated here for the simple case of the rolling cylinder apply to any real mechanical system with a single degree of freedom and with friction neglected. Figure 92 shows a schematic plot of the total potential energy $V = V_e + V_g$ for a hypothetical mechanical system with the four equilibrium conditions described in the foregoing examples. Mathematically the sign of the increment ΔV due to a small displacement from the equilibrium position may be determined by Taylor's expansion for a continuous function in the neighborhood of a certain point. If $x = 0$ is taken to be the equilibrium position and if the subscript zero refers to the conditions at $x = 0$, then the value of V corresponding to a small displacement x away from the equilibrium position is

$$V = V_0 + \left(\frac{dV}{dx}\right)_0 \frac{x}{1!} + \left(\frac{d^2V}{dx^2}\right)_0 \frac{x^2}{2!} + \left(\frac{d^3V}{dx^3}\right)_0 \frac{x^3}{3!} + \left(\frac{d^4V}{dx^4}\right)_0 \frac{x^4}{4!} + \cdots$$

But for equilibrium $(dV/dx)_0 = 0$, so that the change in potential energy $V - V_0$ becomes

$$\Delta V = \left(\frac{d^2V}{dx^2}\right)_0 \frac{x^2}{2!} + \left(\frac{d^3V}{dx^3}\right)_0 \frac{x^3}{3!} + \left(\frac{d^4V}{dx^4}\right)_0 \frac{x^4}{4!} + \cdots$$

where the successive terms diminish rapidly in magnitude. Thus at the equilibrium position the sign of ΔV will be governed by the sign of the lowest order nonzero term that remains. Generally the second derivative will remain, and the equilibrium of the system will be

stable (ΔV positive) if $\quad (d^2V/dx^2)_0$ is positive

unstable (ΔV negative) if $\quad (d^2V/dx^2)_0$ is negative

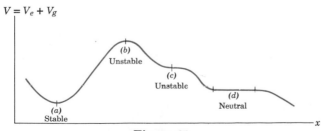

Figure 92

If the second derivative is zero, higher derivatives must be investigated. In general when the order of the lowest remaining nonzero derivative is *even,* the variation of potential energy near the equilibrium position will resemble an even function as shown for conditions (*a*) and (*b*) in Fig. 92, and the equilibrium is stable or unstable according to whether the sign of this lowest even derivative is plus or minus. If the order of the lowest remaining nonzero derivative is *odd,* the variation of potential energy near the equilibrium position will resemble an odd function of the inflection type, similar to that for condition (*c*) in Fig. 92, and the condition is classified as unstable.

Stability is a condition that is difficult to determine always on the basis of mechanical intuition alone, except in the simplest of cases. The foregoing analysis provides a means whereby the engineer can exercise control over the kind and degree of stability that may be involved in the design of structures and mechanisms.

▼   ▼   ▼   ▼   ▼

For a system with two degrees of freedom in the independent coordinates x and y, the stability conditions are derived from Taylor's expansion for V in the neighborhood of the equilibrium position at $x = 0, y = 0$. This expansion for V as a function of two variables is

$$V = V_0 + \left(\frac{\partial V}{\partial x}\right)_0 x + \left(\frac{\partial V}{\partial y}\right)_0 y + \frac{1}{2}\left[\left(\frac{\partial^2 V}{\partial x^2}\right)_0 x^2 + \left(\frac{\partial^2 V}{\partial x\, \partial y}\right)_0 2xy + \left(\frac{\partial^2 V}{\partial y^2}\right)_0 y^2\right] + \cdots$$

where further terms involving higher-order derivatives are somewhat lengthy and are omitted. With the requirement that $(\partial V/\partial x)_0 = 0$ and $(\partial V/\partial y)_0 = 0$ for equilibrium and with the substitutions

$$A = \left(\frac{\partial^2 V}{\partial x^2}\right)_0 \qquad B = \left(\frac{\partial^2 V}{\partial x\, \partial y}\right)_0 \qquad C = \left(\frac{\partial^2 V}{\partial y^2}\right)_0$$

the expression for the increment $\Delta V = V - V_0$ may be approximated by

$$\Delta V = \tfrac{1}{2}(Ax^2 + 2Bxy + Cy^2) \tag{63}$$

where the smaller higher-order terms are dropped.

Equation 63 may be represented by a space surface with ΔV as the vertical coordinate and x and y as horizontal coordinates. The slope of the surface is zero in both directions at the origin or equilibrium position. It may be shown from the geometry of the surface that the system will be

$$
\left.
\begin{array}{lll}
\text{stable } (\Delta V \text{ positive) if} & B^2 - AC < 0 \text{ and } A + C > 0 \\
\text{unstable } (\Delta V \text{ negative) if} & B^2 - AC < 0 \text{ and } A + C < 0 \\
\text{unstable (saddle point) if} & B^2 - AC > 0 \\
\text{undetermined if} & B^2 - AC = 0
\end{array}
\right\} \tag{64}
$$

The potential surfaces for the first three cases are represented in Fig. 93. The stable case, Fig. 93a, requires the surface to be concave upward at the origin regardless of direction. This condition, therefore, ensures that $(\partial^2 V/\partial n^2)_0$ will be positive for any value of θ. The unstable case, Fig. 93b, requires the surface to be concave downward at the origin regardless of direction. The second condition of Eqs. 64 ensures that $(\partial^2 V/\partial n^2)_0$ will be negative for any value of θ. The third case, represented in Fig. 93c, is that of a "saddle," so named because of the reversed curvature. This case is one of instability, since $(\partial^2 V/\partial n^2)_0$ is negative for a certain range of values of θ. If the fourth condition of Eqs. 64 holds, higher-order derivatives in the Taylor expansion for a function of two variables would have to be examined. The condition of neutral equilibrium in two or more variables is given simply by the condition $V = $ constant.

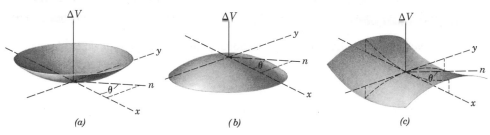

(a)   (b)   (c)

Figure 93

Sample Problems

7/65   The total potential energy $V = V_g + V_e$ of a conservative mechanical system with a single degree of freedom in the coordinate x is given by $V = 3x^3 - 9x^2 + 16$ where x is in feet and V is in foot-pounds. Determine the type of stability at each equilibrium position.

   Solution. Equilibrium requires

$$\left[\frac{dV}{dx} = 0\right]$$

$$\frac{dV}{dx} = 9x^2 - 18x = 0$$

$$x = 0 \qquad x = 2 \text{ ft}$$

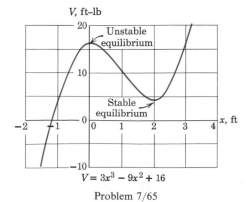

$$V = 3x^3 - 9x^2 + 16$$

Problem 7/65

The stability is determined from the second derivative, which is

$$\frac{d^2V}{dx^2} = 18x - 18$$

For $x = 0$, $d^2V/dx^2 = -18$, and, since the sign is negative, this position is unstable. For $x = 2$ ft, $d^2V/dx^2 = 36 - 18 = +18$, and this position is stable, since the second derivative of V is positive.

The accompanying plot of V versus x discloses at a glance the equilibrium positions and their stability.

7/66 In the mechanism shown the spring of stiffness k is uncompressed when $\theta = 60$ deg. Also the weights of the parts are small compared with the sum W of the two weights. The mechanism is built so that the arms may swing past the vertical as seen in the right-hand projection. Determine the values of θ for equilibrium and investigate the stability of the mechanism in each position. Show graphically the variation with θ of the total potential energy of the system and indicate the stability conditions on the diagram. Neglect friction.

Solution. The zero datum for gravitational potential energy V_g is taken through the fixed point O, so that

$$V_g = -2Wa\cos\theta$$

The compression of the spring is $x = 2a\cos\theta - a$, and the elastic energy of the spring is

$$V_e = \tfrac{1}{2}kx^2 = \frac{ka^2}{2}(2\cos\theta - 1)^2$$

The total potential energy of the mechanism is, then,

$$V = V_e + V_g = \frac{ka^2}{2}(2\cos\theta - 1)^2 - 2Wa\cos\theta$$

The derivatives of V with respect to θ are

$$\frac{dV}{d\theta} = 2a[(W + ka)\sin\theta - 2ka\sin\theta\cos\theta]$$

and

$$\frac{d^2V}{d\theta^2} = 2a[(W + ka)\cos\theta - 2ka(2\cos^2\theta - 1)]$$

For equilibrium

$$\left[\frac{dV}{d\theta} = 0\right] \qquad\qquad [(W + ka) - 2ka\cos\theta]\sin\theta = 0$$

so that

$$\sin\theta = 0 \qquad \text{and} \qquad \cos\theta = \frac{W + ka}{2ka} \qquad\qquad\qquad Ans.$$

give two possible equilibrium values of θ. For the second solution the cosine cannot exceed unity, so this solution is limited to $W \leq ka$.

Stability is determined from the sign of the second derivative. For the solution $\sin\theta = 0$, the second derivative becomes

$$\frac{d^2V}{d\theta^2} = 2a[(W + ka) - 2ka] = 2a(W - ka)$$

If $W > ka$, the second derivative is positive, and the equilibrium is stable for the position $\theta = 0$. If $W < ka$, the second derivative is negative, and the equilibrium is unstable.

For the second solution, $\cos \theta = (W + ka)/2ka$, the second derivative of V becomes

$$\frac{d^2V}{d\theta^2} = 2a\left\{ \frac{(W + ka)^2}{2ka} - 2ka\left(2\left[\frac{W + ka}{2ka}\right]^2 - 1\right)\right\} = \frac{W^2}{k}\left(\frac{3ka}{W} + 1\right)\left(\frac{ka}{W} - 1\right)$$

Since the second solution is valid only if $W < ka$, the second derivative is seen to be positive for this solution, which is therefore stable.

In the right-hand portion of the figure the dimensionless energy

$$\frac{V}{Wa} = \frac{ka}{2W}(2 \cos \theta - 1)^2 - 2 \cos \theta$$

is plotted for two arbitrary values, $ka/W = 2$ and $ka/W = \frac{1}{2}$, over a 90-deg range in θ assuming that the spring remains attached to the collar B beyond $\theta = 60$ deg so that V is expressed by the same continuous function of θ.

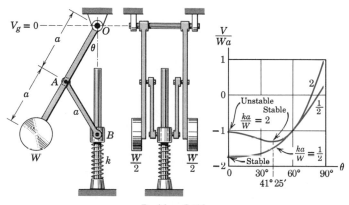

Problem 7/66

7/67   Each of the two uniform hinged links has a weight W and a length a. The combined stiffness of the upper two springs is k, and the bottom two have the same combined stiffness. When both bars are vertical, there is no force in any of the springs. Examine the conditions for stability of the links in the vertical positions and plot the potential-energy surface in the neighborhood of the equilibrium position for the particular value $ka/W = 1$.

Solution. The system possesses two degrees of freedom, since each bar may rotate independently of the other. The independent coordinates that define the configuration of the system are chosen as the angles θ_1 and θ_2 as shown. For small angles the deflections of the bottom and top sets of springs are $a \sin \theta_1$ and $(a \sin \theta_1 + a \sin \theta_2)$, respectively. With the zero datum for gravitational potential energy taken through O, the total potential energy of the system is

$$V = V_e + V_g = \tfrac{1}{2}k(a \sin \theta_1)^2 + \tfrac{1}{2}k(a \sin \theta_1 + a \sin \theta_2)^2$$
$$+ W\frac{a}{2}\cos \theta_1 + W\left(a \cos \theta_1 + \frac{a}{2}\cos \theta_2\right)$$

The pertinent partial derivatives are

$$\frac{\partial V}{\partial \theta_1} = ka^2 \sin 2\theta_1 + ka^2 \sin \theta_2 \cos \theta_1 - \frac{3Wa}{2}\sin \theta_1$$

$$\frac{\partial^2 V}{\partial \theta_1^2} = 2ka^2 \cos 2\theta_1 - ka^2 \sin \theta_2 \sin \theta_1 - \frac{3Wa}{2} \cos \theta_1$$

$$\frac{\partial V}{\partial \theta_2} = \tfrac{1}{2}ka^2 \sin 2\theta_2 + ka^2 \sin \theta_1 \cos \theta_2 - \frac{Wa}{2} \sin \theta_2$$

$$\frac{\partial^2 V}{\partial \theta_2^2} = ka^2 \cos 2\theta_2 - ka^2 \sin \theta_1 \sin \theta_2 - \frac{Wa}{2} \cos \theta_2$$

$$\frac{\partial^2 V}{\partial \theta_1 \, \partial \theta_2} = ka^2 \cos \theta_1 \cos \theta_2$$

For the equilibrium condition at $\theta_1 = \theta_2 = 0$, for which $\partial V/\partial\theta_1 = \partial V/\partial\theta_2 = 0$ also, the values of the second derivatives are

$$A = \left(\frac{\partial^2 V}{\partial \theta_1^2}\right)_0 = 2ka^2 - \frac{3Wa}{2}$$

$$B = \left(\frac{\partial^2 V}{\partial \theta_1 \, \partial \theta_2}\right)_0 = ka^2$$

$$C = \left(\frac{\partial^2 V}{\partial \theta_2^2}\right)_0 = ka^2 - \frac{Wa}{2}$$

Upon substitution of $x = ka/W$ and upon simplification, the quantity $B^2 - AC$ becomes

$$B^2 - AC = \left(\frac{Wa}{2}\right)^2 (-4x^2 + 10x - 3)$$

A plot of $B^2 - AC$ as a function of $x = ka/W$ is included in the figure. There are three regions to be investigated, namely, $x < 0.349$, $0.349 < x < 2.151$, and $x > 2.151$.

Region I: $\dfrac{ka}{W} < 0.349$, $B^2 - AC < 0$

$$A + C = 2ka^2 - \frac{3Wa}{2} + ka^2 - \frac{Wa}{2} = Wa\left(3\frac{ka}{W} - 2\right) = (-)$$

Thus the equilibrium is unstable, and both bars would tend to collapse with a spring stiffness $k < 0.349\ W/a$.

Region II: $0.349 < \dfrac{ka}{W} < 2.151$, $B^2 - AC > 0$

Unstable saddle equilibrium occurs in this region.

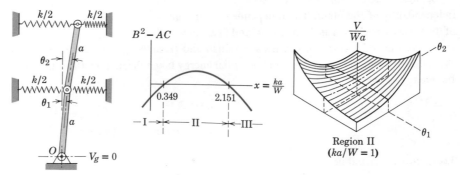

Problem 7/67

Region III: $\frac{ka}{W} > 2.151,$ $B^2 - AC < 0$

$$A + C = Wa\left(3\frac{ka}{W} - 2\right) = (+)$$

Thus the equilibrium is stable. It is apparent, then, that each combination of springs must have a stiffness $k > 2.151\ W/a$ to ensure stability of the bars in the vertical positions.

The right-hand portion of the figure shows a plot of the potential energy expressed in dimensionless form for the value $ka/W = 1$, which is in the region of saddle-type equilibrium. This energy is

$$\left(\frac{V}{Wa}\right)_{ka/W=1} = \sin^2\theta_1 + \sin\theta_1\sin\theta_2 + \tfrac{1}{2}\sin^2\theta_2 + \tfrac{3}{2}\cos\theta_1 + \tfrac{1}{2}\cos\theta_2$$

Other values of ka/W within region II will give other shapes to the saddle surface in the neighborhood of the vertical equilibrium position.

It should be noted that the solution formulated with this example is invalid except for the conditions approaching $\theta_1 = \theta_2 = 0$ by virtue of the geometric assumptions made at the outset.

Problems

7/68 The bar is free to rotate through a complete vertical circle. Verify mathematically the stability conditions that are evident at the two equilibrium positions.

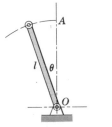

Problem 7/68

7/69 The potential energies of two mechanical systems are given by $V_1 = C_1 x^4$ and $V_2 = C_2 x^3$, where C_1 and C_2 are positive constants and x is the single coordinate expressing the positions of both systems. Specify the stability of each system at the equilibrium position $x = 0$.

7/70 The potential energy of a mechanical system is given by $V = a\cos 2\theta + 2a\sin\theta$, where θ is the coordinate that specifies the position of the system and a is a positive constant. Determine the stability for each position of equilibrium.

Ans. stable for $\theta = \pi/2$ and $\theta = 3\pi/2$
unstable for $\theta = \pi/6$ and $\theta = 5\pi/6$

7/71 Identify the equilibrium conditions represented by the three configurations of the bar of weight W supported by the light links and free to rotate in the vertical plane of the figure.

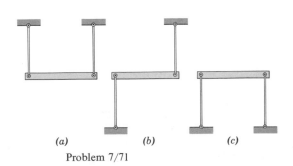

(a)   (b)   (c)

Problem 7/71

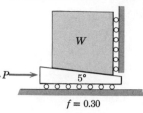

f = 0.30

Problem 7/72

7/72 Calculate the efficiency with which the 5-deg wedge elevates the weight W under the action of the horizontal force P on the wedge. The coefficient of friction between wedge and block is 0.30.

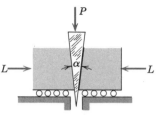

Problem 7/73

7/73 Determine the mechanical efficiency e of the device which uses a force P on the wedge to move the blocks against the forces L. The coefficient of friction is $f = 0.40$, and $\alpha = 5$ deg.

Ans. e = 0.097

7/74 Show that the uniform skylight door of Prob. 7/49 assumes a position of neutral equilibrium for all values of θ if the spring has a stiffness $k = W/(2l)$ where W is the weight of the door. The spring is unstretched when $\theta = 0$.

7/75 Check the stability for the equilibrium position cited for the ventilator door of Prob. 7/48.

Ans. Stable equilibrium for $\theta = 46°10'$

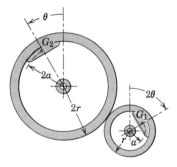

Problem 7/76

7/76 The two gears (teeth not shown) rotate in the vertical plane and carry eccentric weights w with centers of gravity at G_1 and G_2. Both weights are at the tops of their respective arcs when $\theta = 0$. Determine the stability associated with each possible equilibrium configuration of the system.

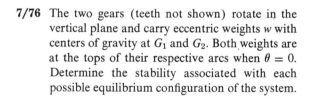

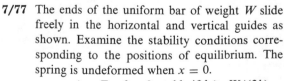

Problem 7/77

7/77 The ends of the uniform bar of weight W slide freely in the horizontal and vertical guides as shown. Examine the stability conditions corresponding to the positions of equilibrium. The spring is undeformed when $x = 0$.

Ans. For $\theta = 0$, stable if $k > W/(2b)$
    unstable if $k < W/(2b)$

For $\theta = \pm\cos^{-1}\dfrac{W}{2kb}$
    unstable if $(k > W/(2b))$

7/78 One of the critical requirements in the design of an artificial leg for an amputee is to prevent the knee joint from buckling under load when the leg is straight. As a first approximation, simulate the artificial leg by the two light links with a torsion spring at their common joint. The spring develops a torque $M = K\beta$, which is proportional to the angle of bend β at the joint. Determine the minimum value of K that will ensure stability of the knee joint for $\beta = 0$.

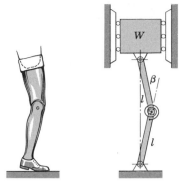
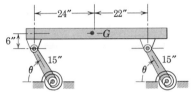

Problem 7/78

7/79 The platform weighs 80 lb with center of gravity at G and is mounted on the two equal spring-loaded cranks of negligible weight. Each torsion spring has a stiffness of 8.2 lb-in. per degree of twist and is untwisted when $\theta = 90$ deg. Determine the stable position of equilibrium.

Ans. $\theta = 22°20'$

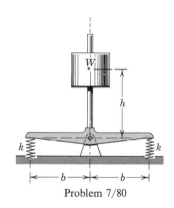

Problem 7/79

7/80 Determine the maximum height h of the weight W for which the inverted pendulum will be stable in the vertical position shown. Each of the springs has a stiffness k, and they have equal precompressions in this position. Neglect the weight of the remainder of the mechanism.

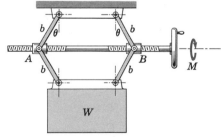

Problem 7/80

7/81 The heavy load W is raised by means of the screw and linkage shown. The screw has left- and right-handed square threads on its respective sections and spreads the threaded blocks A and B as the screw is turned by the action of an applied torque M. The lead (advancement per revolution) of each threaded section of the screw is L, its mean radius is r, and the coefficient of friction for the threads is f. Determine the torque M required to raise the load by equating the efficiency e for the screw thread (Art. 41) to the ratio of the work done by the mechanism in lifting the load to the work done on the mechanism by M.

Ans. $M = 2Wr \tan \theta \tan (\phi + \alpha)$
where $\phi = \tan^{-1} f$

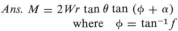

Problem 7/81

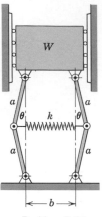

Problem 7/82

7/82 The weight W moves in smooth vertical guides and is supported by the four spring-loaded links of negligible weight. The spring of stiffness k is unstretched in the position for which $\theta = 0$. Specify the stability of the system for its equilibrium positions.

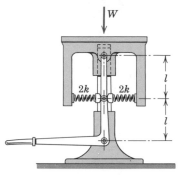

Problem 7/83

7/83 In the figure is shown a small industrial lift with a foot release. There are four identical springs, two on each side of the central shaft. The stiffness of each pair of springs is $2k$. Specify the value of k that will ensure stable equilibrium when the lift supports a load W in the position shown with no force on the pedal. The springs have an equal initial precompression and may be assumed to act in the horizontal direction at all times.

$$Ans. \ k > \frac{W}{2l}$$

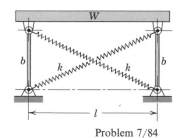

Problem 7/84

7/84 The platform of weight W is supported by equal legs and braced by the two springs as shown. If the weights of the legs and springs are negligible, determine the minimum stiffness k of each spring that will ensure stability of the platform in the position shown. Each spring has a tensile preset deflection Δ.

$$Ans. \ k_{min} = \frac{W}{2b} \frac{l^2 + b^2}{l^2}$$

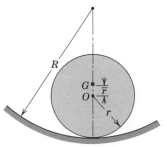

Problem 7/85

7/85 The unbalanced circular disk with center of gravity G a distance $\bar{r}$ from its center O is placed on a concave circular path of radius R. Determine the maximum value that $\bar{r}$ may have and still ensure that the disk remains stable in the bottom position shown.

$$Ans. \ \bar{r}_{max} = \frac{r}{\dfrac{R}{r} - 1}$$

7/86 The uniform garage door *AB* shown in section has a weight *W* and is equipped with two of the spring-loaded mechanisms shown, one on each side of the door. The arm *OB* has negligible weight, and the upper corner *A* of the door is free to move horizontally on a roller. The unstretched length of the spring is $r - a$, so that in the top position with $\theta = \pi$ the spring force is zero. In order to ensure smooth action of the door as it reaches the vertical closed position $\theta = 0$, it is desirable that the door be insensitive to movement in this position. Determine the required spring stiffness *k*.

$$Ans. \ k = \frac{W(r + a)}{8a^2}$$

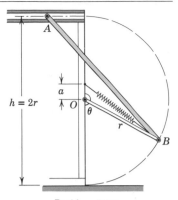

Problem 7/86

7/87 Each of the two links of the double pendulum is free to rotate in the vertical plane through 360 deg. Neglect the weights of the arms compared with *W* and describe all possible equilibrium positions indicating the type of equilibrium for each one.

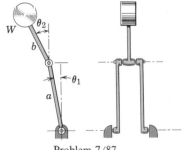

Problem 7/87

7/88 Simulate the cantilever beam of length *l* and end load *P* by a beam of three rigid segments joined with elastic hinges at *A*, *B*, and *C*. The torsional stiffness of each hinge is *K*. Determine the end deflection *y* of the beam for small elastic rotations of the segments.

$$Ans. \ y = \frac{14Pl^2}{9K}$$

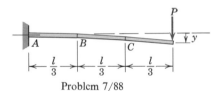

Problem 7/88

7/89 The two uniform bars, each of weight *W* and length *l*, are hinged about their lower ends and supported at their upper ends by three identical springs each of stiffness *k*. If the springs are undeformed when the bars are in the vertical positions, determine the range of values of *l* for which the bars will be stable when vertical.

$$Ans. \ \frac{W}{2k} < l < \infty$$

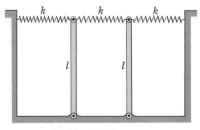

Problem 7/89

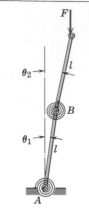

Problem 7/90

◀ 7/90 The two light bars each of length l are hinged with torsion springs at A and B. Each spring has a torsional stiffness of K lb-in. per radian and is untwisted when the bars are vertical. Examine the conditions for stability of the bars in their vertical positions under a vertical load F. (*Note:* The analysis of this problem can be used as a first approximation to the modes of column buckling.)

Ans. For equilibrium at $\theta_1 = \theta_2 = 0$ and with $x = Fl/K$,

stable for $x < 0.382$
saddle instability for $0.382 < x < 2.618$
unstable for $x > 2.618$
buckling modes:
for $x = 0.382$, $\delta\theta_2 = 1.618\,\delta\theta_1$
for $x = 2.618$, $\delta\theta_2 = -0.618\,\delta\theta_1$

8 AREA MOMENTS OF INERTIA

44 Definitions. When forces are distributed continuously over an area upon which they act, it is often necessary to calculate their moment about some desired axis. Frequently the intensity of the force is proportional to the distance of the force from the moment axis, and under these conditions an integral of the form $\int (\text{distance})^2 \, d(\text{area})$ results. This integral is known as the *moment of inertia* of the area. The integral is a function of the geometry of the area and occurs so frequently in the applications of mechanics that it is useful to develop its properties in some detail and to have these properties available for ready use when the integral arises.

Figure 94 illustrates the physical origin of these integrals. In the *a*-part of the figure the surface area $ABCD$ is subjected to a distributed pressure p whose intensity is proportional to the distance y from the axis AB. This situation was covered in Art. 29 of Chapter 5 and describes the action of liquid pressure on a plane surface. The moment about AB that is due to the pressure on the element of area dA is $py \, dA = ky^2 \, dA$. Thus the integral in question appears when the total moment $M = k \int y^2 \, dA$ is evaluated.

In Fig. 94*b* is shown the distribution of stress acting on a transverse section of a simple elastic beam bent by equal and opposite couples applied to its ends. At any section of the beam a linear distribution of force intensity or stress, given by $\sigma = ky$, is present, the stress being positive (tensile) below the axis O–O and negative (compressive) above the axis. The elemental

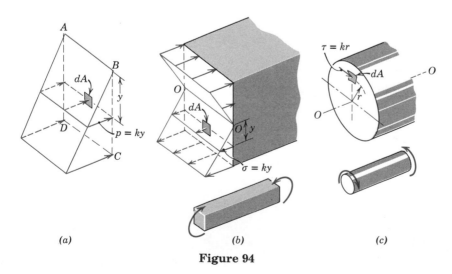

(a) (b) (c)

Figure 94

317

moment about the axis O–O is $dM = y(\sigma\,dA) = ky^2\,dA$. Thus the same integral appears when the total moment $M = k\int y^2\,dA$ is evaluated.

A third example is given in Fig. 94c, which shows a circular shaft subjected to a twist or torsional moment. Within the elastic limit of the material this moment is resisted at each cross section of the shaft by a distribution of tangential or shear stress τ which is proportional to the radial distance r from the center. Thus $\tau = kr$, and the total moment about the central axis is $M = \int r(\tau\,dA) = k\int r^2\,dA$. Here the integral differs from that in the preceding two examples in that the area is normal instead of parallel to the moment axis and in that r is a radial coordinate instead of a rectangular one.

Although the integral illustrated in the preceding examples is generally called the *moment of inertia* of the area about the axis in question, a more fitting term is the *second moment of area,* since the first moment $y\,dA$ is multiplied by the moment arm y to obtain the second moment for the element dA. The word *inertia* appears in the terminology by reason of the similarity between the mathematical form of the integrals for second moments of areas and those for the resultant moments of the so-called inertia forces in the case of rotating bodies. The moment of inertia of an area is a purely mathematical property of the area and in itself has no physical significance.

Consider the area A in the x-y plane, Fig. 95. The moments of inertia of the element dA about the x- and y-axes are, by definition, $dI_x = y^2\,dA$ and $dI_y = x^2\,dA$, respectively. Therefore the moments of inertia of A about the same axes are

$$I_x = \int y^2\,dA$$

$$I_y = \int x^2\,dA$$

(65)

where the integration covers the entire area. The moment of inertia of dA about the pole O (z-axis) is, by similar definition, $dJ_z = r^2\,dA$, and the moment of inertia of the entire area about O is

$$J_z = \int r^2\,dA$$

(66)

The expressions defined by Eqs. 65 are known as *rectangular* moments of inertia, whereas the expression of Eq. 66 is known as the *polar* moment of

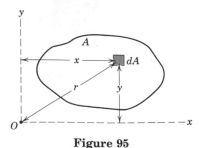

Figure 95

inertia. Since $x^2 + y^2 = r^2$, it is clear that

$$\blacktriangleright \qquad\qquad J_z = I_x + I_y \qquad\qquad (67)$$

A polar moment of inertia for an area whose boundaries are more simply described in rectangular coordinates than in polar coordinates is easily calculated with the aid of Eq. 67.

It should be noted that the moment of inertia of an element involves the square of the distance from the inertia axis to the element. An element whose coordinate is negative contributes as much to the moment of inertia as does an equal element with a positive coordinate of the same magnitude. Consequently the area moment of inertia about any axis is always a positive quantity. In contrast, the first moment of the area, which was involved in the computations of centroids, could be either positive or negative.

The dimensions of moments of inertia of areas are clearly L^4, where L stands for the dimension of length. Thus the units for area moments of inertia are expressed as quartic inches (in.⁴) or quartic feet (ft⁴).

The choice of the coordinates to use for the calculation of moments of inertia is important. Rectangular coordinates should be used for shapes whose boundaries are most easily expressed in these coordinates. Polar coordinates will usually simplify problems involving boundaries that are easily described in r and θ. The choice of an element of area which simplifies the integration as much as possible is also important. These considerations are quite analogous to those discussed and illustrated in Chapter 5 for the calculation of centroids.

Radius of Gyration. The moment of inertia of an area is a measure of the distribution of the area from the axis in question. Assume all the area A, Fig. 96, to be concentrated into a strip of negligible thickness at a distance k_x from the x-axis, so that the product $k_x^2 A$ equals the moment of inertia about the axis. The distance k_x, called the *radius of gyration,* is then a measure of the distribution of area from the inertia axis. By definition, then, for any axis

$$\blacktriangleright \qquad\qquad I = k^2 A \qquad \text{or} \qquad k = \sqrt{\frac{I}{A}} \qquad\qquad (68)$$

When this definition is substituted in each of the three terms in Eq. 67, there results

$$\blacktriangleright \qquad\qquad k_z{}^2 = k_x{}^2 + k_y{}^2 \qquad\qquad (69)$$

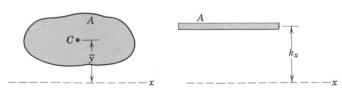

Figure 96

Thus the square of the radius of gyration about a polar axis equals the sum of the squares of the radii of gyration about the two corresponding rectangular axes.

It is imperative that there be no confusion between the coordinate $\bar{y}$ to the centroid C of the area and the radius of gyration k. The square of the centroidal distance, Fig. 96, is $\bar{y}^2$ and is the square of the mean value of the distances y from the elements dA to the axis. The quantity k_x^2, on the other hand, is the mean of the squares of these distances. The moment of inertia is *not* equal to $A\bar{y}^2$, since the square of the mean is less than the mean of the squares.

Transfer of Axes. The moment of inertia of an area about a noncentroidal axis may be easily expressed in terms of the moment of inertia about a parallel centroidal axis. In Fig. 97 the x_0-y_0 axes pass through the centroid C of the area. Let it be desired to determine the moments of inertia of the area about the parallel x-y axes. By definition the moment of inertia of the element dA about the x-axis is

$$dI_x = (y_o + d_x)^2 \, dA$$

Expanding and integrating give

$$I_x = \int y_o^2 \, dA + 2d_x \int y_o \, dA + d_x^2 \int dA$$

The first integral is the moment of inertia $\bar{I}_x$ about the centroidal x_0-axis. The second integral is zero, since $\int y_o \, dA = A\bar{y}_0$ and $\bar{y}_0$ is automatically zero. The third term is simply Ad_x^2. Thus the expression for I_x and the similar expression for I_y become

$$\begin{aligned} I_x &= \bar{I}_x + Ad_x^2 \\ I_y &= \bar{I}_y + Ad_y^2 \end{aligned} \tag{70}$$

By Eq. 67 the sum of these two equations gives

$$J_z = \bar{J}_z + Ad^2 \tag{70a}$$

Equations 70 and 70a are the so-called *parallel-axis theorems.* Two points in particular should be noted. First, the axes between which the transfer is made must be parallel, and, second, one of the axes must pass through the centroid of the area.

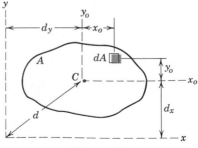

Figure 97

If a transfer is desired between two parallel axes neither one of which passes through the centroid, it is first necessary to transfer from one axis to the parallel centroidal axis and then to transfer from the centroidal axis to the second axis.

The parallel-axis theorems also hold for radii of gyration. With substitution of the definition of k into Eqs. 70, the transfer relation becomes

▶ $$k^2 = \bar{k}^2 + d^2 \tag{70b}$$

where $\bar{k}$ is the radius of gyration about a centroidal axis parallel to the axis about which k applies and d is the distance between the two axes. The axes may be either in the plane or normal to the plane of the area.

A summary of the moment of inertia relations for some of the common plane figures is given in Table C4, Appendix C. The sample problems that follow illustrate the calculation of area moments of inertia in detail. Problems 8/1, 8/2, and 8/3 develop the results for the moments of inertia of three common shapes that find repeated use in mechanics. These results should be understood thoroughly and remembered.

Sample Problems

8/1　Determine the moments of inertia of the rectangular area about the centroidal x_o-y_o axes, the centroidal polar axis z_0 through C, the x-axis, and the polar axis z through O.

　　Solution. For the calculation of the moment of inertia $\bar{I}_x$ about the x_o-axis a horizontal strip of area $b\,dy$ is chosen so that all elements of the strip have the same y-coordinate. Thus

$$[I_x = \int y^2\, dA] \qquad\qquad \bar{I}_x = \int_{-h/2}^{h/2} y^2 b\, dy = \tfrac{1}{12}bh^3 \qquad\qquad\qquad Ans.$$

By interchanging symbols the moment of inertia about the centroidal y_o-axis is

$$\bar{I}_y = \tfrac{1}{12}hb^3 \qquad\qquad\qquad Ans.$$

The centroidal polar moment of inertia is

$$[J_z = I_x + I_y] \qquad\qquad J_z = \tfrac{1}{12}(bh^3 + hb^3) = \tfrac{1}{12}A(b^2 + h^2) \qquad\qquad\qquad Ans.$$

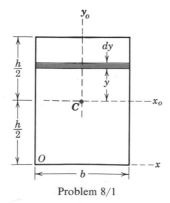

Problem 8/1

By the parallel-axis theorem the moment of inertia about the x-axis is

$$[I_x = \bar{I}_x + Ad_x^2] \qquad I_x = \tfrac{1}{12}bh^3 + bh\left(\frac{h}{2}\right)^2 = \tfrac{1}{3}bh^3 = \tfrac{1}{3}Ah^2 \qquad \textit{Ans.}$$

The polar moment of inertia about O may also be obtained by the parallel-axis theorem. Thus

$$[J_z = \bar{J}_z + Ad^2] \qquad J_z = \tfrac{1}{12}A(b^2 + h^2) + A\left[\left(\frac{b}{2}\right)^2 + \left(\frac{h}{2}\right)^2\right]$$

$$J_z = \tfrac{1}{3}A(b^2 + h^2) \qquad \textit{Ans.}$$

8/2 Calculate the moments of inertia of the area of the circle about a diametral axis and about the polar axis through the center. Specify the radii of gyration.

   Solution. An element of area in the form of a circular ring, shown in the *a*-part of the figure, may be used for the calculation of the moment of inertia about the polar z-axis through O since all elements of the ring are equidistant from O. The elemental area is $dA = 2\pi r_o\, dr_o$, and thus

$$\left[J_z = \int r^2\, dA\right] \qquad J_z = \int_0^r r_o^2(2\pi r_o\, dr_o) = \frac{\pi r^4}{2} = \tfrac{1}{2}Ar^2 \qquad \textit{Ans.}$$

The polar radius of gyration is

$$\left[k = \sqrt{\frac{J}{A}}\right] \qquad k_z = \frac{r}{\sqrt{2}} \qquad \textit{Ans.}$$

   By symmetry $I_x = I_y$, so that from Eq. 67

$$[J_z = I_x + I_y] \qquad I_x = \tfrac{1}{2}J_z = \frac{\pi r^4}{4} = \tfrac{1}{4}Ar^2 \qquad \textit{Ans.}$$

The radius of gyration about the diametral axis is

$$\left[k = \sqrt{\frac{I}{A}}\right] \qquad k_x = \frac{r}{2} \qquad \textit{Ans.}$$

   The foregoing determination of I_x is the simplest possible. The result may also be obtained by direct integration, using the element of area $dA = r_o\, dr_o\, d\theta$ shown in the

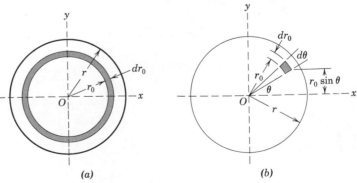

(a)   (b)

Problem 8/2

b-part of the figure. By definition

$$[I_x = \int y^2 \, dA] \qquad I_x = \int_0^{2\pi} \int_0^r (r_0 \sin \theta)^2 r_0 \, dr_0 \, d\theta$$

$$= \int_0^{2\pi} \frac{r^4 \sin^2 \theta}{4} \, d\theta = \frac{r^4}{4} \frac{1}{2} \left[\theta - \frac{\sin 2\theta}{2} \right]_0^{2\pi} = \frac{\pi r^4}{4} \qquad Ans.$$

8/3　Determine the moments of inertia of the triangular area about its base and about parallel axes through its centroid and vertex.

　　Solution. A strip of area parallel to the base is selected as shown in the figure and it has the area $dA = x \, dy = [(h - y)b/h] \, dy$. By definition

$$[I_x = \int y^2 \, dA] \qquad I_x = \int_0^h y^2 \frac{h - y}{h} b \, dy = b \left[\frac{y^3}{3} - \frac{y^4}{4h} \right]_0^h = \frac{bh^3}{12} \qquad Ans.$$

By the parallel-axis theorem the moment of inertia I about an axis through the centroid, a distance $h/3$ above the x-axis, is

$$[\bar{I} = I - Ad^2] \qquad \bar{I} = \frac{bh^3}{12} - \left(\frac{bh}{2} \right) \left(\frac{h}{3} \right)^2 = \frac{bh^3}{36} \qquad Ans.$$

A transfer from the centroidal axis to the x'-axis through the vertex gives

$$[I = \bar{I} + Ad^2] \qquad I_{x'} = \frac{bh^3}{36} + \left(\frac{bh}{2} \right) \left(\frac{2h}{3} \right)^2 = \frac{bh^3}{4} \qquad Ans.$$

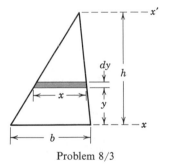

Problem 8/3

8/4　Determine the moment of inertia about the x-axis of the semicircular area shown.

　　Solution. The moment of inertia of the semicircular area about the x'-axis is one half of that for a complete circle about the same axis. Thus from the results of Prob. 8/2

$$I_{x'} = \frac{1}{2} \frac{\pi r^4}{4} = \frac{2^4 \pi}{8} = 2\pi \text{ in.}^4$$

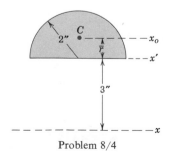

Problem 8/4

The moment of inertia $\bar{I}$ about the parallel centroidal axis x_0 is obtained next. Transfer is made through the distance $\bar{r} = 4r/3\pi = (4)(2)/3\pi = 8/3\pi$ in. by the parallel-axis theorem. Hence

$$[\bar{I} = I - A\bar{d}^2] \qquad \bar{I} = 2\pi - \left(\frac{2^2\pi}{2}\right)\left(\frac{8}{3\pi}\right)^2 = 1.755 \text{ in.}^4$$

Finally, transfer is made from the centroidal x_0-axis to the x-axis, which gives

$$[I = \bar{I} + A\bar{d}^2] \qquad I_x = 1.755 + \left(\frac{2^2\pi}{2}\right)\left(3 + \frac{8}{3\pi}\right)^2$$

$$= 1.755 + 93.08 = 94.8 \text{ in.}^4 \qquad\qquad Ans.$$

8/5 Calculate the moment of inertia about the x-axis of the area enclosed between the y-axis and the circular arcs of radius a whose centers are at O and A.

Solution. The choice of a vertical differential strip of area permits one integration to cover the entire area. A horizontal strip would require two integrations with respect to y by virtue of the discontinuity. The moment of inertia of the strip about the x-axis is that of a strip of height y_2 minus that of a strip of height y_1. Thus, from the results of Sample Prob. 8/1,

$$dI_x = \tfrac{1}{3}(y_2 \, dx)y_2{}^2 - \tfrac{1}{3}(y_1 \, dx)y_1{}^2 = \tfrac{1}{3}(y_2{}^3 - y_1{}^3) \, dx$$

The values of y are obtained from the equations of the two curves, which are $x^2 + y_2{}^2 = a^2$ and $(x - a)^2 + y_1{}^2 = a^2$ and which give $y_2 = \sqrt{a^2 - x^2}$ and $y_1 = \sqrt{a^2 - (x - a)^2}$. Thus

$$I_x = \tfrac{1}{3}\int_0^{a/2} \{(a^2 - x^2)\sqrt{a^2 - x^2} - [a^2 - (x - a)^2]\sqrt{a^2 - (x - a)^2}\} \, dx$$

Solution of the two equations gives the x-coordinate of the intersection of the two curves which, by inspection, is $a/2$. Evaluation of the integrals gives

$$\int_0^{a/2} a^2\sqrt{a^2 - x^2} \, dx = \frac{a^4}{4}\left(\frac{\sqrt{3}}{2} + \frac{\pi}{3}\right)$$

$$-\int_0^{a/2} x^2\sqrt{a^2 - x^2} \, dx = \frac{a^4}{16}\left(\frac{\sqrt{3}}{4} - \frac{\pi}{3}\right)$$

$$-\int_0^{a/2} a^2\sqrt{a^2 - (x - a)^2} \, dx = \frac{a^4}{4}\left(\frac{\sqrt{3}}{2} - \frac{2\pi}{3}\right)$$

$$\int_0^{a/2} (x - a)^2\sqrt{a^2 - (x - a)^2} \, dx = \frac{a^4}{8}\left(\frac{\sqrt{3}}{8} + \frac{\pi}{3}\right)$$

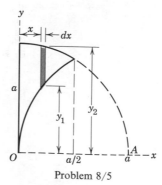

Problem 8/5

Collection of the integrals with the factor of $\frac{1}{3}$ gives

$$I_x = \frac{a^4}{96}(9\sqrt{3} - 2\pi) = 0.0969a^4 \qquad \textit{Ans.}$$

Problems

8/6 Calculate the moment of inertia of the rectangular area about the x-axis and find the polar moment of inertia about point O.

$\qquad$ *Ans.* $I_x = 416$ in.^{4}, $\quad J_O = 704$ in.^{4}

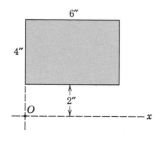

Problem 8/6

8/7 The narrow rectangular strip has an area of 6 in.^{2}, and its moment of inertia about the y-axis is 170 in.^{4} Obtain a close approximation to the radius of gyration about point O.

$\qquad$ *Ans.* $k_O = 7.30$ in.

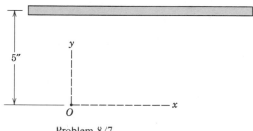

Problem 8/7

8/8 From the results of Sample Prob. 8/1, state without calculation the moment of inertia of the area of the parallelogram about the x-axis through its base and about a parallel axis through its centroid.

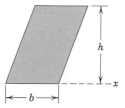

Problem 8/8

8/9 Determine the polar moment of inertia of the semicircular area about point A.

$\qquad$ *Ans.* $J_A = \frac{3}{4}\pi r^4$

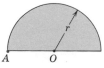

Problem 8/9

8/10 Calculate the polar radius of gyration about the centroid of the quarter-circular area for $r = 4$ in.

8/11 For the quarter-circular area shown with Prob. 8/10, determine the polar radius of gyration about point A. $\qquad$ *Ans.* $k_A = 0.807r$

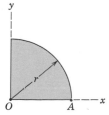

Problem 8/10

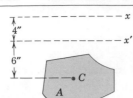

Problem 8/12

8/12 The moments of inertia of the area A about the x- and x'-axes differ by 3650 in.^{4} Compute the shaded area A, which has its centroid at C.

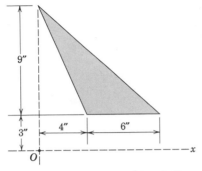

Problem 8/13

8/13 Calculate the moment of inertia of the triangular area about the x-axis. *Ans. $I_x = 1093$ in.^{4}*

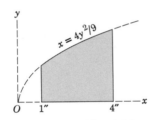

Problem 8/14

8/14 Calculate the moment of inertia of the shaded area about the x-axis. *Ans. $I_x = 13.95$ in.^{4}*

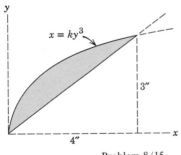

Problem 8/15

8/15 Calculate the moment of inertia of the shaded area about the x-axis.

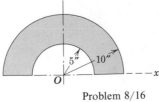

Problem 8/16

8/16 Obtain the polar moment of inertia of the area of the semicircular ring about point O by direct integration and use the result to find the moment of inertia about the x-axis.

  Ans. $I_x = 3682$ in.^{4}

8/17 Calculate by direct integration the moment of inertia of the shaded area about the x-axis. Solve, first, by using a horizontal strip having differential area and, second, by using a vertical strip of differential area.

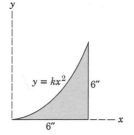

$y = kx^2$ 6″

6″

Problem 8/17

8/18 Determine the polar radius of gyration of the area of the equilateral triangle of side b about its centroid C.

$$Ans.\ \overline{k} = \frac{b}{2\sqrt{3}}$$

Problem 8/18

8/19 Approximate the result for the moment of inertia about the x-axis of the shaded area in Prob. 8/16 by dividing the area into five horizontal strips of equal width. Treat the moment of inertia of each strip as its area times the square of the distance from its center to the axis.

8/20 Determine the moment of inertia of the irregular area of Prob. 5/51, repeated here, about the x-axis. *Ans.* $I_x = 1230$ in.[4]

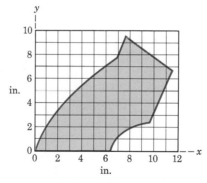

Problem 8/20

8/21 Show that the moment of inertia of the area of the square about any axis x' through its center is the same as that about a central axis x parallel to a side.

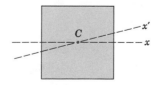

Problem 8/21

8/22 The area of a circular ring of inside radius r and outside radius $r + \Delta r$ is approximately equal to the circumference at the mean radius times the thickness Δr. The polar moment of inertia of the ring may be approximated by multiplying this area by the square of the mean radius. What per cent error is involved if $\Delta r = r/10$?

Ans. Error = 0.226%

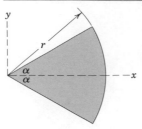

Problem 8/23

8/23 Determine the moments of inertia of the area of the circular sector about the x- and y-axes.

$$\text{Ans. } I_x = \frac{r^4}{4}\left(\alpha - \frac{\sin 2\alpha}{2}\right)$$

$$I_y = \frac{r^4}{4}\left(\alpha + \frac{\sin 2\alpha}{2}\right)$$

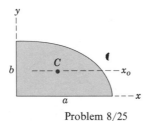

Problem 8/24

8/24 Determine the polar moment of inertia of the shaded area about the origin O.

$$\text{Ans. } J_O = \frac{2ab}{\pi}\left[\frac{2b^2}{9} + a^2\left(1 - \frac{8}{\pi^2}\right)\right]$$

8/25 Determine the moment of inertia of the area of the elliptical quadrant about the centroidal x_o-axis. Also find the polar moment of inertia about the centroid C.

$$\text{Ans. } \bar{J} = \left(\frac{\pi}{16} - \frac{4}{9\pi}\right)(a^2 + b^2)ab$$

Problem 8/25

45 Composite Areas. It is frequently necessary to calculate the moments of inertia for an area that is made up of a number of distinct parts of simple and calculable geometric shape. Since a moment of inertia is the integral or sum of the products of distance squared times element of area, it follows that the moment of inertia of a composite area about a particular axis is simply the sum of the moments of inertia of its individual parts about the same axis. It is often convenient to regard a composite area as being composed of positive and negative parts. The moment of inertia of a negative area is a negative quantity.

When the section is composed of a large number of parts, it is convenient to tabulate the results for the parts in terms of the area A, centroidal moment of inertia $\bar{I}$, the distance d from the centroidal axis to the axis about which the moment of inertia of the entire section is being computed, and the product Ad^2. For any one of the parts the desired moment of inertia is $\bar{I} + Ad^2$, and thus for the entire section the desired moment of inertia may be expressed as $I = \Sigma\bar{I} + \Sigma Ad^2$.

Sample Problem

8/26   Calculate the moment of inertia and radius of gyration about the *x*-axis for the shaded area shown.

Solution. The composite area is composed of the positive area of the rectangle (1) and the negative areas of the quarter circle (2) and triangle (3). For the rectangle the moment of inertia about the *x*-axis, from Sample Prob. 8/1, is

$$I_x = \tfrac{1}{3}Ah^2 = \tfrac{1}{3}(8)(6)(6^2) = 576 \text{ in.}^4$$

From Sample Prob. 8/2, the moment of inertia of the negative quarter-circular area about its base axis *x'* is

$$I_{x'} = -\frac{1}{4}\left(\frac{\pi r^4}{4}\right) = -\frac{\pi}{16}(3^4) = -15.90 \text{ in.}^4$$

Transfer of this result through the distance $\bar{r} = 4r/3\pi = 4(3)/3\pi = 1.273$ in. by the transfer-of-axis-theorem gives for the centroidal moment of inertia of part (2)

$$[\bar{I} = I - Ad^2] \qquad \bar{I}_x = -15.90 - \left[-\frac{\pi(3^2)}{4}(1.273)^2\right] = -4.45 \text{ in.}^4$$

The moment of inertia of the quarter-circular part about the *x*-axis is now

$$[I = \bar{I} + Ad^2] \qquad I_x = -4.45 + \left[-\frac{\pi(3^2)}{4}\right](6 - 1.273)^2 = -162.4 \text{ in.}^4$$

Finally, the moment of inertia of the negative triangular area (3) about its base, from Sample Prob. 8/3, is

$$I_x = -\tfrac{1}{12}bh^3 = -\tfrac{1}{12}(4)(3^3) = -9 \text{ in.}^4$$

The total moment of inertia about the *x*-axis of the composite area is, consequently,

$$I_x = 576 - 162.4 - 9 = 404.6 \text{ in.}^4 \qquad\qquad Ans.$$

The net area of the figure is $A = 6(8) - \tfrac{1}{4}\pi(3^2) - \tfrac{1}{2}(4)(3) = 34.93$ in.^{2} so that the radius of gyration about the *x*-axis is

$$k_x = \sqrt{I_x/A} = \sqrt{404.6/34.93} = 3.40 \text{ in.} \qquad\qquad Ans.$$

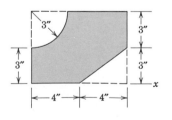
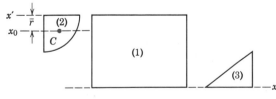

Problem 8/26

Problems

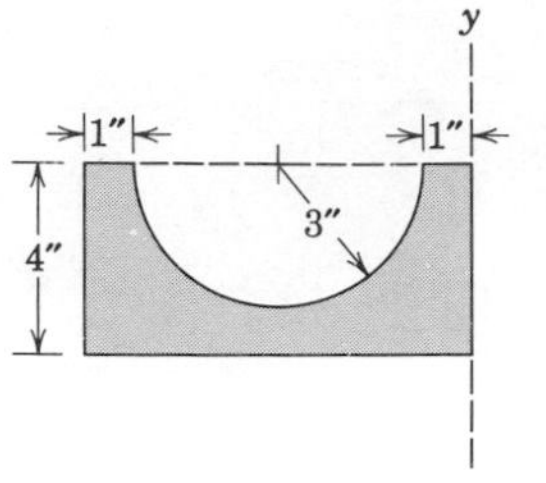

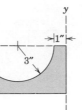

Problem 8/27

8/27 Calculate the moment of inertia of the shaded area about the y-axis.   *Ans.* $I_y = 425$ in.[4]

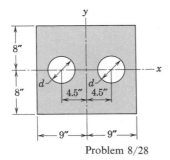

Problem 8/28

8/28 Calculate the diameter d of each of the two holes that will make the moments of inertia of the shaded area about the x- and y-axes equal.

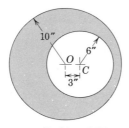

Problem 8/29

8/29 Calculate the polar moment of inertia of the shaded area about point O.
    Ans. $J_O = 12,654$ in.[4]

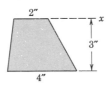

Problem 8/30

8/30 Compute the moment of inertia of the trapezoidal area about the x-axis.

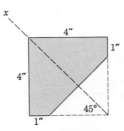

Problem 8/31

8/31 Find the moment of inertia of the shaded area about the 45-deg x-axis of symmetry.
    Ans. $I_x = 17.96$ in.[4]

8/32 Determine the moments of inertia of the *Z*-section about its centroidal x_0- and y_0-axes.

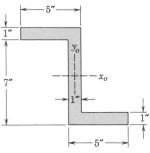

Problem 8/32

8/33 Calculate the moment of inertia of the cross section of the beam about its centroidal x_0-axis.

Ans. $\bar{I}_x = 80.8$ in.^{4}

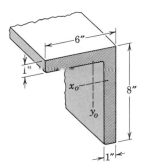

Problem 8/33

8/34 Compute the polar radius of gyration of the U-section about its centroid *C*.

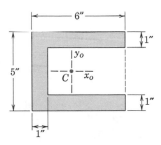

Problem 8/34

8/35 Compute the polar radius of gyration of the shaded area about point *O*.   Ans. $k_O = 4.45$ in.

Problem 8/35

8/36 Develop a formula for the moment of inertia of the regular hexagonal area of side *b* about its central *x*-axis.

8/37 Determine the expression for the radius of gyration of the hexagonal area of Prob. 8/36 about a polar axis through its center *O*.

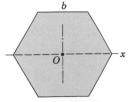

Problem 8/36

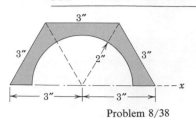

Problem 8/38

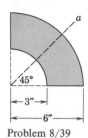

Problem 8/39

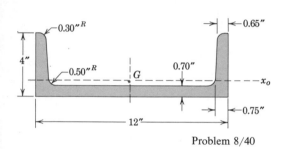

Problem 8/40

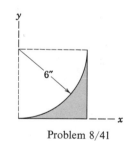

Problem 8/41

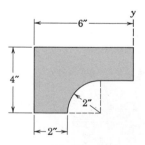

Problem 8/42

8/38 Calculate the moment of inertia of the shaded area about the x-axis. *Ans.* $I_x = 15.64$ in.^{4}

8/39 Calculate the moment of inertia of the quarter-circular ring about the a-axis of symmetry. (*Hint:* Use the results of Prob. 8/23.)
Ans. $I_a = 86.7$ in.^{4}

8/40 Calculate the moment of inertia of the standard 12×4 in. channel section about the centroidal x_o-axis. Neglect the fillets and radii and compare with the handbook value of $I_x = 16.0$ in.^{4}

8/41 Calculate the moment of inertia of the shaded area about the x-axis. *Ans.* $I_x = 23.7$ in.^{4}

8/42 Compute the moment of inertia of the area about the y-axis. *Ans.* $I_y = 256$ in.^{4}

8/43 Compute the moment of inertia of the area about its base. (*Hint:* Use the results of Prob. 8/23.)   *Ans.* $I_x = 933$ in.⁴

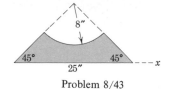

Problem 8/43

◄**8/44** For the H-beam section, determine the flange width b that will make the moments of inertia about the central x- and y-axes equal.   *Ans.* $b = 16.1$ in.

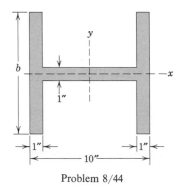

Problem 8/44

46   Products of Inertia and Rotation of Axes.

In certain problems involving unsymmetrical cross sections and in the calculation of moments of inertia about rotated axes, an expression occurs which has the form

$$dI_{xy} = xy\,dA$$

$$I_{xy} = \int xy\,dA \tag{71}$$

where x and y are the coordinates of the element of area dA. The quantity I_{xy} is called the *product of inertia* of the area A with respect to the x-y axes. Unlike moments of inertia, which are always positive for positive areas, the product of inertia may be positive or negative.

The product of inertia is zero whenever either one of the reference axes is an axis of symmetry, such as the x-axis for the area of Fig. 98. Here it is seen that the sum of the terms $x(-y)\,dA$ and $x(+y)\,dA$ due to symmetrically placed elements vanishes. Since the entire area may be considered as composed of pairs of such elements, it follows that the product of inertia for the entire area is zero.

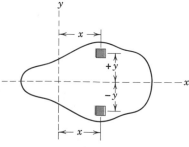

Figure 98

A transfer-of-axis theorem similar to that for moments of inertia also exists for products of inertia. By definition the product of inertia of the area A in Fig. 97 with respect to the x- and y-axes in terms of the coordinates x_o, y_o to the centroidal axes is

$$I_{xy} = \int (x_o + d_y)(y_o + d_x)\, dA$$

$$= \int x_o y_o\, dA + d_x \int x_o\, dA + d_y \int y_o\, dA + d_x d_y \int dA$$

$$I_{xy} = \bar{I}_{xy} + d_x d_y A \tag{72}$$

where $\bar{I}_{xy}$ is the product of inertia with respect to the centroidal x_o-y_o axes which are parallel to the x-y axes.

The product of inertia finds use when it is necessary to calculate the moment of inertia of an area about inclined axes. This consideration leads directly to the important problem of determining the axes about which the moment of inertia is a maximum and a minimum.

In Fig. 99 the moments of inertia of the area about the x'- and y'-axes are

$$I_{x'} = \int y'^2\, dA = \int (y \cos \theta - x \sin \theta)^2\, dA$$

$$I_{y'} = \int x'^2\, dA = \int (y \sin \theta + x \cos \theta)^2\, dA$$

where x' and y' have been replaced by their equivalent expressions as seen from the geometry of the figure.

Expanding and substituting the trigonometric identities,

$$\sin^2 \theta = \frac{1 - \cos 2\theta}{2} \qquad \cos^2 \theta = \frac{1 + \cos 2\theta}{2}$$

and the defining relations for I_x, I_y, I_{xy} give

$$I_{x'} = \frac{I_x + I_y}{2} + \frac{I_x - I_y}{2} \cos 2\theta - I_{xy} \sin 2\theta$$

$$I_{y'} = \frac{I_x + I_y}{2} - \frac{I_x - I_y}{2} \cos 2\theta + I_{xy} \sin 2\theta \tag{73}$$

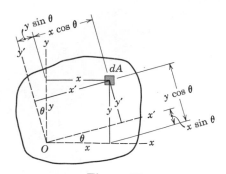

Figure 99

In a similar manner,

$$I_{x'y'} = \int x'y' \, dA = \int (y \cos \theta - x \sin \theta)(y \sin \theta + x \cos \theta) \, dA$$

Expanding and substituting the trigonometric identities

$$\sin \theta \cos \theta = \tfrac{1}{2} \sin 2\theta, \qquad \cos^2 \theta - \sin^2 \theta = \cos 2\theta$$

and the defining relations for I_x, I_y, I_{xy} give

$$I_{x'y'} = \frac{I_x - I_y}{2} \sin 2\theta + I_{xy} \cos 2\theta \qquad (73a)$$

Adding Eqs. 73 gives $I_{x'} + I_{y'} = I_x + I_y = J_z$, the polar moment of inertia about O, which checks the results of Eq. 67.

The angle which makes $I_{x'}$ and $I_{y'}$ a maximum or a minimum may be determined by setting the derivative of either $I_{x'}$ or $I_{y'}$ with respect to θ equal to zero. Thus

$$\frac{dI_{x'}}{d\theta} = (I_y - I_x) \sin 2\theta - 2I_{xy} \cos 2\theta = 0$$

Denoting this critical angle by α gives

$$\tan 2\alpha = \frac{2I_{xy}}{I_y - I_x} \qquad (74)$$

Equation 74 gives two values for 2α which differ by π since $\tan 2\alpha = \tan(2\alpha + \pi)$. Consequently the two solutions for α will differ by $\pi/2$. One value defines the axis of maximum moment of inertia, and the other value defines the axis of minimum moment of inertia. These two rectangular axes are known as the *principal axes of inertia.*

Substitution of Eq. 74 for the critical value of 2θ in Eq. 73a shows that the product of inertia is zero for the principal axes of inertia. Substitution of $\sin 2\alpha$ and $\cos 2\alpha$, obtained from Eq. 74, for $\sin 2\theta$ and $\cos 2\theta$ in Eqs. 73 gives the magnitudes of the principal moments of inertia as

$$I_{\max} = \frac{I_x + I_y}{2} + \frac{1}{2} \sqrt{(I_x - I_y)^2 + 4I_{xy}^2}$$
$$\tag{75}$$
$$I_{\min} = \frac{I_x + I_y}{2} - \frac{1}{2} \sqrt{(I_x - I_y)^2 + 4I_{xy}^2}$$

The relations in Eqs. 73, 73a, 74, and 75 may be represented graphically by a diagram known as Mohr's circle. For given values of I_x, I_y, and I_{xy} the corresponding values of $I_{x'}$, $I_{y'}$, and $I_{x'y'}$ may be determined from the diagram for any desired angle θ. A horizontal axis for the measurement of moments of inertia and a vertical axis for the measurement of products of inertia are first selected, Fig. 100. Next, point A, which has the coordinates (I_x, I_{xy}), and point B, which has the coordinates $(I_y, -I_{xy})$, are located. A circle is drawn with these two points as the extremities of a diameter. The angle from the radius OA to the horizontal axis is 2α or twice the angle from the x-axis of the area in question to the axis of maximum moment of inertia. The angle on the diagram and the angle on the area are both measured in the same sense as shown. The coordinates of any point C are $(I_{x'}, I_{x'y'})$, and those of the corresponding point D are $(I_{y'}, -I_{x'y'})$. Also the angle between OA and OC is 2θ or twice the angle from the x-axis to the x'-axis. Again both angles are measured in the same sense as shown. It may be verified from the trigonometry of the circle that Eqs. 73, 73a, and 74 agree with the statements made.

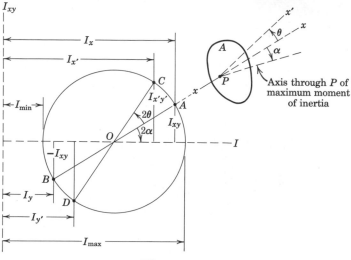

Figure 100

Sample Problems

8/45   Determine the product of inertia I_{xy} for the area under the parabola shown.

Solution. The equation of the curve becomes $x = ay^2/b^2$. The product of inertia for the element $dA = dx\,dy$ is $dI_{xy} = xy\,dx\,dy$ and for the entire area is

$$I_{xy} = \int_0^b \int_{ay^2/b^2}^a xy\,dx\,dy = \int_0^b \frac{1}{2}\left(a^2 - \frac{a^2y^4}{b^4}\right)y\,dy = \tfrac{1}{6}a^2b^2 \qquad Ans.$$

Problem 8/45

8/46   Determine the product of inertia of the semicircular area with respect to the x-y axes.

Solution. The transfer-of-axis theorem, Eq. 72, may be used to write

$$[I_{xy} = \bar{I}_{xy} + d_x d_y A] \qquad I_{xy} = 0 + \left(-\frac{4r}{3\pi}\right)(r)\left(\frac{\pi r^2}{2}\right) = \frac{2r^4}{3} \qquad\qquad Ans.$$

where the x- and y-coordinates of the centroid C are $d_x = +r$ and $d_y = -4r/3\pi$. Since one of the centroidal axes is an axis of symmetry, $\bar{I}_{xy} = 0$.

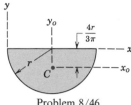

Problem 8/46

8/47   Locate the principal centroidal axes of inertia and determine the corresponding maximum and minimum moments of inertia for the angle section.

    Solution. The centroid C is easily located as shown. The product of inertia for each rectangle about its own centroidal axes parallel to the x- and y-axes is zero by symmetry. Thus the product of inertia for part A is

$$[I_{xy} = \bar{I}_{xy} + d_x d_y A] \qquad I_{xy} = 0 + (-\tfrac{5}{4})(+\tfrac{3}{4})(4) = -3.75 \text{ in.}^4$$

where $d_x = -(\tfrac{3}{4} + \tfrac{1}{2}) = -\tfrac{5}{4}$ in.   and $d_y = +(2 - 1 - \tfrac{1}{4}) = \tfrac{3}{4}$ in.

Likewise for B,

$$[I_{xy} = \bar{I}_{xy} + d_x d_y A] \qquad I_{xy} = 0 + (\tfrac{5}{4})(-\tfrac{3}{4})(4) = -3.75 \text{ in.}^4$$

where $d_x = +(2 - \tfrac{3}{4}) = \tfrac{5}{4}$ in.   and $d_y = -(\tfrac{1}{2} + \tfrac{1}{4}) = -\tfrac{3}{4}$ in.

For the complete angle

$$I_{xy} = -3.75 - 3.75 = -7.50 \text{ in.}^4$$

    The moments of inertia for part A are

$$[I = \bar{I} + Ad^2] \qquad I_x = \tfrac{1}{12}(4)(1^3) + (\tfrac{5}{4})^2(4) = 6.583 \text{ in.}^4$$

$$I_y = \tfrac{1}{12}(1)(4^3) + (\tfrac{3}{4})^2(4) = 7.583 \text{ in.}^4$$

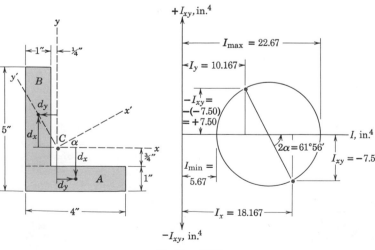

Problem 8/47

In similar manner the moments of inertia for part B are $I_x = 11.583$ in.^{4}, $I_y = 2.583$ in.^{4}
Thus for the entire section

$$I_x = 6.583 + 11.583 = 18.167 \text{ in.}^4$$

$$I_y = 7.583 + 2.583 = 10.167 \text{ in.}^4$$

The inclination of the principal axes of inertia is given by Eq. 74. Therefore

$$\left[\tan 2\alpha = \frac{2I_{xy}}{I_y - I_x}\right] \qquad \tan 2\alpha = \frac{-2(7.50)}{10.167 - 18.167} = 1.875$$

$$2\alpha = 61°56' \qquad \alpha = 30°58' \qquad\qquad\qquad Ans.$$

$$I_{\max} = I_{x'} = \frac{18.167 + 10.167}{2} + \frac{18.167 - 10.167}{2}(0.4705) + (7.50)(0.8824)$$

$$= 22.67 \text{ in.}^4$$

$$I_{\min} = I_{y'} = \frac{18.167 + 10.167}{2} - \frac{18.167 - 10.167}{2}(0.4705) - (7.50)(0.8824)$$

$$= 5.67 \text{ in.}^4 \qquad\qquad\qquad Ans.$$

These results may also be obtained directly from Eqs. 75 or graphically by construction
of the Mohr circle as shown to the right of the angle in the figure.

Problems

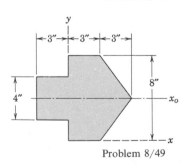

Problem 8/48

8/48 Calculate the product of inertia for the right-
triangular area with respect to the x-y axes.
Ans. $I_{xy} = h^2b^2/24$

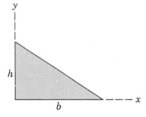

Problem 8/49

8/49 Take advantage of the symmetry of the shaded
area about the x_0-axis and compute the product
of inertia with respect to the x-y axes.
Ans. $I_{xy} = 264$ in.^{4}

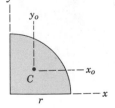

Problem 8/50

8/50 Obtain the product of inertia of the quarter-
circular area with respect to the x-y axes and use
this result to obtain the product of inertia with
respect to the parallel centroidal axes.

8/51 An area has moments of inertia $I_x = 28$ in.^{4} and $I_y = 12$ in.^{4} about a set of x-y axes. The angle measured clockwise from the x-axis to the axis of maximum moment of inertia through the origin O is 20 deg. Determine the minimum moment of inertia of the area about an axis through O.   *Ans.* $I_{\min} = 9.56$ in.^{4}

8/52 Determine the moments and product of inertia of the rectangular area with respect to the x'-y' axes.

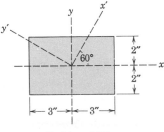

Problem 8/52

8/53 The products of inertia of the shaded area with respect to the x-y and x'-y' axes are 490 in.^{4} and -920 in.^{4}, respectively. Compute the area of the figure whose centroid is C.

Ans. $A = 78.3$ in.^{2}

8/54 Where $I_x = I_y$ for an area which is symmetrical about either the x- or the y-axis, prove that the moment of inertia is the same for all axes through the origin.

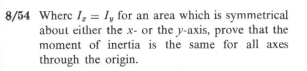

Problem 8/53

8/55 Determine the proportions of the rectangular area for which the moment of inertia about an x'-axis through the center point C of the longer side is a constant value regardless of θ. (See Prob. 8/54.)

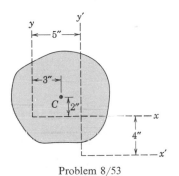

Problem 8/55

8/56 Calculate the base b and the altitude h of an isosceles triangle with an area of 16.0 in.^{2} if the moments of inertia about all axes through the vertex of the triangle are the same. (See Prob. 8/54.)   *Ans.* $b = 10.53$ in., $h = 3.04$ in.

8/57 Calculate the moments and product of inertia of the area of the square with respect to the x'-y' axes.

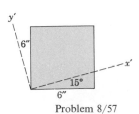

Problem 8/57

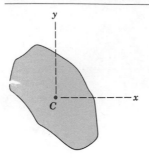

Problem 8/58

8/58 The maximum and minimum moments of inertia of the shaded area are 15 in.⁴ and 5 in.⁴, respectively, about axes passing through the centroid C, and the product of inertia with respect to the x-y axes is $-5\sqrt{3}/2$ in.⁴ From the appropriate equations calculate I_x and the angle α measured counterclockwise from the x-axis to the axis of maximum moment of inertia.

Ans. $I_x = 12.5$ in.⁴, $\alpha = 30$ deg

8/59 Solve Prob. 8/58 by constructing the Mohr circle of inertia.

8/60 The moments of inertia of an area with respect to the principal axes of inertia through a point P are $I_x = 10$ in.⁴ and $I_y = 60$ in.⁴ Determine the moment of inertia $I_{x'}$ and the product of inertia $I_{x'y'}$ for the area with respect to x'-y' axes through P that are rotated 20 deg clockwise from the x-y axes. Also represent the results on a Mohr circle of inertia.

Ans. $I_{x'} = 15.85$ in.⁴
$I_{x'y'} = 16.07$ in.⁴

8/61 Sketch the Mohr circle of inertia for each of the four rectangular areas with the proportions and positions shown. Indicate on each diagram point A with coordinates I_x, I_{xy} and the angle 2α where α is the angle from the x-axis to the axis of maximum moment of inertia.

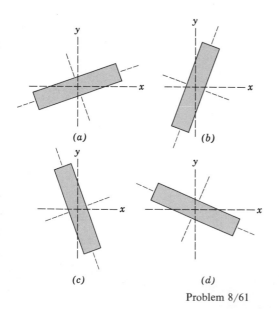

(a)   (b)   (c)   (d)

Problem 8/61

8/62 Prove that the magnitude of the product of inertia can be computed from the relation $I_{xy} = \sqrt{I_x I_y - I_{max} I_{min}}$.

8/63 Calculate the minimum moment of inertia about an axis through point O for the shaded area. Specify the angle α measured clockwise from the x-axis to the axis of minimum moment of inertia.

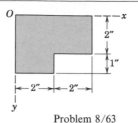

Problem 8/63

8/64 Determine the maximum and minimum moments of inertia with respect to centroidal axes through C for the composite of the four square areas shown. Find the angle α measured from the x-axis to the axis of maximum moment of inertia.   *Ans.* $I_{\max} = a^4(\frac{10}{3} + \sqrt{5})$
$I_{\min} = a^4(\frac{10}{3} - \sqrt{5})$
$\alpha = 76°43'$

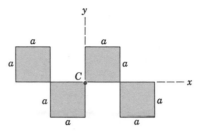

Problem 8/64

8/65 Calculate the maximum and minimum moments of inertia of the structural angle about axes through its corner C and find the angle α measured counterclockwise from the x-axis to the axis of maximum inertia. Neglect the small radii and fillet.

Ans. $I_{\max} = 178.2$ in.^{4}
$I_{\min} = 68.4$ in.^{4}
$\alpha = -13°24'$

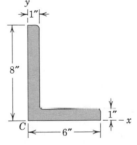

Problem 8/65

8/66 Calculate the product of inertia of the rectangular area with respect to the x-y axes.
Ans. $I_{xy} = 1225$ in.^{4}

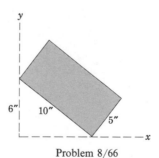

Problem 8/66

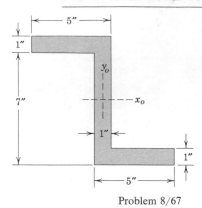

Problem 8/67

8/67 Calculate the maximum and minimum moments of inertia about centroidal axes for the structural Z-section. Indicate the angle α measured counterclockwise from the x_o-axis to the axis of maximum moment of inertia.

Ans. $I_{max} = 181.9$ in.^{4}
$I_{min} = 20.7$ in.^{4}
$\alpha = 30°8'$

Appendices

A REVIEW PROBLEMS

In the preceding chapters the problems included with the various articles illustrate application of the particular topic involved. Thus the problem category and method of solution for these problems are to a great extent indicated automatically by their association with the article. The student of mechanics should develop ability to classify a new problem by recognizing the topic or topics involved and by selecting the appropriate method or methods of solution. The following review problems in Appendix A are included to help the student develop this ability. The problems are arranged without regard for topic or method of solution, although they are arranged approximately in order of increasing difficulty. Some problems include more than one topic or may be worked by more than one method. It is suggested that the student use his time to outline the solution to as many of the problems as possible rather than to concentrate on the complete solution of only a limited number of them. In this way a more comprehensive review can be carried out within a limited period of time. The answers to the problems are included for those who wish them.

A1 A former student of mechanics wishes to weigh himself but has access only to a scale A with capacity limited to 100 lb and a small 20-lb spring dynamometer B. With the rig shown he discovers that when he exerts a pull on the rope so that B registers 18 lb, the scale A reads 67 lb. What is his correct weight?   *Ans.* $W = 157$ lb

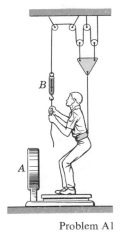

Problem A1

343

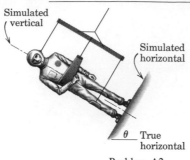

Simulated vertical

Simulated horizontal

θ ___ True horizontal

Problem A2

A2 A simulator for studying human locomotion under reduced gravity conditions consists of a harness on the end of a long wire to support a portion of the weight of the subject. Determine the angle θ which will simulate lunar conditions where gravity is one sixth of that on the earth.

Ans. $\theta = 80°24'$

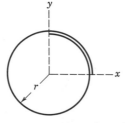

Problem A3

A3 Determine by direct integration the location of the center of gravity of the circular wire of radius r with a quarter-turn overlap.

Ans. $\bar{x} = \bar{y} = \dfrac{2r}{5\pi}$

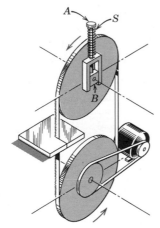

Problem A4

A4 A tension is maintained in the band-saw blade by the spring S which pushes up on the rod A which, in turn, is connected to the bearing block B of the idler pulley. If a 50-lb force is required to shear off a tooth on the blade, determine the spring force F which can be allowed and still ensure that the blade slips on its pulley before a tooth is sheared. Assume that force is applied to only one tooth at a time. The coefficient of friction between the band and the pulley may be taken to be 0.40. Neglect the weight of the upper pulley. *Ans.* $F = 39.8$ lb

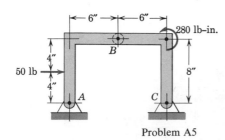

Problem A5

A5 For the two-member frame loaded by the force and couple shown, determine the force supported by the pin at A. *Ans.* $A = 21.1$ lb

A6 All panels of the truss are made of 45-deg right triangles. Determine the forces in members CD, DE, and CE in terms of the load L.

$$Ans. \quad CD = L/\sqrt{2}, \quad C$$
$$DE = L\ \sqrt{2}/8, \quad C$$
$$CE = L/8, \quad T$$

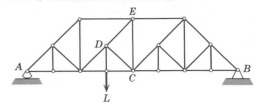

Problem A6

A7 The tension in the hoisting cable for the concrete hopper is 840 lb for a full load. The equilateral frame ABC spreads the cables so that equal vertical forces are applied to the hopper at all times. Calculate the compression P in each leg of the frame. $Ans. \ P = 70$ lb

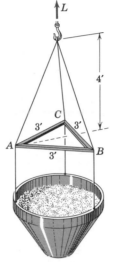

Problem A7

A8 A freeway sign measuring 12 ft by 6 ft is supported by the single mast as shown. The sign and supporting framework, exclusive of mast, together weigh 600 lb with center of gravity 10 ft away from the vertical center line of the mast. If the sign is subjected to the direct blast of a 75 mi/hr wind, an effective average pressure difference between the front and back sides of the sign of 17.5 lb per square foot of sign area is developed. Determine the resultant moment **M** about the base of the mast developed by the wind load and the weight of the structure. Also determine the magnitude M_b of the part of **M** that induces a bending moment in the mast at its base.

$$Ans. \ \mathbf{M} = -100(214.2\mathbf{i} + 60\mathbf{j} + 126\mathbf{k}) \ \text{lb-ft}$$
$$M_b = 22,200 \ \text{lb-ft}$$

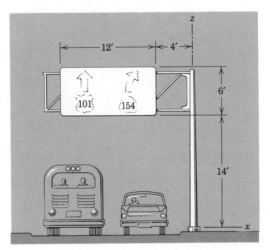

Problem A8

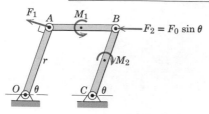

Problem A9

A9 Determine the work done individually by the forces F_1 and F_2 and the couples M_1 and M_2 during a rotation of the parallelogram linkage from $\theta = 0$ to $\theta = \pi$. The force F_1 is applied normal to r and is constant in magnitude. The force F_2 is constant in direction but its magnitude is $F_2 = F_0 \sin \theta$. Both M_1 and M_2 are constant.

$$Ans. \quad U_{F_1} = F_1 r \pi, \quad U_{F_2} = F_0 r \frac{\pi}{2}$$

$$U_{M_1} = 0, \quad U_{M_2} = -M_2 \pi$$

A10 A force $\mathbf{F} = 2\mathbf{i} + 3\mathbf{j} - 4\mathbf{k}$ lb acts at a point A whose position vector from a point O is $\mathbf{r} = 4\mathbf{i} - \mathbf{j} + 2\mathbf{k}$ in. Calculate

(a) the moment $\mathbf{M}$ of $\mathbf{F}$ about O and its magnitude M,

(b) the component $\mathbf{M}_n$ of $\mathbf{M}$ in the direction of the unit vector $\mathbf{n} = 0.5\mathbf{i} + 0.5\mathbf{j} + 0.5\sqrt{2}\mathbf{k}$,

(c) the work U done by $\mathbf{F}$ if A moves a distance of 6 in. in the $\mathbf{n}$-direction.

$$Ans. \quad (a) \ \mathbf{M} = -2\mathbf{i} + 20\mathbf{j} + 14\mathbf{k} \text{ lb-in.}$$
$$(b) \ \mathbf{M}_n = 9.45(\mathbf{i} + \mathbf{j} + \sqrt{2}\mathbf{k}) \text{ lb-in.}$$
$$(c) \ U = -1.970 \text{ in.-lb}$$

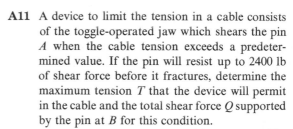

Problem A11

A11 A device to limit the tension in a cable consists of the toggle-operated jaw which shears the pin A when the cable tension exceeds a predetermined value. If the pin will resist up to 2400 lb of shear force before it fractures, determine the maximum tension T that the device will permit in the cable and the total shear force Q supported by the pin at B for this condition.

$$Ans. \ T = 3390 \text{ lb}, \quad Q = 1553 \text{ lb}$$

A12 The gripping mechanism of the logger's hoist is activated by the two hydraulic cylinders. For the particular position shown with a 3000-lb log, the oil pressure acting on each of the 15-in.^{2} pistons in the cylinders is 180 lb/in.^{2} Compute the total force supported by the pin at C. Neglect the weights of the members. $Ans. \ C = 4250$ lb

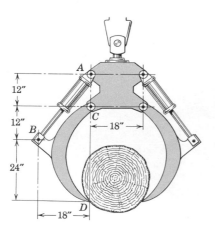

Problem A12

A13 The uniform bar of weight W and length l rests on the horizontal surface and touches the fixed vertical rod at the point shown. Determine the value of the couple M which will cause the bar to slip if the coefficient of friction between the bar and the surface is f. Assume that the weight of the bar is uniformly supported along its length. Specify the least possible value of M to initiate rotation of the bar and the corresponding value of a.

$$Ans.\ M = fWl\left(\frac{1}{2} - \frac{a}{l} + \frac{a^2}{l^2}\right)$$

$$M_{min} = \frac{fWl}{4} \text{ for } a = \frac{l}{2}$$

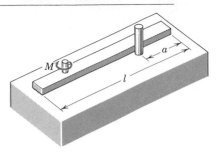

Problem A13

A14 The boom ABC is held in position by the two guy wires BD and BE and supports a vertical load W. The weight of the boom is negligible compared with W. Calculate the minimum coefficient of friction f which can exist between the boom and the horizontal plane so that end A will not slip.   *Ans.* $f_{min} = 0.442$

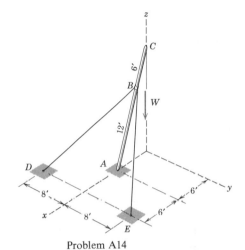

Problem A14

A15 A submarine laboratory to be occupied by marine biologists consists of the shell structure submerged to the depth shown. Water enters the lower portion of the cylindrical structure through an open hatch at the bottom and is maintained at a level of 5 ft by compressed air in the upper portion of the interior space. The entire structure less lead ballast weighs 160,000 lb out of water. Determine the air pressure p (gage) in the interior of the laboratory and the above-water weight L of lead ballast fastened to the bottom of the tank required to produce a 10,000-lb force between the supporting legs and the ocean floor. The specific weight of air at the particular temperature involved and at the atmospheric pressure of 14.7 lb/in.^{2} is 0.0753 lb/ft.^{3}

$$Ans.\ p = 22.2 \text{ lb/in.}^2 \text{ gage}$$
$$L = 423,000 \text{ lb}$$

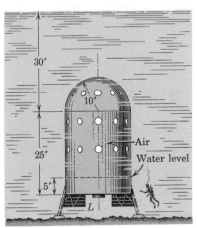

Problem A15

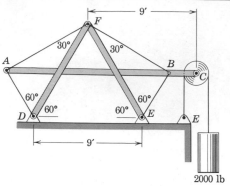

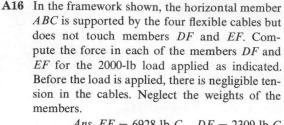

Problem A16

A16 In the framework shown, the horizontal member *ABC* is supported by the four flexible cables but does not touch members *DF* and *EF*. Compute the force in each of the members *DF* and *EF* for the 2000-lb load applied as indicated. Before the load is applied, there is negligible tension in the cables. Neglect the weights of the members.

Ans. EF = 6928 lb *C*,   *DF* = 2309 lb *C*

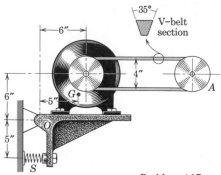

Problem A17

A17 An adjustment in the compression of the spring *S* is provided to accommodate various V-belt tensions. The motor transmits a constant-speed torque of 160 lb-in. to the pulley *A* and to the mechanism it drives, with the belt on the verge of slipping (see Prob. 6/83). Determine the spring force *F* for (*a*) clockwise and (*b*) counter-clockwise rotation. The coefficient of friction between the belt and each pulley is 0.30. The combined weight of the motor and bracket is 75 lb, with center of gravity at *G*.

Ans. (*a*) *F* = 148 lb,   (*b*) *F* = 212 lb

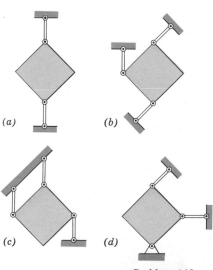

Problem A18

A18 In each of the four cases represented, the plate is confined to the plane of representation and is supported in this plane by the links and connections shown. The plates are subjected to various forces (not shown) in their planes. Identify the plates that correspond to each of the following categories:

(A) Complete fixity in the plane with minimum number of adequate constraints

(B) Partial fixity in the plane with inadequate constraints

(C) Complete fixity in the plane with redundant constraints

Ans. (A) *b*
(B) *a, c*
(C) *d*

A19 A gate for limiting the depth of water in a small reservoir consists of a flat plate which seals a rectangular opening 6 ft high and 4 ft wide. The plate and attached frame have negligible weight and are hinged at C and counterweighted as shown. Determine the counterweight W which will limit the depth of water to the level shown. Also find the force in strut DB.

$$\textit{Ans.} \quad W = 21{,}000 \text{ lb}, \quad DB = 5990 \text{ lb } C$$

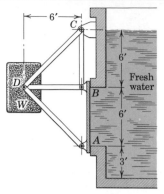

Problem A19

A20 Determine the coordinates of the centroid of the volume obtained by revolving the right triangle through the 90-deg angle about the z-axis.

$$\textit{Ans.} \quad \bar{x} = \bar{y} = \frac{3b}{2\pi}, \quad \bar{z} = \frac{3h}{8}$$

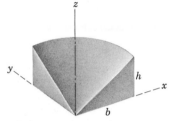

Problem A20

A21 The semicircular beam of uniform section is cantilevered from one end in the vertical plane as shown. If the beam has a total weight W, determine the bending moment M in the beam as a function of θ.

$$\textit{Ans.} \quad M = \frac{Wr}{\pi}\left[(\pi - \theta)\cos\theta + \sin\theta\right]$$

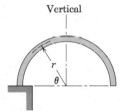

Problem A21

A22 The wedge is inserted into the split ring with a force P. The coefficient of friction between the wedge and the ring is f, and the wedge is self-locking ($\alpha < 2\tan^{-1}f$). If P is removed from the wedge, determine the maximum residual bending moment M in the ring.

$$\textit{Ans.} \quad M = \frac{2Pr}{2f\cos^2\dfrac{\alpha}{2} + \sin\alpha}$$

Problem A22

Problem A23

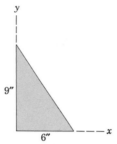

Problem A24

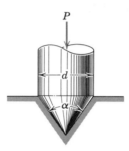

Problem A25

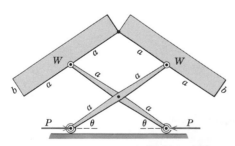

Problem A26

A23 Determine the moment M applied to the reflector dish at the bearing O necessary to prevent rotation of the dish in the vertical plane under the action of its own weight. The dish is a paraboloidal shell of revolution about the z-axis and has a weight μ per unit area.   *Ans.* $M = 2.98\mu a^3$

A24 Calculate the maximum and minimum moments of inertia about axes through the centroid of the triangular area and specify the angle α measured counterclockwise from the x-axis to the axis of maximum moment of inertia. Draw the corresponding Mohr circle of inertia.

$$Ans.\ I_{max} = 140.5\ \text{in.}^4$$
$$I_{min} = 35.0\ \text{in.}^4$$
$$\alpha = 25°6'$$

A25 The conical pivot bearing shown supports a shaft thrust P. Derive an expression for the torque M required on the shaft to overcome friction, assuming (*a*) new bearing surfaces with uniform pressure distribution and (*b*) worn-in surfaces where the constant wear is proportional to the product of the normal pressure and the radial distance from the shaft axis.

$$Ans.\ (a)\ M = \frac{fPd}{3\sin\frac{\alpha}{2}}, \qquad (b)\ M = \frac{fPd}{4\sin\frac{\alpha}{2}}$$

A26 Determine the forces P required to maintain the equilibrium configuration of the hinged weights shown. The center of gravity of each weight is in the center of its respective rectangle. Neglect the weights of the legs. For $b = a$, at what angle θ' will equilibrium occur with $P = 0$?

$$Ans.\ P = W\left(2\operatorname{ctn}\theta - \frac{b}{2a}\right),\quad \theta' = 75°58'$$

A27 Determine the x-coordinate of the centroid of the shaded area and find the radius of gyration of the area about the vertical centroidal axis. The curved boundary is parabolic with zero slope at the y-axis.   *Ans.* $\overline{X} = 3.38$ in., $\overline{k}_y = 1.66$ in.

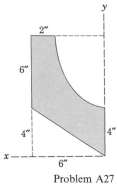

Problem A27

A28 Construct the shear and moment diagrams for the loaded beam shown. Indicate the maximum magnitude $|M|$ of the bending moment and the distance x from the left end of the beam at which this moment occurs.
               Ans. $|M|_{max} = 550$ lb-ft at $x = 7$ ft

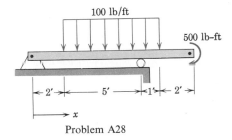

Problem A28

A29 The concrete beam ABC with a uniform cross section weighs 150 lb per foot of length and is supported as shown. Plot the bending moment M in the beam as a function of distance along the beam and determine the maximum magnitude of M and its location on the beam.
            Ans. $|M|_{max} = 4157$ lb-ft at B

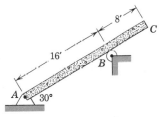

Problem A29

A30 The cable hangs under the action of its own weight and is suspended from the two supports shown. Without solving, outline a procedure for determining the maximum tension T in the cable and the total length S of the cable, using the exact relations.

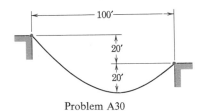

Problem A30

A31 The curved beam is subjected to a force p whose intensity in pounds per foot of arc length varies linearly with θ from zero at the base of the beam to p_0 at the end where $\theta = \pi/2$. Determine the shear force V, bending moment M, and torsional moment T induced in the beam at its base by p.

$$Ans. \quad V = \frac{\pi}{4}p_0 r, \quad M = \frac{2}{\pi}p_0 r^2$$

$$T = p_0 r^2\left(\frac{\pi}{4} - 1 + \frac{2}{\pi}\right)$$

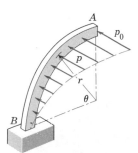

Problem A31

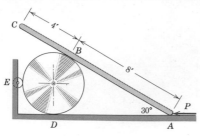

Problem A32

A32 The uniform plank ABC weighs 200 lb and is supported at B by the large 300-lb cylinder and at A by the horizontal plane. The coefficient of friction between the surfaces at A and B is 0.40 and that between the surfaces at D is 0.60. Calculate the force P required to start end A moving to the left. Also compute the corresponding friction force acting at D. The small guiding roller at E turns with negligible friction.

Ans. $P = 155$ lb, $F_D = 52.0$ lb

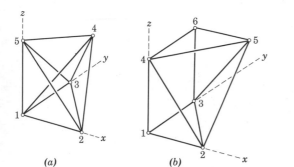

(a)   (b)

A33 For each of the space trusses shown, indicate whether it is rigid with or without redundancy or whether it is unstable. Also indicate the addition or deletion of one or more members that will produce a rigid truss without redundancy.

Ans. (*a*) rigid with redundancy
(*b*) unstable
(*c*) unstable
(*d*) unstable

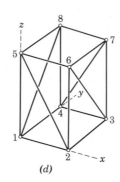

(c)   (d)

Problem A33

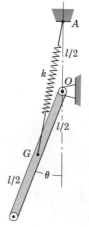

Problem A34

A34 The slender bar of length l and weight W is pivoted freely about a horizontal axis through O. The spring has an unstretched length of $l/2$. Determine the equilibrium positions, excluding $\theta = \pi$, and determine the maximum value of the spring stiffness k for stability in the position $\theta = 0$.

$$Ans.\ \theta = 0,\quad \theta = 2\cos^{-1}\frac{1/2}{1 - \dfrac{2W}{kl}}$$

$$k_{\max} = \frac{4W}{l}$$

A35 The mechanism shown is constrained to swing in the vertical plane, and in the position $\theta = 0$, the spring is unstretched. As θ increases under gradual application of the couple $M = 40$ lb-ft, the rod BC slides through the swivel block at A and extends the spring. If the spring has a stiffness $k = 10$ lb/ft, determine the equilibrium value of θ. Each of the uniform parallel links weighs 15 lb, and the uniform horizontal link weighs 20 lb. Neglect the weight of the rod, spring, and swivel block.   *Ans.* $\theta = 14°7'$

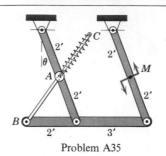

Problem A35

A36 The circular disk with central hole is subjected to a uniform compressive stress around its rim. The thickness t of the disk is a function of r only but is everywhere small compared with r. Derive the differential equation of equilibrium in the radial direction for a differential element of the disk. The radial stress is σ_r and the tangential stress is σ_θ.

$$\text{Ans. } \frac{d\sigma_r}{dr} + \frac{\sigma_r - \sigma_\theta}{r} + \frac{\sigma_r}{t}\frac{dt}{dr} = 0$$

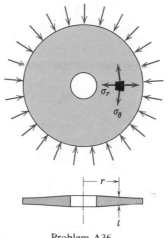

Problem A36

A37 The 1000-lb trailer with center of gravity at G is being towed at a constant speed of 10 mi/hr up the incline when one of its two wheels suddenly locks. If the same speed is maintained with the one wheel dragging, calculate the total force on the ball connection at A. The wheels of the trailer are 5 ft apart, and the trailer is loaded symmetrically about the central vertical plane. Use 0.8 for the coefficient of kinetic friction between the rubber and the pavement.   *Ans.* $A = 480$ lb

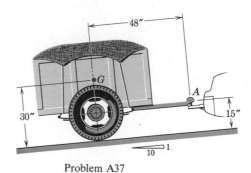

Problem A37

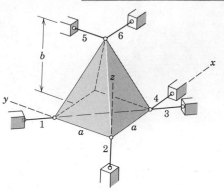

Problem A38

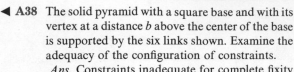

◄ **A38** The solid pyramid with a square base and with its vertex at a distance b above the center of the base is supported by the six links shown. Examine the adequacy of the configuration of constraints.

Ans. Constraints inadequate for complete fixity

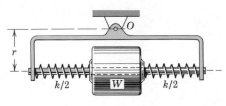

Problem A39

◄ **A39** The weight W slides with negligible friction along the smooth rod mounted in the frame which is pivoted freely about a horizontal axis through O. When the rod is in the horizontal position, the weight is centered below O and the springs are undeformed. Each spring has a stiffness $k/2$ and is secured both to the frame and to the sliding weight. The weights of all parts are negligible compared with W. Determine the equilibrium positions defined by the angle θ of rotation of the frame from the position shown and the displacement x of the weight along the rod measured from the center position. Investigate the stability for each equilibrium position.

Ans. (a) $x = 0, \quad \theta = 0$

saddle equil. for $W > kr$

stable equil. for $W < kr$

(b) $\theta = \cos^{-1} \dfrac{kr}{W}, \quad x = \dfrac{W}{k}\sqrt{1 - \left(\dfrac{kr}{W}\right)^2}$

holds for $k < W/r$, stable

B VECTOR ANALYSIS

B1 Notation. Vector quantities may be described by any notation which accounts properly for both their magnitudes and directions. Scalar algebra may be used to handle the magnitude relationships, and geometric diagrams with appropriate trigonometry may be used to account for the directional properties. Such analysis is generally quite adequate or preferable for problems with one or two geometric dimensions. For much of the analysis in three geometric dimensions, the vector notation invented by Gibbs* is of great advantage and is used widely. The following development is designed to serve as both an introduction to and a concise summary of the relationships of the algebra and calculus of vectors as they are employed in applied mechanics using Gibbs' notation. The material included in Arts. B1 through B5 finds direct use in both statics and dynamics, particularly in three-dimensional problems. The material in Arts. B6 through B11 finds use in dynamics and in other areas of applied mechanics of a more advanced nature and is included for the sake of completeness.

It is absolutely essential to adopt a consistent notation that will always distinguish vector quantities from scalar quantities. In the printed page, boldface type is used for all vector quantities, and lightface italic type is used for scalar quantities. In handwritten work a distinguishing symbol, such as an underline, should be used to designate vector quantities.

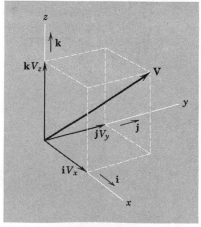

Figure B1

* Josiah Willard Gibbs (1839–1903), a professor of mathematical physics at Yale University.

A vector **V** is represented in three-dimensional space, Fig. B1, in terms of the vector sum of its three mutually perpendicular components

$$\mathbf{V} = \mathbf{i}V_x + \mathbf{j}V_y + \mathbf{k}V_z \tag{B1}$$

where **i**, **j**, **k** are the unit vectors in the *x*-, *y*-, *z*-directions, respectively. A unit vector is a vector with a specified direction and a magnitude of unity. Thus the vector quantity $\mathbf{i}V_x$ has a direction specified by the unit vector **i** in the *x*-direction and a magnitude equal to V_x, the *x*-component of the vector **V**.

The scalar magnitude of **V** is

$$|\mathbf{V}| = V = \sqrt{V_x^2 + V_y^2 + V_z^2} \tag{B2}$$

Any vector **V** may be multiplied by a scalar *a* to give the vector **V***a* or *a***V**, which product has the magnitude *aV* and the direction of **V**.

B2 **Addition.** Two vectors **P** and **Q**, Fig. B2*a*, may be added to obtain their resultant or sum **P** + **Q**, as shown in Fig. B2*b* where the two vectors are the two legs of the parallelogram. For any convenient orientation of reference axes, the sum may be written in component form as

$$\mathbf{P} + \mathbf{Q} = (\mathbf{i}P_x + \mathbf{j}P_y + \mathbf{k}P_z) + (\mathbf{i}Q_x + \mathbf{j}Q_y + \mathbf{k}Q_z)$$
$$= \mathbf{i}(P_x + Q_x) + \mathbf{j}(P_y + Q_y) + \mathbf{k}(P_z + Q_z) \tag{B3}$$

As seen from Fig. B2*b*, the sum may be obtained by combining the vectors head-to-tail in either order in the triangle form of vector addition to obtain their sum. Thus

$$\mathbf{P} + \mathbf{Q} = \mathbf{Q} + \mathbf{P} \tag{B4}$$

and vector addition is seen to be *commutative*.

Vector addition is also *associative* as observed in Fig. B3, where it is clear

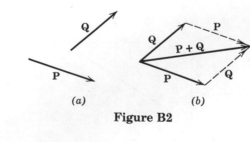

(a)   (b)

Figure B2

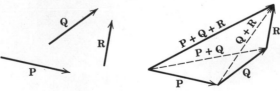

Figure B3

that the sum of the three vectors **P**, **Q**, **R** may be arrived at by adding **P** to the sum of **Q** and **R** or by adding the sum of **P** and **Q** to **R**. Thus

$$\mathbf{P} + (\mathbf{Q} + \mathbf{R}) = (\mathbf{P} + \mathbf{Q}) + \mathbf{R} \tag{B5}$$

The subtraction of a vector is the same as the addition of a negative vector. Thus the vector difference between **P** and **Q** of Fig. B4 is $\mathbf{P} - \mathbf{Q} = \mathbf{P} + (-\mathbf{Q})$.

B3 Dot or Scalar Product. There are two distinctly different ways in which vectors may be multiplied. In the first way the *dot* or *scalar product* of two vectors **P** and **Q**, Fig. B5*a*, is defined as

$$\mathbf{P} \cdot \mathbf{Q} = PQ \cos \theta \tag{B6}$$

where θ is the angle between them. This product may be viewed as the magnitude of **P** multiplied by the component $Q \cos \theta$ of **Q** in the direction of **P**, Fig. B5*b*, or as the magnitude of **Q** multiplied by the component $P \cos \theta$ of **P** in the direction of **Q**, Fig. B5*c*. The commutative law

$$\mathbf{P} \cdot \mathbf{Q} = \mathbf{Q} \cdot \mathbf{P} \tag{B7}$$

holds for the dot product, since the order of the scalar terms in the scalar multiplication may be interchanged without affecting the product.

From the definition of the dot product, it follows that

$$\mathbf{i} \cdot \mathbf{i} = \mathbf{j} \cdot \mathbf{j} = \mathbf{k} \cdot \mathbf{k} = 1$$
$$\mathbf{i} \cdot \mathbf{j} = \mathbf{j} \cdot \mathbf{i} = \mathbf{i} \cdot \mathbf{k} = \mathbf{k} \cdot \mathbf{i} = \mathbf{j} \cdot \mathbf{k} = \mathbf{k} \cdot \mathbf{j} = 0$$

Thus

$$\mathbf{P} \cdot \mathbf{Q} = (\mathbf{i}P_x + \mathbf{j}P_y + \mathbf{k}P_z) \cdot (\mathbf{i}Q_x + \mathbf{j}Q_y + \mathbf{k}Q_z)$$
$$= P_x Q_x + P_y Q_y + P_z Q_z \tag{B8}$$

and

$$\mathbf{P} \cdot \mathbf{P} = P_x{}^2 + P_y{}^2 + P_z{}^2$$

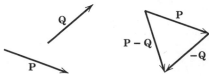

Figure B4

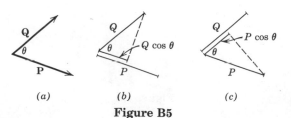

(*a*)   (*b*)   (*c*)

Figure B5

It follows from the definition of the dot product that two vectors $\mathbf{P}$ and $\mathbf{Q}$ are perpendicular when their dot product vanishes, $\mathbf{P} \cdot \mathbf{Q} = 0$.

The angle θ between two vectors $\mathbf{P}_1$ and $\mathbf{P}_2$ may be found from their dot product expression $\mathbf{P}_1 \cdot \mathbf{P}_2 = P_1 P_2 \cos \theta$, which gives

$$\cos \theta = \frac{\mathbf{P}_1 \cdot \mathbf{P}_2}{P_1 P_2} = \frac{P_{1_x} P_{2_x} + P_{1_y} P_{2_y} + P_{1_z} P_{2_z}}{P_1 P_2} = l_1 l_2 + m_1 m_2 + n_1 n_2$$

where l, m, n stand for the respective direction cosines of the vectors. It is also observed that two vectors are perpendicular when their direction cosines obey the relation $l_1 l_2 + m_1 m_2 + n_1 n_2 = 0$.

The distributive law holds for the dot product, as is easily seen from the following expansion.

$$\mathbf{P} \cdot (\mathbf{Q} + \mathbf{R}) = (\mathbf{i}P_x + \mathbf{j}P_y + \mathbf{k}P_z) \cdot (\mathbf{i}[Q_x + R_x] + \mathbf{j}[Q_y + R_y] + \mathbf{k}[Q_z + R_z])$$
$$= P_x(Q_x + R_x) + P_y(Q_y + R_y) + P_z(Q_z + R_z)$$
$$= (P_x Q_x + P_y Q_y + P_z Q_z) + (P_x R_x + P_y R_y + P_z R_z)$$

so that

$$\mathbf{P} \cdot (\mathbf{Q} + \mathbf{R}) = \mathbf{P} \cdot \mathbf{Q} + \mathbf{P} \cdot \mathbf{R} \tag{B9}$$

B4 Cross or Vector Product. The second form of multiplication of vectors is known as the *cross* or *vector product*. For the vectors $\mathbf{P}$ and $\mathbf{Q}$ of Fig. B6 this product is written as $\mathbf{P} \times \mathbf{Q}$ and is defined as a vector whose magnitude equals the product of the magnitudes of $\mathbf{P}$ and $\mathbf{Q}$ multiplied by the sine of the angle θ (less than 180 deg) between them. The direction of $\mathbf{P} \times \mathbf{Q}$ is normal to the plane defined by $\mathbf{P}$ and $\mathbf{Q}$, and the sense of $\mathbf{P} \times \mathbf{Q}$ is in the direction of the advancement of a right-hand screw when revolved from $\mathbf{P}$ to $\mathbf{Q}$ through the smaller of the two angles between them. If $\mathbf{n}$ is a unit vector with the direction and sense of $\mathbf{P} \times \mathbf{Q}$, the cross product may be written

$$\mathbf{P} \times \mathbf{Q} = \mathbf{n}PQ \sin \theta \tag{B10}$$

By using the right-hand rule and reversing the order of vector multiplica-

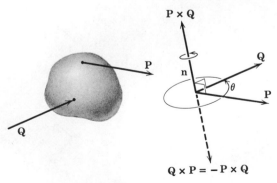

Figure B6

tion, it is seen from Fig. B6 that $\mathbf{P} \times \mathbf{Q} = -\mathbf{Q} \times \mathbf{P}$. Thus the commutative law does not hold for the cross product.

The distributive law does hold for the cross product and may be shown geometrically. In Fig. B7 a triangular prism 1-2-3-4-5-6 is formed from the three arbitrary vectors $\mathbf{P}$, $\mathbf{Q}$, $\mathbf{R}$ as shown. The magnitude of the area of the parallelogram face 1-4-5-2 is the base $|\mathbf{Q}|$ times the altitude, which is $|\mathbf{P}|$ times the sine of the angle between $\mathbf{P}$ and $\mathbf{Q}$. Thus this area is represented by the magnitude of the cross product $\mathbf{P} \times \mathbf{Q}$. The area of this face has a direction normal to its plane, and with its positive sense out from the figure, it may be represented by the vector $\mathbf{P} \times \mathbf{Q}$. In similar fashion the area 2-5-6-3 is represented by $\mathbf{P} \times \mathbf{R}$, and the area 1-3-6-4 by $(\mathbf{Q} + \mathbf{R}) \times \mathbf{P}$. The parallel triangular faces are $\frac{1}{2}(\mathbf{Q} \times \mathbf{R})$ and $-\frac{1}{2}(\mathbf{Q} \times \mathbf{R})$. Since the prism is a closed surface, the sum of its surfaces represented as vectors must vanish. Therefore

$$(\mathbf{Q} + \mathbf{R}) \times \mathbf{P} + \mathbf{P} \times \mathbf{Q} + \mathbf{P} \times \mathbf{R} + \tfrac{1}{2}(\mathbf{Q} \times \mathbf{R}) - \tfrac{1}{2}(\mathbf{Q} \times \mathbf{R}) = 0$$

Reversing the order and the sign of the first term and dropping the last two terms give

$$\mathbf{P} \times (\mathbf{Q} + \mathbf{R}) = \mathbf{P} \times \mathbf{Q} + \mathbf{P} \times \mathbf{R} \qquad (B11)$$

which proves the distributive law for the cross product.

From the definition of the cross product the following relations between the unit vectors are apparent:

$$\mathbf{i} \times \mathbf{j} = \mathbf{k} \qquad \mathbf{j} \times \mathbf{k} = \mathbf{i} \qquad \mathbf{k} \times \mathbf{i} = \mathbf{j}$$
$$\mathbf{j} \times \mathbf{i} = -\mathbf{k} \qquad \mathbf{k} \times \mathbf{j} = -\mathbf{i} \qquad \mathbf{i} \times \mathbf{k} = -\mathbf{j}$$
$$\mathbf{i} \times \mathbf{i} = \mathbf{j} \times \mathbf{j} = \mathbf{k} \times \mathbf{k} = 0$$

With the aid of these identities and the distributive law the vector product may be written

$$\mathbf{P} \times \mathbf{Q} = (\mathbf{i}P_x + \mathbf{j}P_y + \mathbf{k}P_z) \times (\mathbf{i}Q_x + \mathbf{j}Q_y + \mathbf{k}Q_z)$$
$$= \mathbf{i}(P_yQ_z - P_zQ_y) + \mathbf{j}(P_zQ_x - P_xQ_z) + \mathbf{k}(P_xQ_y - P_yQ_x) \qquad (B12)$$

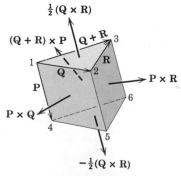

Figure B7

upon rearrangement of terms. This expression may be written compactly as the determinant

$$\mathbf{P} \times \mathbf{Q} = \begin{vmatrix} \mathbf{i} & \mathbf{j} & \mathbf{k} \\ P_x & P_y & P_z \\ Q_x & Q_y & Q_z \end{vmatrix} \qquad \text{(B12a)}$$

which is easily verified by carrying out the expansion.

B5 **Additional Relations.** Two additional relations of vector algebra will be stated without proof, although their validity is not difficult to show geometrically.

The *triple scalar product* is the dot product of two vectors where one of them is specified as a cross product of two additional vectors. This product is a scalar and is given by any one of the equivalent expressions

$$(\mathbf{P} \times \mathbf{Q}) \cdot \mathbf{R} = \mathbf{R} \cdot (\mathbf{P} \times \mathbf{Q}) = -\mathbf{R} \cdot (\mathbf{Q} \times \mathbf{P})$$

Actually the parentheses are not needed since it would be meaningless to write $\mathbf{P} \times (\mathbf{Q} \cdot \mathbf{R})$. It may be shown that

$$\mathbf{P} \times \mathbf{Q} \cdot \mathbf{R} = \mathbf{P} \cdot \mathbf{Q} \times \mathbf{R} \qquad \text{(B13)}$$

which establishes the rule that the dot and the cross may be interchanged without changing the value of the triple scalar product. Furthermore, it may be seen upon expansion that

$$\mathbf{P} \times \mathbf{Q} \cdot \mathbf{R} = \begin{vmatrix} P_x & P_y & P_z \\ Q_x & Q_y & Q_z \\ R_x & R_y & R_z \end{vmatrix} \qquad \text{(B14)}$$

The *triple vector product* is the cross product of two vectors where one of them is specified as a cross product of two additional vectors. This product is a vector and is given by any one of the equivalent expressions

$$(\mathbf{P} \times \mathbf{Q}) \times \mathbf{R} = -\mathbf{R} \times (\mathbf{P} \times \mathbf{Q}) = \mathbf{R} \times (\mathbf{Q} \times \mathbf{P})$$

Here the parentheses must be used since an expression $\mathbf{P} \times \mathbf{Q} \times \mathbf{R}$ would be ambiguous because it would not identify the vector to be crossed. It may be shown that the triple vector product is equivalent to

$$(\mathbf{P} \times \mathbf{Q}) \times \mathbf{R} = \mathbf{R} \cdot \mathbf{P} \mathbf{Q} - \mathbf{R} \cdot \mathbf{Q} \mathbf{P}$$

or

$$\mathbf{P} \times (\mathbf{Q} \times \mathbf{R}) = \mathbf{P} \cdot \mathbf{R} \mathbf{Q} - \mathbf{P} \cdot \mathbf{Q} \mathbf{R} \qquad \text{(B15)}$$

The first term in the first expression, for example, is the dot product $\mathbf{R} \cdot \mathbf{P}$, a scalar, multiplied by the vector $\mathbf{Q}$. The validity of Eqs. B13 and B15 may be checked easily by carrying out the indicated operations with three arbitrary vectors with numerical or algebraic coefficients.

B6 **Derivatives of Vectors.** The derivative of a vector $\mathbf{P}$ with respect to a scalar, such as the time t, is the limit of the ratio of the change $\Delta \mathbf{P}$ in $\mathbf{P}$ to the corresponding change Δt in t as Δt approaches zero. Thus

$$\frac{d\mathbf{P}}{dt} = \lim_{\Delta t \to 0} \frac{\Delta \mathbf{P}}{\Delta t}$$

$$= \lim_{\Delta t \to 0} \left(\mathbf{i} \frac{\Delta P_x}{\Delta t} + \mathbf{j} \frac{\Delta P_y}{\Delta t} + \mathbf{k} \frac{\Delta P_z}{\Delta t} \right)$$

where $\Delta \mathbf{P}$ has been expressed in terms of its components. It follows that

$$\frac{d\mathbf{P}}{dt} = \mathbf{i} \frac{dP_x}{dt} + \mathbf{j} \frac{dP_y}{dt} + \mathbf{k} \frac{dP_z}{dt}$$

and

$$\frac{d^n \mathbf{P}}{dt^n} = \mathbf{i} \frac{d^n P_x}{dt^n} + \mathbf{j} \frac{d^n P_y}{dt^n} + \mathbf{k} \frac{d^n P_z}{dt^n} \tag{B16}$$

The derivative of the sum of two vectors is simply

$$\frac{d(\mathbf{P} + \mathbf{Q})}{dt} = \lim_{\Delta t \to 0} \frac{\Delta(\mathbf{P} + \mathbf{Q})}{\Delta t} = \lim_{\Delta t \to 0} \left(\frac{\Delta \mathbf{P}}{\Delta t} + \frac{\Delta \mathbf{Q}}{\Delta t} \right)$$

$$= \frac{d\mathbf{P}}{dt} + \frac{d\mathbf{Q}}{dt} \tag{B17}$$

since the limit of the sum is the same as the sum of the limits of the terms.

The derivative of the product of a vector $\mathbf{P}$ and a scalar u obeys the same rule as for the product of two scalar quantities and is

$$\frac{d(\mathbf{P}u)}{dt} = \lim_{\Delta t \to 0} \frac{(\mathbf{P} + \Delta \mathbf{P})(u + \Delta u) - \mathbf{P}u}{\Delta t}$$

$$= \lim_{\Delta t \to 0} \frac{\mathbf{P}\Delta u + u\Delta \mathbf{P}}{\Delta t} = \lim_{\Delta t \to 0} \left(\mathbf{P} \frac{\Delta u}{\Delta t} + u \frac{\Delta \mathbf{P}}{\Delta t} \right)$$

$$= \mathbf{P} \frac{du}{dt} + u \frac{d\mathbf{P}}{dt} \tag{B18}$$

The derivatives of the scalar (dot) product and the vector (cross) product of two vectors also obey the same rules as for the product of two scalar quantities. Thus for the dot product

$$\frac{d(\mathbf{P} \cdot \mathbf{Q})}{dt} = \lim_{\Delta t \to 0} \frac{(\mathbf{P} + \Delta \mathbf{P}) \cdot (\mathbf{Q} + \Delta \mathbf{Q}) - \mathbf{P} \cdot \mathbf{Q}}{\Delta t}$$

$$= \lim_{\Delta t \to 0} \frac{\mathbf{P} \cdot \Delta \mathbf{Q} + \mathbf{Q} \cdot \Delta \mathbf{P} + \Delta \mathbf{P} \cdot \Delta \mathbf{Q}}{\Delta t}$$

$$= \lim_{\Delta t \to 0} \left(\mathbf{P} \cdot \frac{\Delta \mathbf{Q}}{\Delta t} + \frac{\Delta \mathbf{P}}{\Delta t} \cdot \mathbf{Q} + \frac{\Delta \mathbf{P} \cdot \Delta \mathbf{Q}}{\Delta t} \right)$$

$$= \mathbf{P} \cdot \frac{d\mathbf{Q}}{dt} + \frac{d\mathbf{P}}{dt} \cdot \mathbf{Q} \tag{B19}$$

The third term drops out in the limit, since it is of a higher order than the ones that remain.

Similarly, for the cross product the derivative is

$$\frac{d(\mathbf{P} \times \mathbf{Q})}{dt} = \lim_{\Delta t \to 0} \frac{(\mathbf{P} + \Delta\mathbf{P}) \times (\mathbf{Q} + \Delta\mathbf{Q}) - \mathbf{P} \times \mathbf{Q}}{\Delta t}$$

$$= \lim_{\Delta t \to 0} \frac{\mathbf{P} \times \Delta\mathbf{Q} + \Delta\mathbf{P} \times \mathbf{Q} + \Delta\mathbf{P} \times \Delta\mathbf{Q}}{\Delta t}$$

$$= \lim_{\Delta t \to} \left(\mathbf{P} \times \frac{\Delta\mathbf{Q}}{\Delta t} + \frac{\Delta\mathbf{P}}{\Delta t} \times \mathbf{Q} + \frac{\Delta\mathbf{P} \times \Delta\mathbf{Q}}{\Delta t} \right)$$

$$= \mathbf{P} \times \frac{d\mathbf{Q}}{dt} + \frac{d\mathbf{P}}{dt} \times \mathbf{Q} \tag{B20}$$

Again the third term disappears in the limit, since it is of a higher order than those that remain. The only special caution to be observed in the differentiation of cross products is that the order of the quantities on the two sides of the cross sign must be preserved, since the quantities in cross-product multiplication are not commutative.

B7   Integration of Vectors. Integration of vectors poses no special problem. If $\mathbf{V}$ is a function of x, y, and z and an element of volume is $d\tau = dx\, dy\, dz$, the integral of $\mathbf{V}$ over the volume may be written as the vector sum of the three integrals of its components. Thus

$$\int \mathbf{V}\, d\tau = \mathbf{i} \int V_x\, d\tau + \mathbf{j} \int V_y\, d\tau + \mathbf{k} \int V_z\, d\tau \tag{B21}$$

B8   Gradient. Consider a scalar function defined by $\phi = f(x, y, z)$. Successive constant values of ϕ such as ϕ_1, ϕ_2, ϕ_3, . . . define adjoining space surfaces, Fig. B8. From any point A on one of the surfaces, there is one path n which goes most directly from surface to surface. The quantity

$$\mathbf{F} = \mathbf{i}\, \frac{\partial \phi}{\partial x} + \mathbf{j}\, \frac{\partial \phi}{\partial y} + \mathbf{k}\, \frac{\partial \phi}{\partial z}$$

called the *gradient* of ϕ, is a vector in the n-direction and represents the maximum space rate of change of ϕ.

The gradient of ϕ may be written in compact notation using the vector

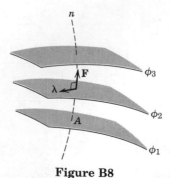

Figure B8

operator ∇ ("del"), so that

$$\mathbf{F} = \nabla\phi$$

where   (B22)

$$\nabla = \mathbf{i}\,\frac{\partial}{\partial x} + \mathbf{j}\,\frac{\partial}{\partial y} + \mathbf{k}\,\frac{\partial}{\partial z}$$

The fact that $\nabla\phi$ is perpendicular to the surface $\phi = $ constant is shown by taking a unit vector $\boldsymbol{\lambda} = \mathbf{i}\,dx + \mathbf{j}\,dy + \mathbf{k}\,dz$ which lies in that surface and noting that $\boldsymbol{\lambda}$ and $\nabla\phi$ satisfy the condition for perpendicularity, namely,

$$\boldsymbol{\lambda} \cdot \nabla\phi = 0$$

Thus

$$dx\,\frac{\partial\phi}{\partial x} + dy\,\frac{\partial\phi}{\partial y} + dz\,\frac{\partial\phi}{\partial z} = d\phi = 0$$

where dx, dy, dz lie in the surface $\phi = $ constant for which $d\phi = 0$

The dot product of the vector ∇ into itself gives the scalar operator

$$\nabla \cdot \nabla = \frac{\partial^2}{\partial x^2} + \frac{\partial^2}{\partial y^2} + \frac{\partial^2}{\partial z^2} \qquad (B23)$$

which is called the *Laplacian.* $\nabla^2\phi = 0$ is known as Laplace's Equation.

B9   Divergence. When the vector operator ∇ is dotted into a vector $\mathbf{V}$, the result is

$$\nabla \cdot \mathbf{V} = \frac{\partial V_x}{\partial x} + \frac{\partial V_y}{\partial y} + \frac{\partial V_z}{\partial z} \qquad (B24)$$

which is known as the *divergence* of $\mathbf{V}$.

B10   Curl. When the vector operator ∇ is crossed into a vector $\mathbf{V}$, the result is

$$\nabla \times \mathbf{V} = \mathbf{i}\left(\frac{\partial V_z}{\partial y} - \frac{\partial V_y}{\partial z}\right) + \mathbf{j}\left(\frac{\partial V_x}{\partial z} - \frac{\partial V_z}{\partial x}\right) + \mathbf{k}\left(\frac{\partial V_y}{\partial x} - \frac{\partial V_x}{\partial y}\right)$$

$$= \begin{vmatrix} \mathbf{i} & \mathbf{j} & \mathbf{k} \\ \dfrac{\partial}{\partial x} & \dfrac{\partial}{\partial y} & \dfrac{\partial}{\partial z} \\ V_x & V_y & V_z \end{vmatrix} \qquad (B25)$$

This expression is known as the *curl* of $\mathbf{V}$.

B11   Other Operations. It may be shown that the curl of the gradient is identically zero

$$\nabla \times \nabla\phi = \mathbf{0}$$

and that the divergence of the curl is also identically zero

$$\nabla \cdot \nabla \times \mathbf{V} = 0$$

C USEFUL TABLES

Table C1   Properties

A. Specific Weight, lb/ft³

Aluminum	168	Iron (cast)	450
Concrete (av.)	150	Lead	710
Copper	556	Mercury	847
Earth (wet, av.)	110	Oil (av.)	56
(dry, av.)	80	Steel	489
Glass	162	Titanium	192
Gold	1205	Water (fresh)	62.4
Ice	56	(salt)	64
		Wood (soft pine)	30
		(hard, oak)	50

B. Coefficients of Friction

(The coefficients in the following table represent typical values under normal working conditions. Actual coefficients for a given situation will depend on the exact nature of the contacting surfaces. A variation of 25 to 100 per cent or more from these values could be expected in an actual application, depending on prevailing conditions of cleanliness, surface finish, pressure, lubrication, and velocity.)

Contacting Surface	Typical Values of Coefficient of Friction, f	
	Static	Kinetic
Steel on steel (dry)	0.6	0.4
Steel on steel (greasy)	0.1	0.05
Teflon on steel	0.04	0.04
Steel on babbitt (dry)	0.4	0.3
Steel on babbitt (greasy)	0.1	0.07
Brass on steel (dry)	0.5	0.4
Brake lining on cast iron	0.4	0.3
Rubber tires on smooth pavement (dry)	0.9	0.8
Wire rope on iron pulley (dry)	0.2	0.15
Hemp rope on metal	0.3	0.2
Metal on ice	—	0.02

	Coefficient of Rolling Friction, f_r
Pneumatic tires on smooth pavement	0.02
Steel tires on steel rails	0.006

Table C2   Solar System Constants

Universal gravitational constant	$K = 6.670(10^{-8})$ cm^3/(gm-sec^2)
	$= 3.442(10^{-8})$ ft^4/(lb-sec^4)
Mass of Earth	$m = 5.976(10^{27})$ gm
	$= 4.089(10^{23})$ lb-sec^2/ft
Period of Earth's rotation (1 sidereal day)	$= 23$ hr 56 min 4 sec
	$= 23.9344$ hr
Angular velocity of Earth	$\omega = 0.7292(10^{-4})$ rad/sec
Angular velocity of Earth-Sun line	$\omega' = 0.1991(10^{-6})$ rad/sec
Mean velocity of Earth's center about Sun	$= 66,600$ mi/hr

Body	Mean Distance to Sun, mi	Eccentricity of Orbit, e	Period of Orbit, solar days	Mean Diameter, mi	Mass Relative to Earth	Surface Gravity, ft/sec^2	Escape Velocity, mi/sec
Sun	—	—	—	865,000	333,000	898	383
Moon	238,854*	0.055	27.32	2160	0.0123	5.32	1.47
Mercury	35.6(10^6)	0.206	87.97	3100	0.054	11.4	2.59
Venus	67.2(10^6)	0.0068	224.70	7700	0.815	27.7	6.36
Earth	92.96(10^6)	0.017	365.26	7917†	1.000	32.22‡	6.95
Mars	141.6(10^6)	0.093	686.98	4200	0.107	12.9	3.20

* Mean distance to Earth (center-to-center)
† Diameter of sphere of equal volume
    polar diameter $= 7900$ mi
    equatorial diameter $= 7926$ mi
‡ For nonrotating spherical Earth, equivalent to absolute value at sea level and latitude 37.5 deg

Table C3   Mathematical Relations

A. Series (expression in bracket following series indicates range of convergence)

$$(1 \pm x)^n = 1 \pm nx + \frac{n(n-1)}{2!}x^2 \pm \frac{n(n-1)(n-2)}{3!}x^3 + \cdots \qquad [x^2 < 1]$$

$$\sin x = x - \frac{x^3}{3!} + \frac{x^5}{5!} - \frac{x^7}{7!} + \cdots \qquad [x^2 < \infty]$$

$$\cos x = 1 - \frac{x^2}{2!} + \frac{x^4}{4!} - \frac{x^6}{6!} + \cdots \qquad [x^2 < \infty]$$

$$\sinh x = \frac{e^x - e^{-x}}{2} = x + \frac{x^3}{3!} + \frac{x^5}{5!} + \frac{x^7}{7!} + \cdots \qquad [x^2 < \infty]$$

$$\cosh x = \frac{e^x + e^{-x}}{2} = 1 + \frac{x^2}{2!} + \frac{x^4}{4!} + \frac{x^6}{6!} + \cdots \qquad [x^2 < \infty]$$

$$f(x) = \frac{a_0}{2} + \sum_{n=1}^{\infty} a_n \cos\frac{n\pi x}{l} + \sum_{n=1}^{\infty} b_n \sin\frac{n\pi x}{l}$$

$$\text{where } a_n = \frac{1}{l}\int_{-l}^{l} f(x) \cos\frac{n\pi x}{l}\, dx, \quad b_n = \frac{1}{l}\int_{-l}^{l} f(x) \sin\frac{n\pi x}{l}\, dx$$

[Fourier expansion for $-l < x < l$]

B. Derivatives

$$\frac{dx^n}{dx} = nx^{n-1}, \qquad \frac{d(uv)}{dx} = u\frac{dv}{dx} + v\frac{du}{dx}, \qquad \frac{d\left(\frac{u}{v}\right)}{dx} = \frac{v\frac{du}{dx} - u\frac{dv}{dx}}{v^2}$$

$$\lim_{\Delta x \to 0} \sin \Delta x = \sin dx = \tan dx = dx$$

$$\lim_{\Delta x \to 0} \cos \Delta x = \cos dx = 1$$

$$\frac{d \sin x}{dx} = \cos x, \qquad \frac{d \cos x}{dx} = -\sin x, \qquad \frac{d \tan x}{dx} = \sec^2 x$$

$$\frac{d \sinh x}{dx} = \cosh x, \qquad \frac{d \cosh x}{dx} = \sinh x, \qquad \frac{d \tanh x}{dx} = \operatorname{sech}^2 x$$

C. Integrals

$$\int x^n \, dx = \frac{x^{n+1}}{n+1}$$

$$\int \frac{dx}{x} = \log x$$

$$\int \sqrt{a + bx} \, dx = \frac{2}{3b} \sqrt{(a+bx)^3}$$

$$\int x\sqrt{a + bx} \, dx = \frac{2}{15b^2} (3bx - 2a)\sqrt{(a+bx)^3}$$

$$\int \frac{dx}{\sqrt{a+bx}} = \frac{2\sqrt{a+bx}}{b}$$

$$\int \frac{x \, dx}{a + bx} = \frac{1}{b^2}[a + bx - a \log (a + bx)]$$

$$\int \frac{x \, dx}{(a + bx)^n} = \frac{(a+bx)^{1-n}}{b^2}\left(\frac{a+bx}{2-n} - \frac{a}{1-n}\right)$$

$$\int \frac{dx}{a + bx^2} = \frac{1}{\sqrt{ab}} \tan^{-1}\frac{x\sqrt{ab}}{a} \qquad \text{or} \qquad \frac{1}{\sqrt{-ab}} \tanh^{-1}\frac{x\sqrt{-ab}}{a}$$

$$\int \sqrt{x^2 \pm a^2} \, dx = \tfrac{1}{2}[x\sqrt{x^2 \pm a^2} \pm a^2 \log (x + \sqrt{x^2 \pm a^2})]$$

$$\int \sqrt{a^2 - x^2} \, dx = \tfrac{1}{2}\left(x\sqrt{a^2 - x^2} + a^2 \sin^{-1}\frac{x}{a}\right)$$

$$\int x\sqrt{a^2 - x^2} \, dx = -\tfrac{1}{3}\sqrt{(a^2 - x^2)^3}$$

$$\int x^2\sqrt{a^2 - x^2} \, dx = -\frac{x}{4}\sqrt{(a^2 - x^2)^3} + \frac{a^2}{8}\left(x\sqrt{a^2 - x^2} + a^2 \sin^{-1}\frac{x}{a}\right)$$

$$\int x^3\sqrt{a^2 - x^2} \, dx = -\tfrac{1}{5}(x^2 + \tfrac{2}{3}a^2)\sqrt{(a^2 - x^2)^3}$$

$$\int \frac{dx}{\sqrt{a + bx + cx^2}} = \frac{1}{\sqrt{c}} \log\left(\sqrt{a + bx + cx^2} + x\sqrt{c} + \frac{b}{2\sqrt{c}}\right) \text{ or } = \frac{-1}{\sqrt{-c}} \sin^{-1}\left(\frac{b + 2cx}{\sqrt{b^2 - 4ac}}\right)$$

$$\int \frac{dx}{\sqrt{x^2 \pm a^2}} = \log (x + \sqrt{x^2 \pm a^2})$$

$$\int \frac{dx}{\sqrt{a^2 - x^2}} = \sin^{-1}\frac{x}{a}$$

$$\int x\sqrt{x^2 \pm a^2} \, dx = \tfrac{1}{3}\sqrt{(x^2 \pm a^2)^3}$$

$$\int x^2\sqrt{x^2 \pm a^2} \, dx = \frac{x}{4}\sqrt{(x^2 \pm a^2)^3} \mp \frac{a^2}{8}x\sqrt{x^2 \pm a^2} - \frac{a^4}{8} \log (x + \sqrt{x^2 \pm a^2})$$

$$\int \frac{x \, dx}{\sqrt{x^2 - a^2}} = \sqrt{x^2 - a^2}$$

$$\int \frac{x\,dx}{\sqrt{a^2 \pm x^2}} = \pm\sqrt{a^2 \pm x^2}$$

$$\int \sin x\,dx = -\cos x$$

$$\int \cos x\,dx = \sin x$$

$$\int \sec x\,dx = \frac{1}{2}\log\frac{1 + \sin x}{1 - \sin x}$$

$$\int \sin^2 x\,dx = \frac{x}{2} - \frac{\sin 2x}{4}$$

$$\int \cos^2 x\,dx = \frac{x}{2} + \frac{\sin 2x}{4}$$

$$\int \sin x \cos x\,dx = \frac{\sin^2 x}{2}$$

$$\int \sin^3 x\,dx = -\frac{\cos x}{3}(2 + \sin^2 x)$$

$$\int \cos^3 x\,dx = \frac{\sin x}{3}(2 + \cos^2 x)$$

$$\int x \sin x\,dx = \sin x - x \cos x$$

$$\int x \cos x\,dx = \cos x + x \sin x$$

$$\int x^2 \sin x\,dx = 2x \sin x - (x^2 - 2)\cos x$$

$$\int x^2 \cos x\,dx = 2x \cos x + (x^2 - 2)\sin x$$

$$\int \sinh x\,dx = \cosh x$$

$$\int \cosh x\,dx = \sinh x$$

$$\int \tanh x\,dx = \log\cosh x$$

$$\int \log x\,dx = x\log x - x$$

$$\int e^{ax}\,dx = \frac{e^{ax}}{a}$$

$$\int x e^{ax}\,dx = \frac{e^{ax}}{a^2}(ax - 1)$$

$$\int e^{ax}\sin px\,dx = \frac{e^{ax}(a \sin px - p \cos px)}{a^2 + p^2}$$

$$\int e^{ax}\cos px\,dx = \frac{e^{ax}(a \cos px + p \sin px)}{a^2 + p^2}$$

$$\int e^{ax}\sin^2 x\,dx = \frac{e^{ax}}{4 + a^2}\left(a\sin^2 x - \sin 2x + \frac{2}{a}\right)$$

$$\int e^{ax}\cos^2 x\,dx = \frac{e^{ax}}{4 + a^2}\left(a\cos^2 x + \sin 2x + \frac{2}{a}\right)$$

$$\int e^{ax}\sin x \cos x\,dx = \frac{e^{ax}}{4 + a^2}\left(\frac{a}{2}\sin 2x - \cos 2x\right)$$

Radius of curvature

$$\rho_{xy} = \frac{\left[1 + \left(\dfrac{dy}{dx}\right)^2\right]^{3/2}}{\dfrac{d^2y}{dx^2}}$$

$$\rho_{r\theta} = \frac{\left[r^2 + \left(\dfrac{dr}{d\theta}\right)^2\right]^{3/2}}{r^2 + 2\left(\dfrac{dr}{d\theta}\right)^2 - r\dfrac{d^2r}{d\theta^2}}$$

Table C4. Properties of Plane Figures

Figure	Centroid	Area Moments of Inertia
Arc Segment	$\bar{r} = \dfrac{r\sin\alpha}{\alpha}$ *for small α* $\bar{r} = r$	—
Quarter and Semicircular Arcs	$\bar{y} = \dfrac{2r}{\pi}$	—
Triangular Area	$\bar{x} = \dfrac{a+b}{3}$ $\bar{y} = \dfrac{h}{3}$	$I_x = \dfrac{bh^3}{12}$ $\bar{I}_x = \dfrac{bh^3}{36}$ $I_{x_1} = \dfrac{bh^3}{4}$
Rectangular Area	—	$I_x = \dfrac{bh^3}{3}$ $\bar{I}_x = \dfrac{bh^3}{12}$ $\bar{J} = \dfrac{bh}{12}(b^2 + h^2)$
Area of Circular Sector	$\bar{x} = \dfrac{2}{3}\dfrac{r\sin\alpha}{\alpha}$	$I_x = \dfrac{r^4}{4}\left(\alpha - \tfrac{1}{2}\sin 2\alpha\right)$ $I_y = \dfrac{r^4}{4}\left(\alpha + \tfrac{1}{2}\sin 2\alpha\right)$ $J = \tfrac{1}{2}r^4\alpha$
Quarter Circular Area	$\bar{x} = \bar{y} = \dfrac{4r}{3\pi}$	$I_x = I_y = \dfrac{\pi r^4}{16}$ $J = \dfrac{\pi r^4}{8}$
Area of Elliptical Quadrant Area $A = \dfrac{\pi ab}{4}$	$\bar{x} = \dfrac{4a}{3\pi}$ $\bar{y} = \dfrac{4b}{3\pi}$	$I_x = \dfrac{\pi ab^3}{16}$ $I_y = \dfrac{\pi a^3 b}{16}$ $J = \dfrac{\pi ab}{16}(a^2 + b^2)$

Table C5.   Properties of Homogeneous Solids

(m = mass of body shown)

Body	Mass Center	Moments of Inertia
Circular Cylindrical Shell	—	$I_{xx} = \frac{1}{2}mr^2 + \frac{1}{12}ml^2$ $I_{x_1x_1} = \frac{1}{2}mr^2 + \frac{1}{3}ml^2$ $I_{zz} = mr^2$
Half Cylindrical Shell	$\bar{x} = \dfrac{2r}{\pi}$	$I_{xx} = I_{yy}$ $\quad = \frac{1}{2}mr^2 + \frac{1}{12}ml^2$ $I_{x_1x_1} = I_{y_1y_1}$ $\quad = \frac{1}{2}mr^2 + \frac{1}{3}ml^2$ $I_{zz} = mr^2$
Circular Cylinder	—	$I_{xx} = \frac{1}{4}mr^2 + \frac{1}{12}ml^2$ $I_{x_1x_1} = \frac{1}{4}mr^2 + \frac{1}{3}ml^2$ $I_{zz} = \frac{1}{2}mr^2$
Semicylinder	$\bar{x} = \dfrac{4r}{3\pi}$	$I_{xx} = I_{yy}$ $\quad = \frac{1}{4}mr^2 + \frac{1}{12}ml^2$ $I_{x_1x_1} = I_{y_1y_1}$ $\quad = \frac{1}{4}mr^2 + \frac{1}{3}ml^2$ $I_{zz} = \frac{1}{2}mr^2$
Rectangular Parallelepiped	—	$I_{xx} = \frac{1}{12}m(a^2 + l^2)$ $I_{yy} = \frac{1}{12}m(b^2 + l^2)$ $I_{zz} = \frac{1}{12}m(a^2 + b^2)$ $I_{y_1y_1} = \frac{1}{12}mb^2 + \frac{1}{3}ml^2$

Table C5. *Continued*
(m = mass of body shown)

Body	Mass Center	Moments of Inertia
 Spherical Shell	—	$I_{zz} = \frac{2}{3}mr^2$
 Hemispherical Shell	$\bar{x} = \dfrac{r}{2}$	$I_{xx} = I_{yy} = I_{zz} = \frac{2}{3}mr^2$
 Sphere	—	$I_{zz} = \frac{2}{5}mr^2$
 Hemisphere	$\bar{x} = \dfrac{3r}{8}$	$I_{xx} = I_{yy} = I_{zz} = \frac{2}{5}mr^2$
 Uniform Slender Rod	—	$I_{yy} = \frac{1}{12}ml^2$ $I_{y_1y_1} = \frac{1}{3}ml^2$

Table C5. *Continued*

(m = mass of body shown)

Body	Mass Center	Moments of Inertia
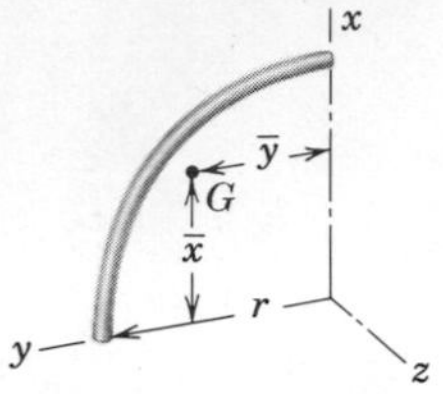 Quarter Circular Rod	$\bar{x} = \bar{y}$ $= \dfrac{2r}{\pi}$	$I_{xx} = I_{yy} = \frac{1}{2}mr^2$ $I_{zz} = mr^2$
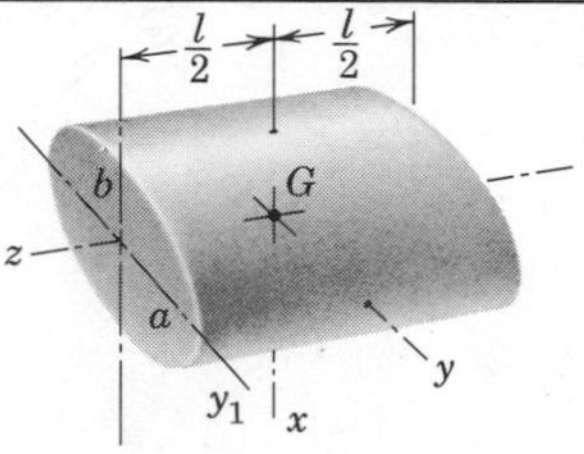 Elliptical Cylinder	—	$I_{xx} = \frac{1}{4}ma^2 + \frac{1}{12}ml^2$ $I_{yy} = \frac{1}{4}mb^2 + \frac{1}{12}ml^2$ $I_{zz} = \frac{1}{4}m(a^2 + b^2)$ $I_{y_1y_1} = \frac{1}{4}mb^2 + \frac{1}{3}ml^2$
Conical Shell 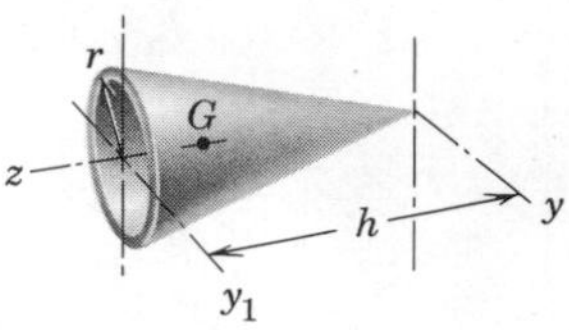	$\bar{z} = \dfrac{2h}{3}$	$I_{yy} = \frac{1}{4}mr^2 + \frac{1}{2}mh^2$ $I_{y_1y_1} = \frac{1}{4}mr^2 + \frac{1}{6}mh^2$ $I_{zz} = \frac{1}{2}mr^2$
Half Conical Shell 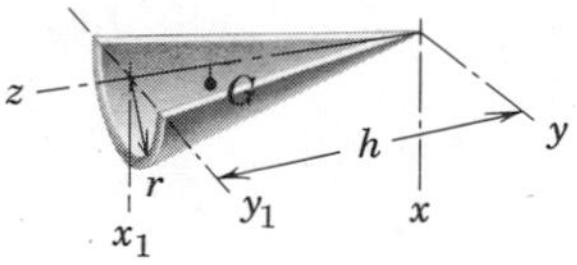	$\bar{x} = \dfrac{4r}{3\pi}$ $\bar{z} = \dfrac{2h}{3}$	$I_{xx} = I_{yy}$ $\quad = \frac{1}{4}mr^2 + \frac{1}{2}mh^2$ $I_{x_1x_1} = I_{y_1y_1}$ $\quad = \frac{1}{4}mr^2 + \frac{1}{6}mh^2$ $I_{zz} = \frac{1}{2}mr^2$
Right Circular Cone 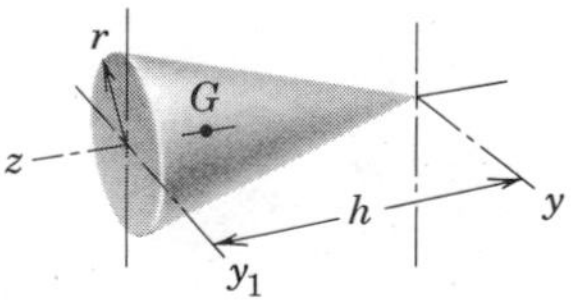	$\bar{z} = \dfrac{3h}{4}$	$I_{yy} = \frac{3}{20}mr^2 + \frac{3}{5}mh^2$ $I_{y_1y_1} = \frac{3}{20}mr^2 + \frac{1}{10}mh^2$ $I_{zz} = \frac{3}{10}mr^2$

Table C5.   *Continued*
(m = mass of body shown)

Body	Mass Center	Moments of Inertia
Half Cone	$\bar{x} = \dfrac{r}{\pi}$ $\bar{z} = \dfrac{3h}{4}$	$I_{xx} = I_{yy}$ $\quad = \frac{3}{20}mr^2 + \frac{3}{5}mh^2$ $I_{x_1x_1} = I_{y_1y_1}$ $\quad = \frac{3}{20}mr^2 + \frac{1}{10}mh^2$ $I_{zz} = \frac{3}{10}mr^2$
$\dfrac{x^2}{a^2} + \dfrac{y^2}{b^2} + \dfrac{z^2}{c^2} = 1$ Semiellipsoid	$\bar{z} = \dfrac{3c}{8}$	$I_{xx} = \frac{1}{5}m(b^2 + c^2)$ $I_{yy} = \frac{1}{5}m(a^2 + c^2)$ $I_{zz} = \frac{1}{5}m(a^2 + b^2)$
$\dfrac{x^2}{a^2} + \dfrac{y^2}{b^2} = \dfrac{z}{c}$ Elliptic Paraboloid	$\bar{z} = \dfrac{2c}{3}$	$I_{xx} = \frac{1}{6}mb^2 + \frac{1}{2}mc^2$ $I_{yy} = \frac{1}{6}ma^2 + \frac{1}{2}mc^2$ $I_{zz} = \frac{1}{6}m(a^2 + b^2)$
Rectangular Tetrahedron	$\bar{x} = \dfrac{a}{4}$ $\bar{y} = \dfrac{b}{4}$ $\bar{z} = \dfrac{c}{4}$	$I_{xx} = \frac{1}{10}m(b^2 + c^2)$ $I_{yy} = \frac{1}{10}m(a^2 + c^2)$ $I_{zz} = \frac{1}{10}m(a^2 + b^2)$
Half Torus	$\bar{x} =$ $\dfrac{a^2 + 4R^2}{2\pi R}$	$I_{xx} = I_{yy} = \frac{1}{2}mR^2 + \frac{5}{8}ma^2$ $I_{zz} = mR^2 + \frac{3}{4}ma^2$

INDEX